Linear Dichroism and Circular Dichroism
A Textbook on Polarized-Light Spectroscopy

Linear Dichroism and Circular Dichroism
A Textbook on Polarized-Light Spectroscopy

Bengt Nordén
Department of Physical Chemistry, Chalmers University of Technology, Gothenburg, Sweden

Alison Rodger
Department of Chemistry, University of Warwick, Coventry, UK

Timothy Dafforn
School of Biosciences, University of Birmingham, Edgbaston, Birmingham, UK

RSCPublishing

ISBN: 978-1-84755-902-9

A catalogue record for this book is available from the British Library

Published by The Royal Society of Chemistry,
Thomas Graham House, Science Park, Milton Road,
Cambridge CB4 0WF, UK

Registered Charity Number 207890

For further information see our web site at www.rsc.org

10 0621494 7

Preface

Many of the questions we ask about chemical and biological systems relate to what shapes the molecules have and how they interact with each other. New or modified methods are constantly being developed to address these questions, either in response to new problems or as the result of technical advances. The aim of this book is to show how linear dichroism (*LD*) and circular dichroism (*CD*) spectroscopies can be used to provide information about molecular structure and about interactions between molecules.

The methods and applications we present include ways, based on the use of polarized ultra violet-visible light spectroscopy, to study the three dimensional structures of molecules in solution or in membranes or other pertinent environments and also study the interactions between molecules. The molecules may be small or large, even molecules of a few atoms may be studied by both *LD* and *CD* and there is no upper size limit for macromolecules, such as proteins, nucleic acids, and even assemblies of molecules.

For several reasons we shall put emphasis on *LD*, in front of *CD*, in this successor of our *CD* and *LD* textbook from 1997. The main reason is that since that time there has been a continuously increasing spectrum of applications of *LD* because *LD* provides structural information not obtainable by any other technique. In addition, *LD* is easier than *CD* to explain and interpret in terms of structural information. *LD* and *CD* can be measured in any molecular environment that transmits light and there is no molecular size limitation. Both techniques can be used to determine dynamic and kinetic information. As such they are particularly useful in situations where techniques such as crystallography and NMR are inappropriate.

However, neither *LD* nor *CD* give unambiguous atomic level structures, meaning that if a certain structure of a modeled molecular system is found to be consistent with the observed spectrum, it does not prove that this is the true and only possible structure. Often the application of several structural methods in parallel need to be used to improve the reliability of a structural or mechanistic assignment. *LD* and *CD* are a particulary effective complementary pair, one giving orientation data and one giving information on the asymmetry of the system. Thus we have included *CD* methods in parallel with *LD*. While excellent textbooks on *CD* have appeared recently, covering a variety of applications to chiral molecules, there have been no up-to-date textbooks on the range of use of *LD* for studying oriented molecules and little emphais on how to complement *CD* data with *LD* data (and conversely).

This book is designed like a ramp, the text gradually getting more advanced: one may start reading at the beginning and continue until sufficient information has been covered for the purpose at hand. Chapter 1 contains all that is necessary for understanding the basic concepts of *LD* and *CD*. Chapter 2 continues with the experimental and analytical techniques and how they can be implemented in the laboratory. The two spectroscopies are interleaved in the first two chapters for the reasons mentioned above, and also because they depend on the same spectroscopic principles. In addition they have some experimental aspects in common, including the fact that the same kind of spectrophotometric instrumentation may be used to perform both types of experiment. Subsequent chapters focus on either *LD* or *CD* and are presented with increasingly detailed levels of interpretation and analysis of the data collected. For example, in Chapter 2 orientation phenomena and the various methods we may use to macroscopically align the samples are discussed as these are essential for any *LD* experiment. The emphasis of Chapters 3–6 is further into methodology and the analysis of experimental data, and in the final chapters the theoretical foundations for *LD* and *CD* are laid.

We have endeavoured to illustrate the text throughout the book. The illustrations—whether recorded spectra or structures or sketches—are primarily typical examples or illustrations included to assist the understanding of the text material. Many of the figures have been discussed in more detail in the original literature to which the reader is referred.

Finally, this book would not have been possible without our many colleagues, students and friends. Most of them are acknowledged specifically at the appropriate place in the text where we have described their research. They have been the community that has made this book possible and we are extremely grateful to them.

To our families

Contents

**Chapter 9 Analysis of circular dichroism: magnetic dipole allowed transitions   211
and magnetic *CD***

Chapter 10 Circular dichroism formalism   229

Chapter 11 Further derivations and definitions

1 Linear and circular dichroism spectroscopy: basic principles

Measurement of linear or circular dichroism involves shining two polarizations of respectively linearly or circularly polarized light onto a sample. Any resultant difference in absorption of the light polarizations gives information about the environment and the structures of the molecules contained in the sample. This chapter contains an introduction to both linear and circular dichroism.

Linear Dichroism and Circular Dichroism: A Textbook on Polarized-Light Spectroscopy
By Bengt Nordén, Alison Rodger and Timothy Dafforn
© B. Nordén, A. Rodger and T. Dafforn, 2010
Published by the Royal Society of Chemistry, www.rsc.org

1.1   Introduction

The aim of this book is to show how linear dichroism and circular dichroism spectroscopies can be used to provide information about molecular structure and about interactions between molecules.

The somewhat odd word 'dichroism' is derived from the Greek διχρωμα meaning 'two colours'[2] since for some samples two colours are seen when they are viewed from two different directions. Dichroism was first observed for crystals: when light that has passed through a polarizer shines on a dichroic crystal, the colour you will see depends on the orientation of the crystal relative to the polarizer.

Amethyst is an example of a dichroic crystal. It is a form of crystalline quartz with transition metal 'impurities' which shows distinct dichroism, *e.g.*, bluish violet and reddish violet for the ordinary and extraordinary rays, respectively.[1]

Chiral is derived from a Greek word χειρ meaning hand (hence the alternate term 'handedness'). Two molecules that are non-identical mirror images of each other are often referred to as enantiomers.

Superposed technically means: to place one geometric figure over another so that all like parts coincide. The more commonly used 'superimposed' does not require the coincidence of all like parts.

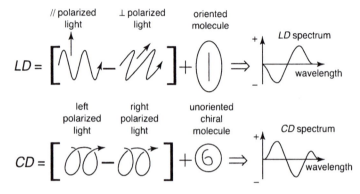

Figure 1.1 Schematic illustration of *LD* and *CD*.

Although polarized light has not yet been defined in this book, it is still appropriate to begin with the definitions of linear and circular dichroism (Figure 1.1). Linear dichroism (*LD*) is the difference in absorption, *A*, of light linearly polarized parallel (//) and perpendicular ($\perp$) to an orientation axis:

$$LD = A_{//} - A_{\perp} \tag{1.1}$$

LD is used with systems that are either intrinsically oriented or are oriented during the experiment. Circular dichroism (*CD*) is the difference in absorption of left and right circularly polarized light:

$$CD = A_{\ell} - A_r \tag{1.2}$$

CD is particularly useful for studying solutions of chiral molecules, by which we mean ones that cannot be superposed on their mirror images.[2]

The justification for our interest in *LD* and *CD* is that both oriented systems and chiral systems are intrinsic features of the world in which we live. Oriented molecular systems are key components of biological cells as well as leading to macroscopic effects such as crystals, liquid crystals, membranes, and muscles; and all biological systems and many non-living ones are chiral. Sometimes the oriented system is itself built up of chiral molecules and the whole system may exhibit a macroscopic handedness.

This book is designed like a ramp with the text getting gradually more advanced: one starts to read at the beginning and continues until sufficient information has been covered for the purpose at hand. Chapter 1 contains all that is necessary to understand the basic concepts of *LD* and *CD* spectroscopies; Chapter 2 continues with what is required to enable the techniques to be implemented in the laboratory. Subsequent chapters are presented with increasingly detailed levels of interpretation and analysis of data. The emphasis of Chapters 3–6 is on the analysis of experimental data, and in the final chapters the theoretical foundations are laid. A selection of general references may be found in references [3-16].

1.2   Electromagnetic radiation and spectroscopy

Electromagnetic radiation

Electromagnetic radiation, as its name implies, has an electric field, **E**, and a magnetic field, **B**; these oscillate at right-angles to one another and to the propagation direction. It may be described by a transverse wave whose *polarization* is defined by the direction of its electric field (Figure 1.2). Electromagnetic radiation also has particle character. It is impossible to really understand wave-particle duality, we just have to accept the dual character of light. If we do an experiment that would probe wave-like behaviour, it is apparent that light has it, and if we look for particle character we find that too. Thus we refer to the *wavelength*, λ, (or *frequency*, v) of radiation, but acknowledge that if a molecule absorbs radiation energy it is as discrete units or particles called *photons*.

Vectors, denoted throughout by bold letters, are a convenient way of summarizing information about physical phenomena that have magnitude (the vector's length) and a direction in space (indicated by where the arrow points). See §11.1.

Electromagnetic radiation is referred to as light if the energy range of interest is near the visible or easily accessible ultraviolet (UV) wavelength range (~ 170–800 nm).

(a) Linearly polarized   (b) Right circularly polarized

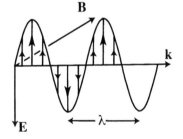
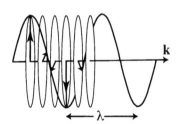
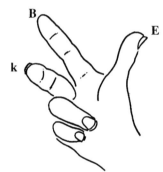

Figure 1.2 (a) Linearly and (b) circularly polarized electromagnetic radiation. Arrows denote direction of **E**. The hand on the right indicates the relative orientations of **E**, **B**, and **k** with the so-called 'right-hand rule'. **E**, **B** and **k** (the direction of propagation of the radiation) are related by equations derived by James Clerk Maxwell in 1864. **B** = $(1/c)$ **k** x **E** with c the velocity of light and x denoting the vector cross product (see §11.1 for a brief discussion of vector products).

Two types of polarized light are mentioned above: linear and circular. In a linearly polarized light beam all photons have their electric field, **E**, oscillating in the same plane, whereas in a circularly polarized light beam the electric field vector retains constant magnitude in time but traces out a helix about the propagation direction. Following the optics convention, we take the end of the electric field vector of right circularly polarized light to form a right-handed helix in space at any instant of time. At each point in space or time, the magnetic field, **B**, is perpendicular to the electric field such that **k**,

If we view right circularly polarized light sitting at a fixed point in space looking towards the light source, down the direction of propagation **k**, then the electric field vector traces out a clockwise circle.

E, and **B** form a right-handed system as illustrated in Figure 1.2.

While linearly polarized (plane-polarized) light is characterized by the electric field vector oscillating in a plane, circularly polarized light is a form of polarization in which the field vector describes a cork-screw motion, completing one turn of a helix after having travelled a distance equal to the wavelength of light. The two forms of light polarizations obviously have quite different 'textural' features, one looking symmetric, the other looking asymmetric, like a screw. This textural difference illustrates the basis for quite different applications of the two forms of dichroism in the study of molecular structure: linear dichroism using linearly polarized light probes macroscopically oriented molecules, whereas circular dichroism using circularly polarized light probes 'helical' molecules. Thus, while *macroscopic orientation* is a prerequisite for non-zero *LD*, *chirality* of the molecules in the sample is a prerequisite for non-zero *CD*.

When is chirality important?
Consider a pair of shoes, a left foot, and a shoe box. The shoes and the foot are chiral and the left shoe only fits comfortably about the left foot, whereas the achiral shoe box cannot distinguish the two shoes. So chirality is important if and only if two chiral species interact.

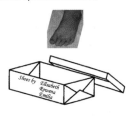

1.3   Normal absorption spectroscopy

Most spectroscopic phenomena arise from the interaction of a molecule with one photon at a time. The molecule may either absorb or scatter the photon; it may also emit a photon. The simplest form of spectroscopy is absorption, where we measure how much light of a given frequency is absorbed by a collection of molecules. If a molecule absorbs a photon of frequency v, it increases its energy by

$$\Delta E = hv = hc / \lambda \tag{1.3}$$

where h is Planck's constant, λ is wavelength, and c is speed of light. In this book the focus is on what happens when molecules absorb photons of visible or ultraviolet (UV) light, since the magnitude of the energy of these photons $((2\text{–}12) \times 10^{-19}$ J/molecule, $\lambda = 170\text{–}800$ nm) is that required to rearrange the electron distribution of a molecule.

In practice, in a collection of molecules, the photons absorbed by different molecules will be of slightly different energies so what we measure is a curve such as the one in Figure 1.3, where the signal that is plotted is a measure of the probability that a transition will occur at that energy (or wavelength). Such a plot of the absorbance of light verses λ or v is known as an *absorption spectrum*.

Absorbance is defined in terms of the intensity of incident, I_o, and transmitted, I, light:[17-19]

$$A = \log_{10}(I_o / I) \tag{1.4}$$

The bandwidth of a transition (usually defined as the width at half the maximum height) is not infinitely narrow but broadened due to:
- the uncertainty principle

$$\Delta E \Delta t \geq \frac{h}{4\pi} = \frac{\hbar}{2}$$

for ΔE the uncertainty in energy, Δt the lifetime of the excited state, and h Planck's constant
- doppler broadening
- intermolecular interactions
- rotational transitions
- vibrational transitions.

The Beer-Lambert law for the absorption, A, of light by a sample of concentration C is (*cf.* §10.2 for derivation)

$$A = \varepsilon C \ell \tag{1.5}$$

where ℓ is the length of the sample through which the light passes and ε is known as the extinction coefficient. If C is measured in mol dm^{-3} and ℓ is measured in cm, then ε has units of mol^{-1} dm^3 cm^{-1}. ε depends on the identity of the molecules in the sample, their environment, and the wavelength, λ. Therefore A is also a function of wavelength. The Beer-Lambert law is valid as long as: the spectrometer can measure *I* and there are no concentration-dependent intermolecular interactions.

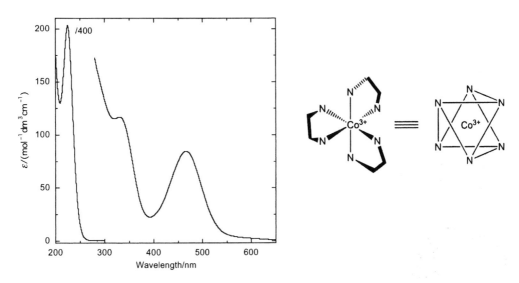

Figure 1.3 Normalized absorption spectrum, ε, of the chiral molecule [Co(en)$_3$]$^{3+}$, en = ethylenediamine. This is the spectrum that would be measured for a 1 M solution in a 1 cm path length cuvette. Since most absorbance spectrometers do not perform well above absorbances of 2, spectra for 100 mM and 250 µM solutions in a cell with ℓ = 0.1 cm were measured and the above spectrum constructed using equation (1.5). $\varepsilon_{466\ nm}$ = 84 mol^{-1} dm^3 cm^{-1}.

Absorption can be pictorially viewed as either the electric field or the magnetic field (or both) of the radiation pushing the electron density from a starting arrangement to a higher energy final one. The electric field is far more effective than the magnetic field in achieving the push that gets a photon absorbed. The direction of net linear displacement of charge is known as the *polarization of the transition*. Some examples of transition polarizations for DNA bases are given in Figure 1.4. The polarization and intensity of a transition are characterized by its *electric dipole transition moment*, which is a vectorial property (*cf.* §11.1) having a well-defined direction (the transition polarization) within each molecule and a well-defined length (which is proportional to the square root of the absorbance). The transition moment may be regarded as an antenna by which the molecule absorbs light. Each transition thus has its own antenna and the maximum probability of absorbing light is obtained when the molecular antenna and the electric field of the light are parallel. Conversely, the absorption is zero when the light polarization and antenna are perpendicular to one another.

Another way to think of absorbance is to visualize a key on a piano. If your finger pushes the key down then it makes a sound. Your finger is the analogue of the electric field, the transition moment of the key is its vertical motion, and the sound produced is the analogue of the absorbance. If your finger strikes at an angle, it is less effective at making sound and if your finger moves horizontally, it has no effect on the key and no sound is generated.

Many of the linear and circular dichroism applications described in this book relate to DNA and proteins. To understand them it is essential that key features of their absorbance spectroscopy are understood.

Spectroscopy of DNA bases

The absorption, and hence *LD* and *CD*, spectra of nucleic acids in the easily accessible region of the spectrum (down to ~180 nm) are dominated by transitions involving electrons redistributing between π and π^* orbitals of the purine and pyrimidine bases.[20] These $\pi \rightarrow \pi^*$ transitions are, by symmetry, all polarized in the plane of the bases, so approximately perpendicular to the helix axis in B-DNA (*cf.* Figure 1.11). The absorbance spectra of the DNA bases thymine (T), adenine (A), guanine (G), and cytosine (C) and the current best estimates of their transition polarizations are illustrated in Figure 1.4. The UV spectra of the bases look as if there are two or three simple Gaussian shaped bands, however, each 'simple' band observed is generally composed of more than one transition.

(a)   (b)

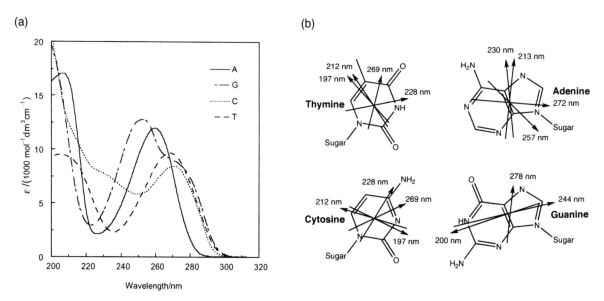

Figure 1.4 (a) UV spectra of the DNA nucleotides deoxyadenosine 5'-monophosphate (A), deoxyguanosine 5'-monophosphate (G), deoxycytidine 5'-monophosphate (C), and thymidine 5'-monophosphate (T) (that of uracil is very similar). (b) Probable transition polarizations for UV transitions of adenine, guanine, cytosine, and thymine from crystal data [21,22,23,24,25] and *LD* measurements on monomeric nucleobases in stretched films.[26]

Protein UV spectroscopy

UV spectra of proteins are usually divided into the 'near' and 'far' UV regions. The near UV in this context means 250–300 nm and is also referred to as the aromatic region, though transitions of disulfide bonds (cystines) also contribute to the total absorption intensity in this region. The far UV (< 250 nm) is dominated by transitions of the peptide backbone of the protein; transitions from some side chains also contribute in this region.

The aromatic side chains of Figure 1.5 and Figure 3.19 all have transitions in the near UV region of the spectrum: the indole of tryptophan has three transitions absorbing in the 240–290 nm region with maximum extinction coefficient $\varepsilon_{279\,nm} \sim 5,000$ mol^{-1} dm^3 cm^{-1}; tyrosine has one transition with $\varepsilon_{274\,nm} \sim 1,400$ mol^{-1} dm^3 cm^{-1}; phenyl alanine also has one transition with $\varepsilon_{258\,nm} \sim 190$ mol^{-1} dm^3 cm^{-1}; and a cystine disulfide bond absorbs from 250–270 nm with $\varepsilon \sim 300$ mol^{-1} dm^3 cm^{-1}.

The aromatic region is so-called because aromatic chromophores absorb in this region. Cystine is classed as an honorary aromatic chromophore only in this context as it enables one to refer to the aromatic and backbone regions of the spectrum.

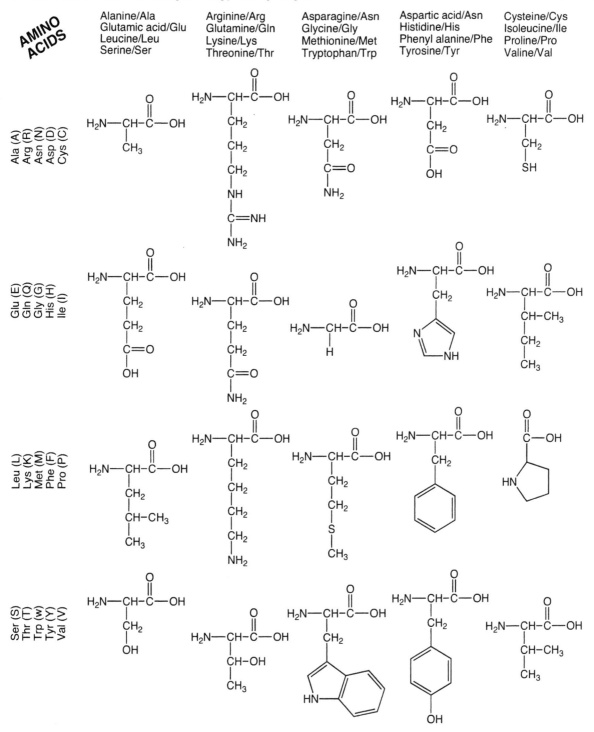

Figure 1.5 The 20 commonly occurring amino acids.

Although tryptophans have by far the most intense transitions in the aromatic region (characterized by the largest extinction coefficients), many proteins have few tryptophans compared with the other aromatic groups, so the region is not necessarily dominated by tryptophan transitions.

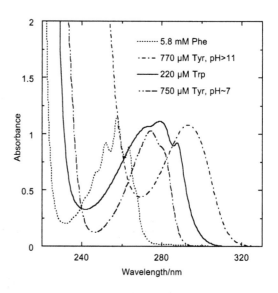

Figure 1.6 Aromatic absorption spectra of tryptophan, tyrosine and phenylalanine. Note all spectra have similar magnitudes as illustrated, however, the concentrations to achieve this are different. Note the pH dependence of the tyrosine spectrum.

The peptide chromophore (Figure 1.7) which gives rise to the transitions observed in the far UV region (180–240 nm) has, of course, σ-bonding electrons but it also contains non-bonding electrons on the oxygen and on the nitrogen atoms, and π-electrons which are delocalized to some extent over the carbon, oxygen, and nitrogen atoms. The lowest energy transition of the peptide chromophore is one from a non-bonding orbital, n, to an anti-bonding π orbital: an $n{\rightarrow}\pi^*$ transition analogous to that in ketones (Chapter 9). The next lowest energy transition is a $\pi{\rightarrow}\pi^*$ transition. As in the carbonyl case, the $n{\rightarrow}\pi^*$ transition is of low intensity ($\varepsilon{\sim}100$ mol^{-1} dm^3 cm^{-1}), though it is not as low as for a simple ketone; it occurs at about 210–230 nm (depending mainly upon the extent of hydrogen bonding of the oxygen lone pairs) and its electric character is polarized more or less along the carbonyl bond. The $\pi{\rightarrow}\pi^*$ transition ($\varepsilon{\sim}7000$ mol^{-1} dm^3 cm^{-1}) is dominated by the carbonyl π-bond but is also affected by the involvement of the nitrogen in the π orbitals; its electric dipole transition moment is polarized somewhere near the line between the oxygen and the nitrogen and is centred at ~190 nm. These transitions are schematically illustrated in Figure 1.7a and b and in more detail in Figure 3.20. When combined together in a protein, they appear at the wavelengths indicated in Figure 1.7c, d, and e.

As noted above, a number of amino acid side chains also have transitions in the peptide region. Although these transitions are often stronger than the peptide $\pi\to\pi^*$ transitions, since the peptide chromophores are usually in significant excess, the side chain transitions are generally nearly impossible to detect. However, the presence of these side chain transitions can be sufficient to confuse attempts to empirically determine the percentage of a given structural unit from *CD*. This is particularly true for proteins with a low α-helical content.

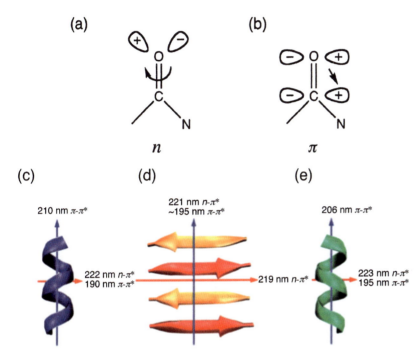

Chromophores with far UV transitions include the aromatic side chains, disulfide cystines, arginine, asparagine, aspartic acid, glutamine, glutamic acid, and histidine.

Figure 1.7 The peptide chromophore and the *n* and *π* orbitals with predominant movement of electron density for the (a) $n\to\pi^*$ and (b) $\pi\to\pi^*$ transitions indicated by the arrows. The net transition moments in (c) an α-helix, (d) β-sheet, and (e) polyproline II helix are illustrated.[27]

Some proteins, for example, metallo-proteins, have so-called prosthetic or extrinsic groups with additional chromophores that can be used in analysis; the transitions often occur in the visible region and the groups can be analyzed in the same way as the molecules that are discussed in Chapter 8. It should be noted that if a protein has an extrinsic chromophore, such as a haeme group, then that chromophore's spectroscopy may also interfere with the peptide region of the spectrum.

1.4   Linear dichroism

We are now in a position to understand *LD* a little better. Imagine we have a linearly polarized light beam (Figure 1.2) and a sample of molecules all oriented in exactly the same way, such as in the dichroic crystal mentioned above. $A_{//}$ denotes the absorbance when the light polarization is parallel to some reference direction in the crystal; turning the polarization at right

The 'perpendicular' direction is determined by its requirement to be at right angles to both 'parallel' and the propagation direction of the light.

angles will give a different colour and hence a different absorption spectrum, $A_\perp$. A quantitative measure of the difference in absorption at a given wavelength is the linear dichroism of equation (1.1). The sign and amplitude of this difference depends on the direction along which the light passes through the crystal and, of course, which direction is denoted as 'parallel'. Permuting 'parallel' and 'perpendicular' inverts the sign of *LD*, producing a mirror image *LD* spectrum. For a more general case than a crystal, a prerequisite for a sample to display non-zero *LD* is that its molecules are somehow macroscopically oriented.

A simple example of such an oriented sample is a stretched sheet of polyethylene into which some substance, such as the linear small molecule S=C=S has been allowed to solubilize. As discussed in §2.3, such a polymer host is a convenient matrix for producing *LD* samples of small molecules. Assume that 'parallel' is the direction of stretch of the film and that we measure with light polarized parallel and perpendicular to this direction. Consider the longest wavelength transition of CS_2. Since the *LD* of this band is positive (Figure 1.8) we know that $A_{//}$ for this transition is larger than $A_\perp$.

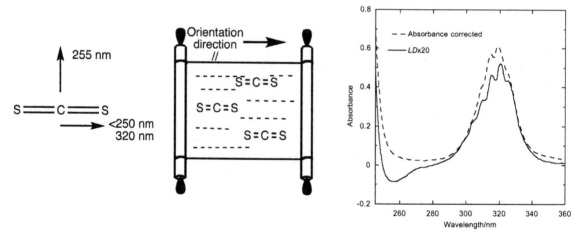

Figure 1.8 Transition polarizations and stretched film orientation of CS_2 together with its PE film absorbance and *LD* spectrum (measured using the solution cell film stretcher unit of Figure 2.4).[28]

Steric forces are repulsive forces between crowded molecules. *Dispersion forces* (London forces) are longer range forces that attract molecules to one another. They depend on the polarizability of the molecules. Polarizabilities generally increase with size so dispersion force orientation tends to favour the molecules orienting with their long-axes parallel to each other.

Qualitatively what does the observation that $A_{//} > A_\perp$ at 320 nm tell us? The positive *LD* of the longest wavelength band of CS_2 tells us that its antenna for absorbing light at 320 nm is sitting along the stretch direction of the film. Hence, the observed *LD* indicates that the CS_2 molecules are oriented with their long axes preferentially parallel with the stretch direction of the polyethylene film. Since stretching the polyethylene film leads to the polymeric chains becoming aligned parallel to the stretching direction, we may infer that CS_2 is itself somehow aligned parallel with the polymer chains. This is precisely the orientation mode to be expected if either steric or dispersive forces (or both) between solute and polymer host molecules are the main determinants for the orientation. Experience indeed shows that all elongated (rod-like) molecules become aligned in this way when inside stretched polymer matrices.

The 255 nm *LD* of CS_2 in Figure 1.8 indicates the presence of a relatively weak short axis polarized transition. Below this band, the positive *LD* suggests that long axis polarized transitions are dominant.

In general, if molecules are perfectly oriented and we measure their absorbance spectrum we have three possible situations illustrated in Figure 1.9. Which one is operative in a given situation depends on how the electric transition dipole moment (pictured above as an antenna having a fixed orientation in the molecular frame and shown as μ in Figure 1.9) is oriented with respect to the polarization of the light. The three situations are as follows.

(1)   If the polarization (*i.e.* the direction of net electron displacement) of the transition is perfectly *parallel* to the orientation direction, as would be the case for a transition polarized along the long-axis of a rod-shaped molecule when the orientation comes from putting the molecules in a film and stretching it (Figure 1.8), then the photons of light may be absorbed in the parallel direction and

$$LD = A_{//} - A_{\perp} = A_{//} > 0 \qquad (1.6)$$

These three situations can be summarized using the piano key analogy given above. (1) would be represented by the piano standing normally on the floor and someone hitting the key of the piano using a vertical motion of the finger. If the piano is replaced by an accordian, in which the key board is oriented vertically, then situation (2) exists and a vertical motion of the finger gives no sound whereas a horizontal motion does. Situation (3) exists for all other orientations of the keyboard.

(2)   If the polarization of the transition we are probing is *perpendicular* to the orientation direction, as would be the case for a pure short-axis polarized transition of the molecule in the stretched film of Figure 1.8, then absorption occurs only in the perpendicular direction

$$LD = A_{//} - A_{\perp} = -A_{\perp} < 0 \qquad (1.7)$$

(3)   Between these two extreme orientations, the absorption intensity varies as the cosine square of the angle between the electric field of light and the transition moment as illustrated in Figure 1.9.

If S=C=S were perfectly aligned in a film, $A_{\perp}$ at 320 nm would be zero (since no absorption of light can occur if all antennae are perpendicular to the light polarization) and $A_{//}$ would be three times the magnitude of the isotropic absorbance. Therefore a large positive quantity would be expected. However, $A_{//}$ is typically only 5% larger than $A_{\perp}$ (depending how much the film is stretched, *cf.* §2.3) for CS_2 in a stretched polyethylene film. This means that the molecules are not perfectly oriented. The magnitude of the *LD* signal therefore also depends on how well the molecules are oriented in the sample. If there is no orientation then there is no *LD*. The orientation parameter, *S*, which equals zero for unoriented samples and equals 1 for perfectly oriented samples, is used to summarise this for many situations as discussed in more detail in Chapter 7.

With sensitive phase-modulation techniques, an *LD* absorbance differential as small as 10^{-6} may be detected.

The converse of the above discussion is that if we know how the transition moments are oriented in the molecular coordinate framework, we can, from the observed *LD*, deduce something about how the molecules themselves are aligned in the sample. This illustrates one of the two major applications of *LD* spectroscopy, namely the structural one—how to obtain structural information from the polarized light spectrum. The second major application of *LD* spectroscopy is to exploit a known orientation of molecules to deduce spectroscopic information, such as how transition moments of various transitions are oriented in the molecular frame. This then enables observed peaks in a spectrum to be assigned to various excited states. This information is in turn needed for structural applications.

As discussed above, a molecule only absorbs light when its electric dipole transition moment aligns with the electric field of the radiation. Thus for an isotropic (randomly oriented) sample, the absorption of a sample is one third of the maximum possible, due to rotational averaging (*cf.* Chapter 6). Hence the factor of 3 in the Figure 1.9 equation.

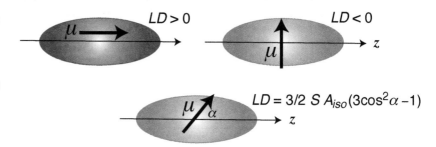

Figure 1.9 Schematic diagram of an *LD* experiment. z is the molecular orientation direction which is the same as the // direction of equation (1.1) if the sample is perfectly oriented. μ is the electric dipole transition moment vector which points along the transition's polarization direction. The formula refers to a uniaxial sample such as a stretched polymer. S is the so-called orientation parameter; $S = 1$ for perfect orientation and $S = 0$ for an isotropic sample. A_{iso} is the absorbance of the corresponding unoriented (isotropic) sample.

A simple and sometimes very useful application of *LD* is for detecting interactions between a small molecule and some biological macromolecule *via* a resulting indirect orientation. Thus, a sample of DNA aligned by laminar (smooth) fluid flow may be titrated with various dye molecules and *LD* measured at the wavelength where the dyes absorb. Complete absence of *LD* is generally a safe indication that the dye does not interact with DNA, while the appearance of an *LD* signal indicates some interaction. The sign and relative amplitude of the measured *LD* further tells us something about the binding geometry in terms of angles between the absorbing dye transition moments and the DNA helix axis (*cf.* formula in Figure 1.9 and Chapter 3).

A molecule has a reflection plane if its mirror image can be rotated round to look exactly like the starting molecule. Solutions of such molecules have zero *CD*.

It is tempting to imagine the helix traced out by the photon's electric field pushing the electrons in a the helix during a transition. However, the wavelength of the light is ~1000 times the size of a molecule, so the helical pitches of photon and molecule are too different to make significant discrimination.

1.5   Circular dichroism

Now consider chiral molecules. Once one has a sample containing a net excess of one enantiomer, collecting *CD* data is simpler than performing an *LD* experiment for which one has to orient the sample. Because a chiral molecule has no reflection plane, any rearrangement of its electrons will not have one either, so the electrons may be thought to move in some kind of helix. From the fact that in circularly polarized light the electric field vectors trace out helices (Figure 1.2), the interaction between a chiral molecule and left and right handed photons will be different. This is the idea behind the definition of *CD*. However, even in an 'obviously' helical molecule, different electronic transitions involve electron redistributions of different handedness so any one molecule will have both positive and negative *CD* signals (Figure 1.10). The challenging task is then to relate the helical motions of the electrons to the arrangement of the atoms and bonds in space. This forms the subject matter of Chapters 8, 9 and 10.

CD is now a routine tool in many laboratories. The most common applications include proving that a chiral molecule has indeed been synthesized or resolved into pure enantiomers and probing the structure of biological macromolecules, in particular determining the secondary structure content of proteins. It is also useful for probing the binding of molecules to a chiral molecule as the *CD* spectrum is perturbed by the interaction. In particular, if an achiral molecule binds to a chiral molecule, an induced *CD* signal will appear in the absorption bands of the achiral molecule. Most *CD* applications involve randomly oriented samples. In §4.6 brief consideration is given to some applications of *CD* with oriented samples.

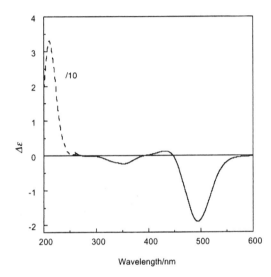

CD measurements on anisotropic (*i.e.* not randomly oriented) samples require meticulous care including correction for sample imperfections as well as instrument optics.[29, 30]

Figure 1.10 *CD* spectrum of Δ-[Co(en)$_3$]$^{3+}$. $\Delta\varepsilon_{490 \text{ nm}} = -1.89 \text{ mol}^{-1} \text{ dm}^3 \text{ cm}^{-1}$. [Co(en)$_3$]$^{3+}$ is illustrated in Figure 1.3.

CD data are more complex to interpret in structural terms than are *LD* data. This complexity has its roots in circular dichroism (and optical activity in general) being a phenomenon involving both electric and magnetic field interactions—thus intrinsically more complex spectroscopy. *LD* and normal absorption are, to a first approximation, simply related to the square of the electric transition dipole moment, and to $\cos^2$ of the angle between the light polarization and the transition moment. *CD*, however, is related to the scalar product between the electric and magnetic transition dipole moments as discussed in Chapter 8. In most cases either the electric or magnetic dipole moment is small which makes simple calculations difficult.

Examples of molecules in which different dominant mechanisms give rise to most of the measured *CD* intensity are illustrated in Figure 1.11. In later chapters of this book we will discuss how the measured *CD* signal from such molecules can be analyzed to give structural data. *CD* is particularly useful where the observed *CD* can be expressed empirically as sum of component parts. The most widely used example of this is with the backbone spectra of proteins being analyzed to give protein secondary structure content as discussed in §4.5.

The focus of this book is on electronic spectroscopy. However, vibrational circular dichroism and Raman optical activity are being increasingly widely used since the combination of improved instrumentation and improved theoretical methods mean that accurate comparisons between experimental data and the spectrum predicted for a given molecular structure can now be made.[3]

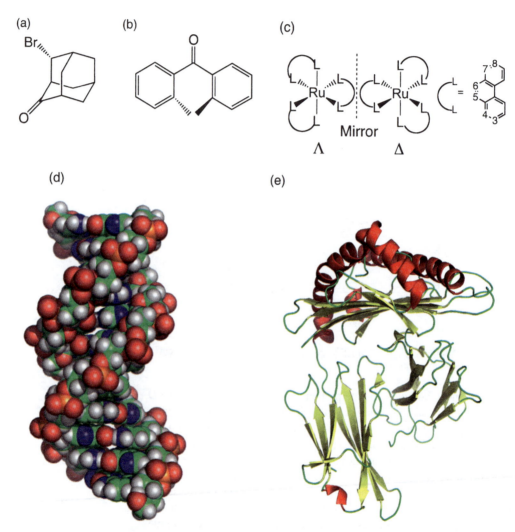

Figure 1.11 Chiral molecules: (a) R-β-equatorial bromoadamantanone, (b) a chiral benzophenone conformation, (c) Λ- and Δ- [Ru(phen)$_3$]$^{2+}$, phen = 1,10-phenanthroline, (d) B-DNA (carbon: green; nitrogen: blue; oxygen: red; phosphorous: orange; hydrogen: grey) and (e) human leukocyte antigen protein where red denotes α-helices and yellow denotes β-strands (*cf.* Figure 3.18).

2   Spectroscopic practicalities

2.1 Introduction

2.2 Measuring linear dichroism spectra

2.3 Molecular alignment techniques

2.4 Experimental considerations for CD and LD spectroscopy

2.5 Sample preparation for LD and CD spectroscopy

2.6 Spectral artifacts

2.7 Instrumentation considerations

Although it is fairly easy to load a sample into a cuvette and record an LD or CD spectrum, a little care is required to ensure that the data are meaningful. The contents of this chapter include the basics of how to collect LD and CD data and also some more advanced considerations about how to avoid artifacts and to perform instrument calibrations.

Linear Dichroism and Circular Dichroism: A Textbook on Polarized-Light Spectroscopy
By Bengt Nordén, Alison Rodger and Timothy Dafforn
© B. Nordén, A. Rodger and T. Dafforn, 2010
Published by the Royal Society of Chemistry, www.rsc.org

2.1   Introduction

Polarized spectroscopy techniques such as linear and circular dichroism need to be implemented carefully to ensure that one ends up with reliable data—not artifacts—to analyze. This chapter contains the basic practical methodologies to enable good data to be collected. The focus of §2.2 is on methods for measuring *LD* spectra. §2.3 contains a summary of a range of sample orientation methods. Instrument and parameter considerations are covered in the context of *CD* spectra in §2.4. The chapter concludes with consideration of spectral artifacts to be avoided and instrument calibration.

A common form of polarizer is a polaroid, a plastic sheet in which molecules, typically long iodine chains, have been aligned mechanically by extruding the sheet at elevated temperature. A more commonly used alternative for *LD* spectroscopy is a quartz photoelastic modulator acting as a half wave plate (*cf.* §10.3).

2.2   Measuring linear dichroism spectra

Rationale behind *LD* spectroscopy

If one places one polarizer, such as a polaroid film, on top of another and rotates it in front of a light source, the light coming through the polarizers goes from maximum to minimum every 90° of rotation (Figure 2.1). This is an *LD* experiment if one thinks of one polarizer as the sample. In other words, the oriented sample in an *LD* experiment has the properties of a wavelength-dependent polarizer.

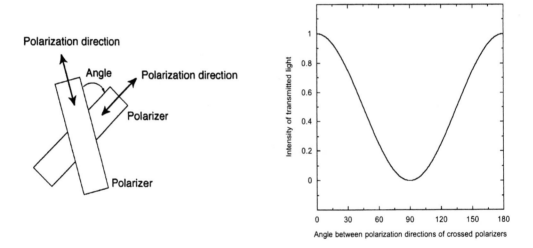

Figure 2.1 Light intensity transmitted through crossed polarizers as a function of rotation angle.

Malus' law[31] states that the intensity of the light between crossed polarizers scales as cosine square of the angle between their polarization directions—hence the function of Figure 2.1. If the polarizers are perfect, no light would pass at the 90° angle between the polarizers, while maximum light intensity is let through when they are set in parallel orientation. In practice, in an *LD* experiment, while we hope that the polarizer used in the manufacture of our spectrometer is perfect, the second 'polarizer'—the sample—only absorbs a fraction of the incident photons and is seldom perfectly oriented, so even at wavelengths where it absorbs and thus operates as a polarizer, it is not a perfect polarizer.

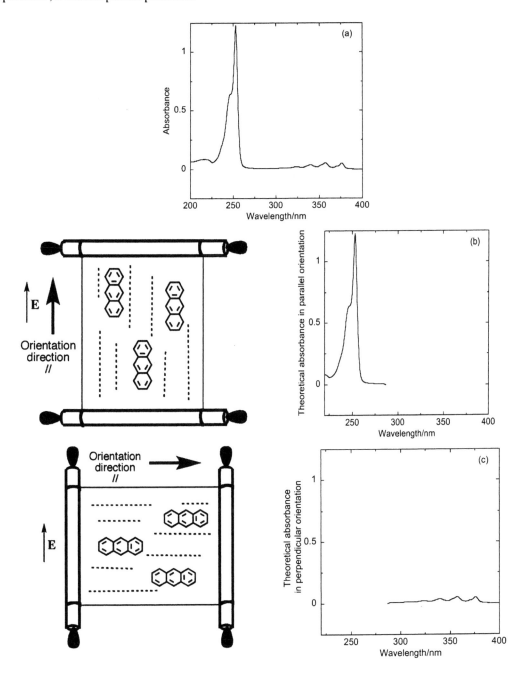

Figure 2.2 Principle of the two-spectra method for measuring *LD* of anthracene. (a) Absorbance in cyclohexane. (b) $A_{//}$ (constructed assuming the 254 nm band of (a) is long axis polarized). (c) $A_{\perp}$ (constructed assuming the 380 nm region of (a) is short axis polarized). The difference between (b) and (c) gives the *LD* spectrum of Figure 2.3a. **E** is the electric field of the light.

The polarized spectra of Figure 2.2 and the theoretical spectrum of Figure 2.3a are simplified in two ways:
- perfect orientation of the molecules has been assumed, and
- the long wavelength band has been assumed to be of pure short-axis polarization.

In reality perfect orientation is never achieved and there is a significant long-axis polarized component at ~320 nm due to coupling of this transition with the 250 nm transition.

Alternatively, the polarizer may be rotated, in which case one must consider any effect of internal polarization of the light by the optics of the spectrometer.

The requirements of an *LD* experiment

The component parts of an *LD* experiment are:
- a source of linearly polarized light,
- a method of orienting the sample,
- a means of detecting how much light is absorbed, and
- a way to change the relative orientations of sample and light beam polarization.

There are two main methods for measuring *LD* spectra. The one requiring less specialized equipment is the two-spectra method. The second method, the differential method, is easier for the user to implement and a wide range of sample orientation techniques may be used. However, much more sophisticated instrumentation is required with this method.

Two-spectra method for measuring LD

For measurement of strongly dichroic samples the simplest method of measuring *LD* is to use a standard double beam normal absorption spectrophotometer equipped with a polarizer. The sample is first oriented parallel to the polarization direction of the polarizer to obtain $A_{//}$. It is then oriented perpendicular to the polarization direction of the polarizer to obtain $A_\perp$. It is advisable to put identical polarizers into both the sample and the reference beam, if a double-beam spectrometer is used, to improve the baseline and measurement sensitivity. This approach is illustrated in Figure 2.2 and Figure 2.3.

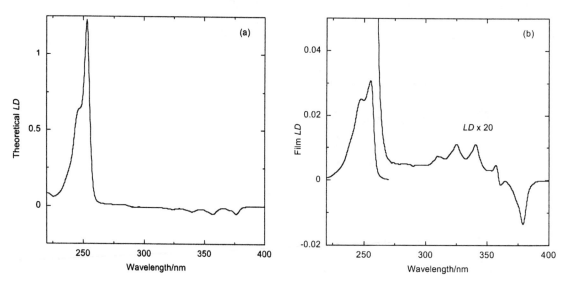

Figure 2.3 (a) The theoretical *LD* spectrum made from the idealized component spectra of Figure 2.2. (b) An *LD* spectrum of anthracene collected by the differential method. The base line was the stretched PE film to which dichloromethane had been added dropwise and allowed to evaporate between additions. The sample was dissolved in dichloromethane and added drop-wise to a 5x stretched polyethylene film (*cf.* §2.3).

Differential method for measuring LD

LD spectra may also be measured directly as a differential absorption spectrum using a *CD* spectropolarimeter (*cf.* §2.4) whose phase-modulation technique produces alternating left and right circularly polarized light. To turn a *CD* instrument into an *LD* instrument, one may insert a quarter-wave device (*e.g.* a Fresnel rhomb or an Oxley prism) into the light beam to produce alternating beams of horizontal and vertical linearly polarized light instead of left and right circularly polarized light. Alternatively, some *CD* spectropolarimeters have a software option which doubles the driving voltage across the photoelastic modulator (Figure 10.3) to achieve, instead of alternating left and right circularly polarized light, alternating orthogonal forms of linearly polarized light at double the beat frequency. The lock-in amplifier is adjusted to detect the higher frequency and the spectrum is produced directly (Figure 2.3).

The *LD* spectra of Figure 2.3a and Figure 2.3b are not the same, which tells us that our assumption that we could take the 350 nm band as being a simple independent short-axis polarized transition is wrong. This is because it is quite weak and as the molecule vibrates, the 350 nm transition couples strongly with the neighbouring much more intense long-axis polarized 250 nm transition. It therefore 'borrows' intensity from the 250 nm band with the result that the shorter wavelength vibronic components of the 350 nm band have net parallel polarization and thus positive *LD*.

> Most transitions have changes in their vibrational energy levels happening at the same time as the electronic changes. This is sometimes seen as regularly spaced fairly sharp peaks in the spectrum. The 350 nm band of anthracene shows this effect particularly clearly (Figure 2.2 and Figure 2.3).

The differential method of measuring *LD* has the advantage of being extremely sensitive, thus not needing high sample orientations and concentrations. However, for any quantitative analysis of the spectrum an independent measurement is also required, either of the unpolarized absorption, A, of the same sample or of the isotropic absorption, A_{iso}, of an unoriented sample. Since this measurement is often performed using a different instrument, care has to be taken to avoid artifacts due to the different monochromators of the instruments and/or due to the probability that the light irradiates different parts of the sample (or even a different sample). The results can be variations in wavelengths of peaks, stray light, scattering, *etc*.

> Isotropic means that there is no directional preference for what we are measuring. In other words, the molecules are randomly distributed to measure A_{iso}.

2.3   Molecular alignment techniques

LD can be used to give relative orientations of subunits of a system, often in a situation where no other technique can be used. However, this is only possible if the sample can be macroscopically oriented in a manner that does not perturb the molecules of interest. Which orientation method one should use depends on the sample and the question being asked. Long and relatively rigid polymers, such as double-stranded DNA or RNA, or molecular assemblies of micrometer dimension, may be oriented in a fluid by shear flow whereas small molecules require a stronger orienting force. Some molecules which cannot themselves be oriented may be oriented when they bind to another molecule that is oriented.

A selection of orientation methods is described below. The sensitivity of bench-top instruments based on *CD* spectropolarimeters is of the order of one part in a million, so the orientation method does not have to be particularly efficient to give a good *LD* signal. More details of the underlying principles of molecular alignment and its significance for *LD* are given in Chapter 7 and references [11,32,28]. Some applications may be found in Chapters 3 and 5.

Stretched polymers as matrices in which to orientate small molecules

In 1934 A. Jablonski was the first to propose a method for orienting molecules by adsorption in anisotropic matrices.[33]

Small molecules can often be absorbed into polymer films. When the film is mechanically stretched, either before or after the small molecules are added, the included molecules align their long axes preferentially along the stretch direction (Figure 2.2). For a molecule to be aligned in a stretched polymer film it must be either an integral part of the film material or associated sufficiently strongly with the polymer chains that when the film is stretched the molecule follows the alignment of the polymer. The best baseline for such an experiment is that collected for a piece of the same film that has been stretched but does not contain the molecule of interest: either the same film after washing out the analyte molecules or before adding them. In practice, one of two types of films enable one to prepare aligned samples of most molecules: polyethylene for non-polar molecules and polyvinyl alcohol for polar molecules. Polyvinyl chloride is also valuable in some applications.

(a)

(b)

Figure 2.4 (a) A mechanical film stretcher with oppositely threaded screws to ensure even stretching of the film. The data of Figure 2.3 were collected with this holder. (b) A mechanical film stretcher with IR cell attachment to provide a liquid environment for the film. If one first fills the cell with the sample in a water immiscible solvent (*e.g.* chloroform or cyclohexane), then flushes the cell through with ethanol and finally water, the result is that all analyte not absorbed into the film is removed. In this way it is guaranteed that the unpolarized absorbance spectrum is due only to sample in the film. The data of Figure 1.8 were collected with this cell and holder.

Alignment in polyethylene films

Ideally the stretching force is applied uniformly across the film using a film stretcher such as those of Figure 2.4. One can obtain reasonable *LD* spectra with almost any degree of stretching. The arms of a micrometer can be used to stretch a film affixed to it. Alternatively, light weight PE can be stretched by hand.

Polyethylene (PE) is microcrystalline and when it is mechanically stretched along the manufacturer's stretch direction a molecular orienting environment is produced. PE is well suited for orienting non-polar molecules for spectroscopy as it has transparency in UV (above 200 nm), in the visible, and in the infrared regions. The key to success with PE film *LD* is the choice of PE and the degree of stretching. Fairly thick transparent plastic bags are usually a good source of PE. However, to stretch this weight of PE a mechanical stretcher (Figure 2.4) into which the film can be fixed is required. The assembly then needs to be placed in the light beam in a reproducible manner.

By convention the parallel direction of the polarized light is usually taken to be horizontal, *i.e.* parallel to the floor, so the stretch direction of the film should be aligned horizontally. It is advisable not to stretch too close to the breaking point of the polymer since the film has a tendency to become opaque and to rip suddenly. With a film stretcher, a stretch factor of 5–10 is fairly straightforward. One should always endeavour to collect the sample spectrum on the same part of the film as the baseline.

There has been no evidence that a solute is better oriented in a stretched PE film when introduced into the film before stretching compared with it being added after stretching. The orientation of the solute molecules, hence, appears to be caused by their adsorption under equilibrium conditions to aligned polymer chains or crystallites or by occupying anisotropic cavities. Thus the most efficient protocol is to first stretch the film, then measure the baseline spectrum (see below). Since the solvent may also enter and swell the PE and is itself aligned when the film is stretched, the baseline must be measured on a film that has been treated with solvent in the same way as the sample film will be (*cf.* Figure 2.3b). After the baseline has been collected, then introduce the analyte into the stretched film by adding droplets of a solution containing the analyte in cyclohexane or chloroform or dichloromethane to the surface of the film. Allow the solvent to evaporate between drops. The 1,10-phenanthroline spectrum of Figure 5.15 was collected using this methodology.

To study small molecules, such as the volatile liquid CS_2 (Figure 1.8) or the gases SO_2 and NO_2, a standard liquid cell for IR spectroscopy containing the stretched PE sample may be used (Figure 2.4b).[28]

Polyvinyl alcohol films

Polyvinylalcohol (PVA) is a near universal host for polar molecules; the film is transparent in the UV (above 200 nm) and in the visible region of the spectrum, though it has a strong absorption over large regions of the infrared.[26] Small molecules inserted in a dry (less than a few percent water) PVA film may be oriented by stretching under low heat. PVA can also be used for *LD* when it is equilibrated in a humid atmosphere to form an elastic gel containing approximately 50% water. *LD* spectra of oriented B- and A-DNAs, as well as denatured polynucleotides have been obtained using such an elastic gel (*cf.* Figure 5.26).[26, 34, 35]

PVA films are more difficult to prepare than PE films, however, the quality of the resulting data is often better. To prepare a PVA film, one mixes well-hydrolysed low molecular weight commercial PVA powder in cold water (10% w/v) to make a slurry which is then heated to near boiling to form a viscous solution. The sample solution (typically the sample is prepared at a concentration of ~5 mM in water, however, the aim is to have the final film with a maximum absorbance of between 0.1 and 1) is then added to half of the PVA solution, and the mixture is cast onto a glass plate and left to dry. The same volume of water is added to the remainder of the solution which is also cast onto a glass plate and left to dry to make a baseline film. Finally the films are stretched by the same factor (typically 2–5) at an elevated temperature (~80 °C) by holding the films in the hot air from a hair dryer as they are being stretched. PVA films easily rip if the

The half-life of orientation of a dry PVA stretched film has been estimated to be greater than 500 years. There are no data to support this statement, but the orientation parameter of a sample of the three-ring aromatic dye methylene blue in a PVA film has remained at $S = 0.95$ for over 30 years.

PVA films typically take one to three days to dry in a well-ventilated dust-free place.

temperature gets too high and the film gets soft; one really needs to use a well-engineered film stretcher that stretches the film firmly and evenly. It is also advisable to stretch a little bit first and measure the spectra before trying a larger stretch on a precious sample.

As with PE, the greater the stretch factor of the PVA, the greater the *LD* signal magnitude will be, until of course the film breaks. At the limit of extreme stretch, the PVA film suddenly becomes hydrophobic (spherical water droplets form on the film surface) indicating a very tight packing of the polymer chains and the squeezing out of remaining solvent water.

If the solute is added to PVA films after stretching, the alignment is usually poor. This observation together with the very efficient orientation of samples that can be obtained when they are added prior to polymerization, suggests that orientation in PVA may not be an equilibrium orientation but one due to the strong crowding and high tension built into the glassy matrix—a tension that will eventually relax, like a pane of glass.

Polyvinyl chloride films

Many substances that are too polar to dissolve in PE, but insoluble in water and thus not able to be included in PVA films, may be oriented in stretched PVC films. The method used to prepare the films is similar to that of PVA with the difference that DMF or chloroform are used as the PVC solvent instead of water. The films are cast on a glass plate. In contrast to PVA, one has to reduce the ventilation so that the solvent does not evaporate too fast as this causes the PVC to become opaque. The solid PVC film may be stretched in the hot air from a hair dryer. As the film is even more brittle than PVA, it is therefore advisable to practice before stretching a real sample.

Flow orientation of macromolecules

If a rigid or semi-flexible polymer, such as DNA, is dissolved in a solvent and then flowed past a stationary surface at $0.1-3$ m s^{-1}, then the molecules experience sufficient shear forces to give a net orientation of the long axis of the polymer along the flow direction. If light is incident on the sample perpendicular to the flow direction, then the absorbance parallel to the flow $A_{//}$, and the absorbance perpendicular to the flow direction, $A_{\perp}$, are different so an *LD* signal may be measured. If the cell components are quartz then data in the visible and UV regions may be collected.

Even a flow rate of 100 mm s^{-1} gives reasonable orientation for most samples.

Other geometries of a flow *LD* experiment are also possible including measuring *LD* at ±45° to the flow direction.[36]

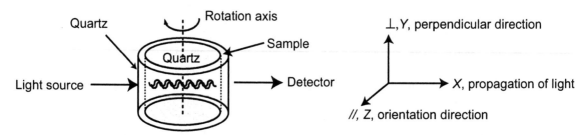

Figure 2.5 Schematic diagram of a Couette flow cell showing flow orientation in a coaxial flow cell with radial incident light. $\{X,Y,Z\}$ denotes the laboratory fixed axis system.

The obvious method for producing the required flow rate is to use a linear flow-through system such as provided by an HPLC pump or a pair of syringes. However, such an open system has an inherently large sample requirement and tends not to be completely stable. Another problem is that any air bubble in the system will multiply as the sample circulates through the tubing. Wada in 1964[37, 38] solved these problems with the invention of a Couette flow cell for *LD* where the sample is endlessly flowed between two cylinders one of which rotates and one of which is stationary. This is schematically illustrated in Figure 2.5.

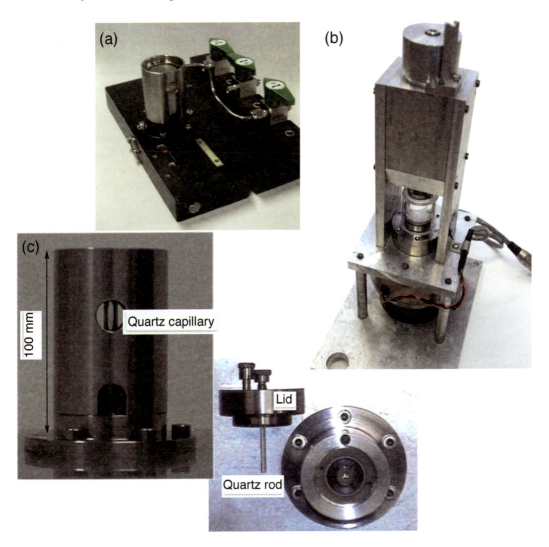

Figure 2.6 (a) Large volume (2–3 mL) inner rotating cylinder Couette flow cell with 500 μm annular gap.[11] (b) Medium volume (0.6 mL) outer rotating Couette cell with 500 μm annular gap. The box under the cell contains high-precision ball bearings to minimize wobbling of the rotating cylinder. (c) Microvolume (25–60 μL) outer rotating[39, 40] Couette flow cell showing the outer quartz capillary (3 mm inner diameter) and inner quartz rod (2.5 mm outer diameter) which when assembled results in an annular gap of 250 μm.

Some Couette flow cells are illustrated in Figure 2.6. The most recent developments include microvolume Couette flow cells, which require 25–60 μL of sample rather than the mLs of the previous Couette cells, and Peltier temperature control.[39, 40] These developments have increased both the range of samples and types of experiments that can be undertaken.

As is the case for *CD* spectroscopy (§2.4), a baseline spectrum must always be subtracted from an *LD* spectrum. With a Couette flow system a number of possible baselines can be used.

(1)   The simplest option is to stop the flow and measure the spectrum. The validity of such a baseline relies on the non-rotating spectrum being independent of the position of the rotor, and hence of the part of the rotating quartz that happens to have stopped in the light beam.

(2)   A baseline of the same cell rotating slowly, causing no measurable orientation of the sample, may be used if the motor is stable at low rotation speeds and the scan speed is slow compared with the rotation speed.

(3)   A spectrum of only the solvent/buffer with higher rotation speed may be measured as the baseline. However, if the sample scatters light significantly then the contribution of scattering to the spectrum needs to be taken into account. See §2.6 for more discussion of this.

Flow orientation of liposomes

Membrane proteins and peptides are peripherally associated with or embedded within the phopholipid bilayer (or membrane) that surrounds the cell and its organelles. They play a number of key roles in cells including acting as ion channels, adhesion molecules and cell signalling systems. The malfunctioning of these units is directly linked to diseases, such as cancer, diabetes, and arteriosclerotic disease, *etc*. Membrane proteins are therefore important drug targets and peptides are also drug candidates; however, we have only a very limited understanding of their structure, function, and intermolecular interactions in their membranes.

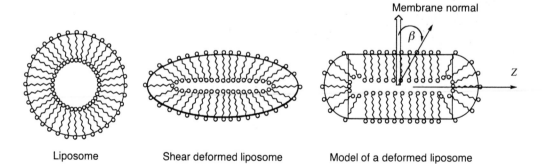

Liposome   Shear deformed liposome   Model of a deformed liposome

Figure 2.7 Schematic diagram of liposomes distorted in shear flow.

Membrane proteins have been studied by *LD* for a long time using more-or-less dried films and squeezed gel methodologies (see below). The possibilities for lipid/membrane protein *LD* were opened up to the more biologically relevant solution phase when it was shown that unilamellar lipid vesicles (liposomes)—used as model membrane systems with a single bilayer of lipid enclosing a central space—were distorted in shear flow and could be aligned (Figure 2.7).[41, 42] Anything (small molecule or proteins or peptides) bound to the liposome is also aligned. Examples of how *LD* measured on these systems may provide information about the structures formed upon interaction between the membrane and membrane-binding proteins are given in §3.7 and §5.4. One should be aware, however, that due to the size of unilamellar lipid vesicles being near the wavelength of light, significant light scattering below 250 nm may occur. It should be noted that the equation for the *LD* magnitude given in Figure 1.9 needs to be modified for this geometry as discussed in §3.7 and given by equation (3.7).

The effect of shear rate, sample viscosity and particle length on flow LD signals

One of the key experimental parameters for flow *LD* is how fast the Couette cell spins. This is directly related to shear rate by

$$G/s^{-1} = \frac{dv_z}{dx} = \frac{2\pi R_o \Omega}{60(R_o - R_i)} \qquad (2.1)$$

where the rate of rotation Ω is revolutions per minute (rpm) for an outer rotating Couette cell with outer cylinder radius R_o, and inner cylinder radius R_i. The dependence of *LD* on shear rate for some DNAs is illustrated in Figure 2.8 (note the plots are $-LD$ *vs* G^2 or G). To a first approximation, most DNAs with which one works have a linear dependence on G. Stiffer polymers reach saturation at lower G than the semi-flexible DNA. Others (often including DNA) do not reach saturation before the flow becomes turbulent.

Viscosity of water at 60° is half that at 20°. Viscosity of water at $G = 1000\ s^{-1}$ is one third that at $0\ s^{-1}$; higher shear rates have less effect.[43]

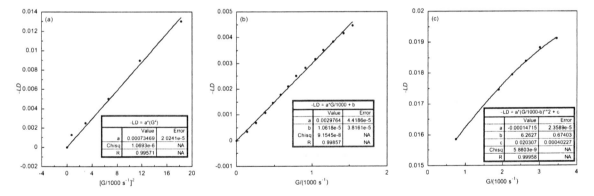

Figure 2.8 (a) $-LD$ (at 260 nm) *versus* G^2 (G=shear rate) for a 450 base pair DNA (130 μM, 10 mM NaCl, 23 °C). (b) $-LD$ (at 260 nm) *versus* G for a linearized plasmid DNA (pC3.1 variant, 6882 base pairs) (100 μM, 0.1 M NaCl, 30 °C) and (c) $-LD$ (at 260 nm) *versus* G for calf thymus DNA (200 μM base, 10 mM salt, 23 °C). *LD* cell has $R_o = 3.00$ mm and $R_i = 2.75$.[43]

Flow *LD* magnitudes are also dependent on solution viscosity, η, as illustrated in Figure 2.9a. Small *LD* signals can thus be enhanced by adding, *e.g.*, glycerol to the solution to increase its viscosity.[44, 45] It must be noted that viscosity itself is a function of both temperature and shear rate as illustrated in Figure 2.9b and c. Thus T, *G* and η dependence of *LD* cannot be considered in isolation.

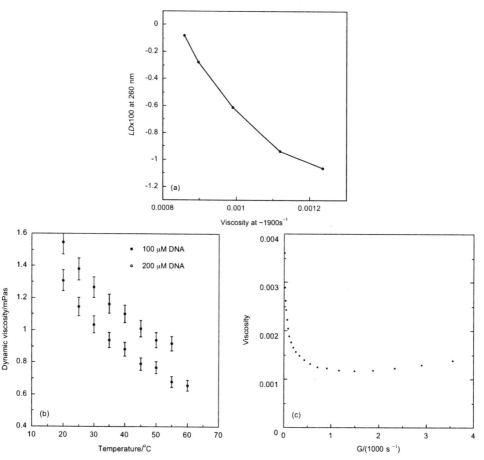

Figure 2.9 (a) Viscosity dependence of Couette flow *LD* (measured on a Jasco J-815 spectropolarimeter) at 260 nm for a solution of DNA (200 μM calf thymus DNA in water). The viscosity was determined on the same solution as the *LD* using an Advanced Rheometer AR100. The viscosity was varied by changing the temperature from 20–60 °C. The shear rate was 1900 s[−1]. (b) Temperature dependence of the dynamic viscosity of DNA (200 μM calf thymus DNA in water). (c) Shear rate dependence of the dynamic viscosity of DNA (200 μM calf thymus DNA in water). Viscosities for (b) and (c) were determined using a Cannon-Manning 25 E50 semi-micro viscometer.[46]

Electric field orientation

Effective orientation for polar or polarizable molecules may be achieved by setting up an electric field between two parallel plates. This produces uniaxial orientation of the molecules (Figure 2.10). Electric field orientation is conceptually simpler than flow orientation, and, when performed in non-conducting solvents, is an equilibrium orientation methodology whose effect can be predicted theoretically. It is usual to measure $A_\perp$ and A_{iso} (the unoriented isotropic absorbance) in electric field *LD*, rather than $A_{//} - A_\perp$ directly. The lower sensitivity of this method compared with *e.g.* the Couette

DNA may be oriented by an electric field, although it is charged rather than polar, since its ionic environment is polarizable and establishes a net dipole in the presence of the field.

flow method is often compensated by strong orientation effects. In addition, at least for DNA, the *LD* magnitude for perfect orientation ($S = 1$) may be determined by plotting *LD versus* inverse field-strength.

In conducting solvents (*e.g.* aqueous solutions), heating effects may be circumvented by using short-pulsed field techniques, which additionally enable the study of relaxation phenomena. The pulsed nature of the field, in early studies, required each electric dichroism experiment to be performed at a fixed wavelength while today a CCD (charge-coupled device) system may be used to collect a full spectrum as a function of time.

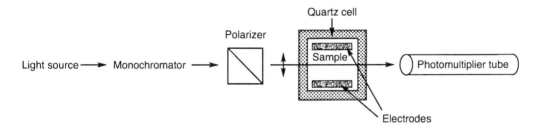

Figure 2.10 Schematic illustration of an electric dichroism experiment.

Magnetic field orientation

Magnetic fields may be used to orient molecules in much the same way as electric fields, although the effect is much smaller. This method is not widely used unless the particles are large (such as chloroplasts or carbon nanotubes, *cf.* Figure 5.20) and carry substantial magnetic dipole moments or have very high magnetic susceptibility anisotropy yielding large induced magnetic dipole moments. There is also always some concern about the effect of magnetic fields on the electron cascades in the photomultiplier tube that is used as the detector in most experiments. However, an advantage of magnetic over electric field orientation in conducting media is the lack of dissipated heat which makes detailed spectral recordings over longer time periods possible. This type of data has been used to probe the thermal randomization decay of carbon nanotubes after switching off the magnetic field. When done as a function of solution viscosity, the data allow an estimate of the length of the oriented particles.[47]

Gel orientation methods

Squeezed gel orientation

A method of orientation that has proved successful for photosynthetic membrane proteins is to embed the protein in a gel and then physically squeeze the gel.[48] Figure 2.11 shows the principles of uni-directional squeezing (giving biaxial orientation) and bidirectional squeezing (giving uniaxial orientation). Such a cell typically has quartz windows and may be operated electro-magnetically. The gap between the windows is changed by a well defined amount (typically compressing by ~10%) by the fast activation of an electromagnet. Light propagated through the gel slab at inclined incidence (*cf.* equation (6.42)) to the axis of applied pressure is used to probe

To make a polyacrylamide gel to orient, *e.g.*, a photosynthetic pigment, a buffer suspension of the photosynthetic reaction centre particles is mixed with acrylamide (0–15% w/v); N,N'-methylene-bisacrylamide (0.3–0.5% w/v); and glycerol (50% v/v). The mixture is then polymerized by adding 0.03% (v/v) N,N,N',N'-tetramethylethylene-diamine and 0.05% (w/v) ammonium persulfate.[49]

the *LD* and its temporal relaxation.

The main shortcomings of the gel orientation technique are its limitation to wavelengths longer than about 250 nm due to light scattering by the gel, and its limited dynamic range of deformation which results from the to the risk of rupturing the gel.

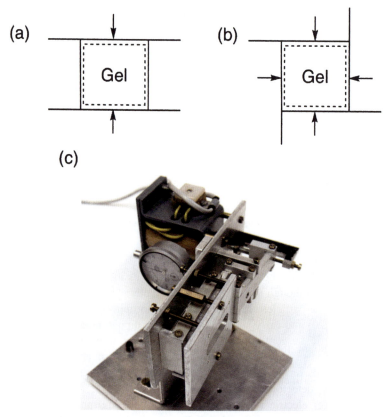

Figure 2.11 Schematic illustration of gel deformation perpendicular to the plane of the paper to achieve a thin slab of gel compressed (a) biaxially and (b) uniaxially.[11] The uniaxial gel requires the light to be incident at less than 90° in order to give an *LD* signal as discussed in Chapter 6. (c) A gel compression device which produces uniaxial deformation.[50]

Migrative orientation through gels

By migrative orientation we mean the alignment due to translational motion of nonspherical particles through a viscous or porous medium. The most studied case of migrative orientation is the electrophoretic orientation that DNA displays in porous media[52] as well as in free water solution.[51]

Free water migrative orientation of DNA for *LD* has been performed in a sounding rocket to avoid convective disturbance from dissipating heat.[51]

Studies of electrophoretic orientation of DNA has given valuable information about the mechanisms underlying the success of pulsed fields to separate large 'reptating' DNA molecules in gel electrophoresis.[52,9] Figure 2.12 shows the non-monotonic behaviour of the *LD* of double-stranded T2 DNA (166,000 base pairs) in an electrophoresis agarose gel experiment as a long electric-field pulse (~20 s) is applied. The *LD* starts from zero (random orientation of the DNA coil) and then becomes negative at 260 nm, as expected for B-form DNA aligned parallel with the field. The orientation is several orders of magnitude higher than the applied electric field would directly induced, given that it is weak (9 V/cm).

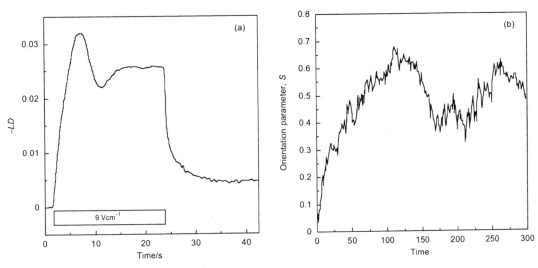

Figure 2.12 (a) Electrophoretic orientation of T2 DNA (1.1×10^{-4} mol dm^{-1} DNA base) in a 1% agarose gel. *LD* as a function of time before, during and after a rectangular electric field pulse (9 V cm^{-1}) is applied.[9] (b) Brownian dynamics simulation in a 3 dimensional rigid fibre gel model.[9]

The behaviour of the *LD* signal as a function of time is as follows: the magnitude first goes to a maximum and then decreases, then increases, but to a lesser amount than the first maximum, after which it asymptotically approaches a steady-state (non-zero) value. As has been shown in single-molecule experiments and also been inferred from Brownian dynamics simulations,[9] the oscillating behaviour of the experimental *LD* signal as well as its magnitude can be explained by the fact that the migrating DNA undergoes several phases of extension and contraction during its path through the gel. As the random coil DNA chain begins to move, the friction from surrounding obstacles (gel fibres) leads to a stretching, giving an *LD* signal as a result of the net orientation of the DNA helix along the electric field direction (which is also the migration direction). Since the polarization of the counterions (cations) surrounding the DNA is similar for the two ends of DNA, both ends move towards the anode leaving the middle behind. The two ends are essentially independent and will, with high probability, take different paths (pores) through the gel and thereby sooner or later end up in a strained hairpin-like configuration. The arm of the hairpin which has the longest part along the field, and hence the biggest accumulated force, will win the tug of war and pull the shorter arm backwards until it is free of the gel fibres and can relax back to form a random-like configuration (this is the point where the orientation drops to a minimum). As indicated by the dynamics simulation, each DNA molecule undergoes cyclic behaviour oscillating between stretched and collapsed configurations.

This example illustrates a limitation of *LD* namely that, despite the high sensitivity with which it can be measured, it is an average that may conceal both strong spatial (distribution function) as well as temporal (local dynamics) variations.

The fact that only the first overshoot in Figure 2.12 is clear in the experimental *LD* data is a consequence of the fact that the *LD* (in contrast to the simulations or single-molecule microscope observations) is the ensemble average over a large number of molecules which, as the cycles are repeated, soon get out of phase with each other.

Crystalline samples

Crystalline samples can provide perfect orientation if the molecules in the unit cell are aligned, however, for transmission spectroscopy very thin slices are usually required because of high absorptivity. Alternatively, a polarized reflection spectrum of the thick crystal may be converted to the polarized absorption spectrum using a Kramers–Kronig transform.[22]

Information from an *LD* experiment may conversely be helpful for solving the crystal structure, either assisting the phase-angle determination or by providing angular data for moieties that are not distinctly resolved by the diffraction.[53] Such cases include situations where a small drug molecule may diffuse into a pure protein crystal. *LD* may also be used to investigate the subtle atomic displacements of a piezo-electric crystal (*e.g.* voltage appearing across a barium titanate crystal upon application of pressure), or the birefringence observed in pressure-deformed plastic and glass materials.

Liquid crystal orientation on flat surfaces

There are two kinds of liquid crystals (LC) made from organic molecules: thermotropic and lyotropic, both of which are prone to easy alignment either spontaneously (in domains or homogeneously when confined to a surface) or by application of an external electric field. Thermotropic liquid crystals are polar non-ionic molecules that in pure form, or at high concentrations in organic solvent, form oriented aggregates—these are the components of liquid crystal displays and their alignment and associated optical birefringence are crucial parameters for their function. Lyotropic liquid crystals are composed of ionic (sometimes together with non-ionic) amphiphilic molecules in aqueous phases. They are less well studied in orientation contexts, although biological lipid bilayer membranes are examples of this kind of structure with generally a high degree of internal order.

Thermotropic nematic liquid crystals (orientationally but not positionally ordered molecules) are excellent hosts for most molecules that dissolve in organic solvents and pioneering work on aromatic chromophores is due to Maier, Saupe and Sackmann (Saupe introduced the *S* orientation tensor formalism mentioned in Chapter 6). In contrast to their lyotropic counterparts, which usually contain reasonably large amounts of water, thermotropic liquid crystals are also well-suited for infrared *LD* studies.

Solute molecules incorporated within a liquid crystal are probably oriented by similar forces to those within the stretched polymer films discussed above, although dispersive and dipole forces seem more important in these soft structures than the steric forces that presumably dominate in the nm-size cavities between super-stretched polymer chains. The liquid crystal and solute molecules within a liquid crystal can then be macroscopically oriented

---

Sidebar notes:

Incidence of the light along an axis about which the sample has three-fold or higher symmetry will give *LD* = 0 as discussed below for liquid crystals. An effective infinite rotational axis may also be given to a sample due to the symmetry of the orientation method.

The inverted piezo-electric effect (pressure from voltage) as well as pressure-induced birefringence are both exploited in the photoelastic modulators used in modern *LD* and *CD* spectropolarimeters.

Thermotropic LCs exhibit a phase transition into being a LC phase as temperature is changed. Lyotropic LCs exhibit phase transitions as a function of both temperature and concentration of the LC molecules in a solvent (typically water).

In the absence of an electric field, the thermotropic LC molecules in a LC display are aligned parallel with a glass surface which has been prepared by, *e.g.*, polishing parallel scratches into it. Between crossed polaroids its birefringence then yields a bright image. When an electric field is put across the sample gap, the liquid crystal becomes oriented perpendicular to the surface (so-called homeotropic orientation) and the image becomes dark.

in a laboratory axis system by sandwiching the liquid crystal between two quartz plates. This makes a technically very simple means of obtaining an oriented sample for *LD* spectroscopy.

A lamellar phase liquid crystal (extended sheets of amphiphiles separated by thin layers of water) that may be used for *LD* spectroscopy may be formed from a mixture of sodium *n*-octanoate (18.2% w/w)/*n*-decanol (35.5% w/w)/water (46.3% w/w). If this sample is sandwiched between quartz plates, then upon gentle heating (or leaving for a few hours incubation), it forms a system with alternating layers as indicated in Figure 2.13. In this case small hydrophobic solute molecules are oriented parallel to the long hydrocarbon chains just as in a polyethylene film. Ionic species may also become oriented, either parallel to the hydrocarbon chains, for an amphiphilic molecule, or perpendicular to the orientation axis but in the hydrophilic water layer.[54] Since there is no single direction of preferred orientation within the plane of the sandwiched quartz plates, the unique axis is perpendicular to the quartz plates.

A wide range of different molecules can be oriented in such a system: diphenylacetylene (rod-like molecule); biaxial molecules such as anthracene and pyrene behave nearly exactly as in stretched polyethylene; retinal and retinoic acid are anchored in the polar head region, *etc*.[54] Double-stranded DNA oligonucleotides may also be aligned this way if water is added to make the water layers just wide enough.[56]

This lamellar phase was one of the first such system studied and is known as 'Ekwall's system'.[55] It is typical of the structures formed by almost all lipid or lipid-like systems in water where 'infinite' parallel bilayers are separated by layers of water.

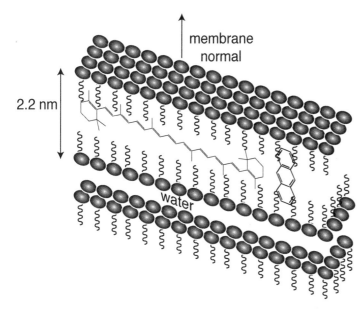

Figure 2.13 Liquid crystal orientation for hydrophobic molecules. Molecules smaller than the bilayer thickness of the illustrated liquid crystal align parallel to the hydrocarbon chains.[54]

Many of liquid crystal systems that have been used for orienting samples are lamellar. Since the orientation in the lamellar system is uniaxial, with the unique axis parallel to the normal of the planes of the bilayers, no preferred direction of orientation is observed within the planes. Therefore, the *LD* is zero when propagating the light through the sample at normal incidence. Instead one has to send in the light at inclined incidence — by tilting the sample — to the light beam of the spectrometer as discussed in Chapter 6, summarized in equation (6.42) and described in reference [52]. The apparent *LD* from solubilized probe molecules at inclined incidence follows a cosine square dependence on the angle of incidence.

Orientation by evaporation or assembly

Many molecules may be oriented simply by evaporating them onto a surface that is transparent to the radiation. This method works particularly well for planar aromatic molecules. However, the plate may need to be tilted for *LD* measurements (as discussed above for lamellar liquid crystals and crystalline samples) since the molecules usually orient preferentially parallel to the surface of the quartz resulting in the unique axis being perpendicular to the plate and all in-plane directions having the same distribution of molecular orientations. Orientation by evaporation has probably been most extensively used for infra-red studies of lipid bilayers and molecules bound to the bilayers.[57] It is also used for oriented *CD* studies of membrane proteins (*cf.* §4.6). Mono-molecular layers of surface-assembled dye molecules may also be studied in this way.[8]

The extent to which the solvent needs to be removed depends on the sample. Most work with lipids usually proceeds by drying the sample then adding a salt solution chosen to give the reqired humidity.[58] However, one can proceed with lipid bilayer membranes and systems that mimic them by depositing them onto a flat surface (usually quartz or glass), often with a second surface placed above them. The lipids spontaneously orient as the humidity is reduced, resulting in bi-layers of lipid with thin water interstitia separating the bilayers. These methods of orienting lipids complement the solution-phase flow-method that both distorts and orients liposomes that is discussed above.

Some molecules such as Alzheimer's fibres are sufficiently large and rigid that simply pipetting them onto a surface from solution may produce a significant degree of orientation.[59]

Classification of orientation methods

The selection of orientation methods described above may be divided into two categories: *equilibrium* and *dissipative*. Equilibrium orientation means that no external energy has to be supplied to the sample and there is no dissipation of the orientation with time. To this category belong crystals, liquid crystals and orientation in (non-dissipating) electric or magnetic fields.

The stretched polymers for film orientation themselves are not at equilibrium, but if the small solute molecules hosted by the polymer may be added afterwards with the same result as when added before — which is the case with stretched polyethylene — their orientation can be considered to fulfil the criterion of local equilibrium.

Dissipative orientation on the other hand means that the orientation is due to some 'dissipative structures' of a relaxing system, generally a system where orientation is induced by addition of energy such as hydrodynamic flow or an electric current. Dissipative orientations therefore decay in the absence of the applied force.

2.4   Experimental considerations for *CD* and *LD* spectroscopy

As with many other analytical techniques, the quality of the data from *CD* and *LD* spectroscopy is directly influenced by two elements: firstly the quality of the instrumentation used to make the measurements and secondly the quality of the sample. *CD* spectra are very easily collected when one has access to a *CD* spectropolarimeter, however, it is not hard to find nonsensical *CD* spectra in the literature. What follows is a description of the instrumentation and how to optimise its performance. In the following sections, outlines of good practice that will help ensure the collection of reliable data are given. An extremely valuable supporting reference is that of Kelly *et al.*[60]

The essential features of a *CD* spectropolarimeter are: a source of (more-or-less) monochromatic left and right circularly polarized light and a means of detecting the difference in absorbance of the two polarizations of light. *CD* spectra could in principle be measured by the analogue of the two-spectra *LD* method, however, *CD* signals are an extremely small (typically one thousandth those from *LD*) difference between two large signals, so this method is no longer used. Thus, the normal method of measuring *CD* is to implement a polarization phase-modulation technique as briefly mentioned above.

A photoelastic modulator (in older instruments a Pockels' cell) produces alternatively right and left circularly polarized light with a switching frequency of typically 50 kHz (in the ultraviolet to near-infrared region of the spectrum). The light intensity is constant, but, upon passage through a sample exhibiting *CD*, an intensity fluctuation (corresponding to the different absorptions of left and right circularly polarized light) that is in phase with the modulator frequency appears. The unabsorbed photons hit a photomultiplier tube which produces a current whose magnitude depends on the number of incident photons. This current is detected by a lock-in amplifier. Thus, while the DC (direct current) component of the photomultiplier current only depends on the total absorption of light by the sample (and on lamp intensity and monochromator characteristics), it is the AC (alternating current) component that relates to the *CD* (the lock-in amplifier is needed to assign the phase of the AC component, *i.e.* the sign of the *CD*). In most instruments the photomultiplier voltage is automatically increased to compensate for variations in the DC and keep it a constant level (*cf.* §10.3).

Most *CD* spectropolarimeters, although they measure differential absorbance, produce a *CD* spectrum in units of ellipticity, θ, in millidegrees, *versus* λ, rather than ΔA *versus* λ. The conversion between

Friedrich Pockels in 1893 found that non-centrosymmetric crystals exhibited birefringence when subjected to an electric field. If one applies an AC voltage to such a crystal, with peak values corresponding to a birefringence that gives quarter-wave retardation, then linearly polarized light will be made to oscillate sinusoidally between opposite forms of circular polarization. In the 1960's J. Badoz[61] invented the photoelastic modulator (PEM), a piece of fused silica in which birefringence is induced by pressure from a piezoelectric crystal.

these two is (*cf.* §11.5):

$$CD = \Delta A \,/\text{(absorbance units)} = A_\ell - A_r$$

$$= \frac{4\pi\theta \,/\text{(degrees)}}{180\ln 10} = \frac{\theta \,/\text{(millidegrees)}}{32{,}982} \tag{2.2}$$

The *CD* version of the Beer-Lambert law (equation (1.5)) is:

$$\Delta A = (\Delta\varepsilon)C\ell \tag{2.3}$$

Equation (2.3) was used to give the $\Delta\varepsilon$ plot of Figure 1.10 since the data were measured with concentrations and path lengths lower than 1 M and 1 cm and rescaled.

'Ellipticity' (and hence millidegrees) dates from the time when the *CD* was measured in terms of the change of polarization of linearly polarized light into elliptically polarized light when it passes through the sample. This was measured by a compensator technique. Today one measures instead changes of intensity, *i.e.*, the true differential absorption of light. Millidegrees remain in wide use simply because the measured numbers are conveniently above unity rather than requiring many decimal places.

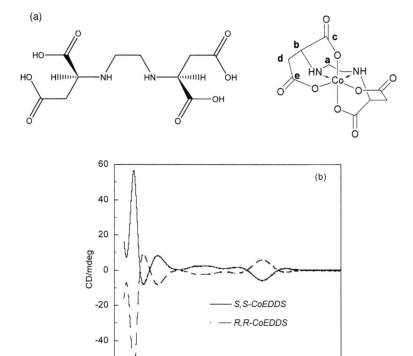

Figure 2.14 (a) The chiral ligand EDDS and its Co(III) complex Co(EDDS) with distinct protons indicated; (b) overlaid *CD* spectra of *R,R*- and *S,S*-Co(EDDS) (0.072 mM, 1 cm path length) illustrating the equal magnitude and opposite sign of the two enantiomers.[62]

Instrument calibration for CD spectroscopy

Wavelength calibration

A simple check of instrument performance is to check its wavelength and intensity accuracy. Traditionally wavelength has been determined using a solid neodynium filter. A scan from say 580–590 nm with a 1 nm bandwidth and data pitch of less than 0.1 should show a maximum in the high tension voltage trace at 586 nm with an error of less than 1 nm. A single wavelength calibration at 586 nm is not entirely convincing, particularly if one only wishes to work in the UV region of the spectrum. Another option is to use holmium which has bands in a more appropriate region of the spectrum. Alternatively one can use the fairly sharp bands of a sample such as the Co(EDDS) complex whose spectra are given in Figure 2.14.

Intensity calibration

Intensity has traditionally been calibrated on *CD* instruments using either d-10-camphor sulfonic acid or the less hygroscopic but spectroscopically equivalent ammonium d-10-camphor sulfonate (ACS). In a 1 cm path length cuvette (after a water baseline has been subtracted, see below) the wavelength and intensity of the longest wavelength peak of a 0.06% w/v ACS in water sample should be 190.4 ± 1 mdeg at 290.5 nm. There is a second negative band with maximum intensity at 192.5 nm, however, there is still some debate as to what its magnitude is (the consensus is that the 192.5 nm:290.5 nm ratio should be about −2.1).[63]

A range of other *CD* standards are available. The Co(EDDS) compound of Figure 2.14 has a set of bands from the visible into the UV region of the spectrum that can be measured with one solution (Table 2.1). Both enantiomers of this compound are available which helps make it obvious if an instrument is out of calibration.

Table 2.1 Extinction coefficients and delta epsilon values for both enantiomers of Co(EDDS).[62] Error quoted is standard deviation.

λ/nm	ε /(mol^{-1} dm^3 cm^{-1})	$\Delta\varepsilon_{S,S}$ /(mol^{-1} dm^3 cm^{-1})	$\Delta\varepsilon_{R,R}$ /(mol^{-1} dm^3 cm^{-1})
545	210±0.7	−2.40±0.05	+2.33±0.05
515	298±1.0		
480		+0.68±0.015	−0.67±0.015
382	103.5±0.5	−0.95±0.025	+0.93±0.025
274	6,136±25	+3.35±0.10	−3.30±0.10
237		+3±0.3	+4±0.3
221.5	18,991±30	+0.665±0.10	−0.675±0.10
210	17,495±35	+24.0±0.3	−22.5±0.3

Path length calibration

A final calibration that is required if short path length cells are being used is to determine their path length (the manufacturer's value is usually only nominal). To assign the path length of a 0.01 mm cuvette, *e.g.*, one prepares a 0.2 M potassium chromate solution. Longer path lengths cuvettes require appropriately diluted samples. The path length is then calculated using the Beer-Lambert Law (equation (1.5)) and the extinction coefficient of potassium chromate which is 4830 mol^{-1} dm^3 cm^{-1} at 372 nm. Interference patterns may also be used to determine cell path length. However, unless the cell is pre-assembled and then the sample loaded it is unlikely that the empty cell path length is the same as the loaded cell path length.

1.942 g potassium chromate and a pellet of potassium hydroxide (to keep the solution stable) in a 50 mL volumetric flask give the required 0.2 M solution.

Parameters for CD spectroscopy

If $b = 2$ nm and $\lambda = 350$ nm, then the light ranges from approximately 348 to 352 nm.

CD spectropolarimeters give the operator considerable control over the time constant, τ, (time over which the machine averages data), scan speed, s, and bandwidth, b (the wavelength range of incident light). The choice of parameters is usually determined by deciding what quality of data is required. If the perfect spectrum for a data base is being collected then operator and instrument time is not a consideration. However, if, for example, a series of 50 spectra is required then minimizing run-time is attractive. Some issues are as follows.

- Signal to noise ratio increases roughly as the *square root* of: the number of scans, the time constant, and the intensity of the light beam.

Some instruments use response time, $\rho = 2\tau$.

- τ should be as large as possible subject to $\tau \times s \le b/2$. If τ is too long for the chosen s and b, then the maxima of peaks (both positive and negative) will be cut off and their wavelengths shifted. A control scan using $\tau' = \tau/2$ (or $s' = s/2$) should be used to check that spectra are not being distorted by the chosen parameters.

The high tension voltage must not get too high, which means that concentration and/or path length must be considered as discussed below.

- A fast preliminary scan will indicate whether there is any advantage in collecting an accurate spectrum and whether the sensitivity scale has been chosen correctly to appropriately display the *CD* at all wavelengths of the spectrum.

- The data interval determines how often a data point is collected. If the instrument works in a stopped-scan mode (it stops at each point to collect data) then this parameter determines the scan speed. Some instruments let one deal with scan speed and data collection time more or less independently, in which case one needs to ensure the spectrum is not distorted by scanning too quickly. The advantage of independent scan speed and time constant parameters is that with broad bands, the user can save time without collecting distorted spectra. However, care must be taken to ensure this is indeed happening.

Titration experiments where spectra are collected as a function of concentration, ionic strength, pH *etc.* often involve adding solution to the cuvette. A simple way to avoid dilution effects is as follows. Consider a starting sample that has concentration x M of species X. Each time $y\,\mu$L of Y is added, also add $y\,\mu$L of a $2x$ M solution of X. The concentration of X remains constant at x M. Other ratios may be readily be devised.

- Sometimes the sample concentration or the time available means the spectra collected are very noisy. There are many options for smoothing the data (some of them within the instrument software). However, unless it is obvious 'to the eye' what the result should be, avoid smoothing the data as you will probably either introduce 'structure' into or remove structure from the spectrum. To ensure the result of smoothing is reasonable, always overlay the noise-reduced result on the original data set and use your eye to decide the validity of the transformed spectrum. If it does not 'look right', reject the result.

Light beam and cells

All of the light beam incident upon the cell must pass through the sample and not be clipped or reflected by the walls or base of the cell or the meniscus of the solution (otherwise the measured spectrum is affected by scattered light). The narrow cells often used to minimize sample volume in normal absorption spectrophotometers *cannot* be used for *CD* unless the light beam is chopped or focused prior to hitting the cell—otherwise light may be reflected by the side walls of the cell. Black quartz masking on cells usually avoids the problem, but should be checked on standards and a sample of similar light scattering propensity as the one of interest. Business cards are ideal for inserting in the light beam at ~550 nm (the green light most easily detected by our eyes) to see the beam width and height. However, note that the beam width is dependent on the instrument slit width, which in turn may be designed to depend on the lamp energy so may be (much) larger in the UV region of the spectrum than at 550 nm. Interestingly, if the beam dimensions in the UV are also required then the same white business card can be used as the card itself will often fluoresce where the beam irradiates the card. If this method does not work at the wavelength of interest then painting the card with a fluorecent dye *e.g.* fluorescein, which produces visible fluorescence when excited in the UV is an option.

Either cylindrical or rectangular cuvettes may be used for *CD*. Cylindrical cells are deemed to have lower birefringence (less distorted baseline *CD*) and may require slightly less sample. Rectangular cells are cheaper, may be used in standard absorption spectrophotometers (so *CD* and normal absorption data may be collected on exactly the same sample), and may be used for serial titration experiments as ~60% of a rectangular cell can be empty for the first spectrum and gradually filled.

If path lengths of 0.1 mm or less are required, it is probably best to use demountable cuvettes where the sample is dropped onto a quartz disk or plate that is etched to a predefined depth and then another quartz disk carefully placed on top. In this case sample recovery is very difficult. The cell holder for demountable cells must be positioned so that the cell is perpendicular to the light beam with the whole incident light beam passing through the sample windows when data are being collected. The path length of demountable cells must be determined independently (see above) and by each user as the volume loaded and the pressure used to assemble them affects the path length.

> Removing the sample holder and checking what the light beam will 'see' is good practice if one is trying to minimize sample volume.

> With care, *CD* data can be collected in an *LD* Couette cell. It is essential that the optical quality of the system is confirmed with a good standard compound, preferably one such as Co(EDDS) (*cf.* Figure 2.14) with multiple positive and negative bands.

Baselines, wavelength ranges, and numbers of accumulations

The baseline in a *CD* experiment is rarely flat as it depends on the intrinsic birefringence of the spectropolarimeter optics and the cuvette. So a baseline spectrum of the solvent/buffer under the same conditions as the sample spectrum using the same cuvette in the same orientation with respect to the light beam should be collected and then the baseline spectrum subtracted from the sample spectrum to produce the final *CD* plot. *CD* spectra often also need to be zeroed so it is essential that data at least 20 nm beyond the normal absorption envelope are available. When the baseline spectrum is subtracted from the sample spectrum, the region outside the absorption envelope should be flat. If it is not then there is some artifact in the spectrum (*cf.* §2.6) or perhaps a transition with low absorbance but large *CD*.

Most *CD* spectropolarimeters have both short timescale (millisecond to minutes) and long timescale (minutes to hours) baseline variations. If the *CD* signal is large both can be ignored. To avoid any problem from short timescale variations, if possible, collect data averaged over a number of

> It is better actively to measure a baseline spectrum for each sample rather than automatically to use a baseline stored in the instrument's memory since any problems with the baseline will adversely affect the *CD* spectrum but may be hidden.

> The magnitudes of baseline variations are usually small on modern instruments and can often be dealt with simply by zeroing the spectra outside an absorption band.

faster scans rather than one slow one. Longer timescale fluctuations can be more problematic and are usually dealt with by alternating collection of sample and baseline spectra.

2.5   Sample preparation for *LD* and *CD* spectroscopy

The quality of the sample used in *CD* and *LD* experiments is something over which the user has significant control. It is true that, no matter how good the instrument is, data collected on a low quality sample will produce low quality data. This part of the chapter will cover the aspects of sample preparation that will effect the quality of *LD* and *CD* data. Some additional issues associated with biological samples are also discussed.

Sample quality

It is essential that every effort is made to ensure only samples of the highest attainable quality are used for *CD* and *LD* measurements. But, what is meant by sample quality? The key objective is to know what is in the sample. Ideally it is just the molecules of interest. Spectrometers are not able to discriminate between the molecule you may be interested in and any other molecules that may be in the solution: the data reported is that of the sample as a whole complete with contributions from contaminants.

Solvent and buffers, for example, may not contribute to the expected *LD* or *CD* (though if they bind to the analytes they may). However, if they absorb light they may cause the high tension voltage to be too high. In this case, the spectrum measured will not be real since not enough photons will reach the photomultiplier tube.

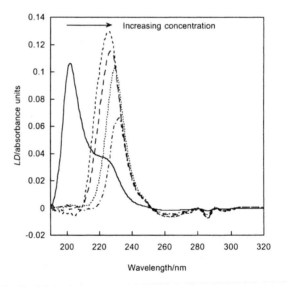

Figure 2.15 Far UV *LD* spectra showing the apparent shift to shorter wavelength of the maximum signal as the concentration of strongly light scattering F-actin fibres (which exhibit strong light scattering) is reduced. F-actin concentrations 93, 74, 62, 53 and 12 *μ*M. The true spectrum is the 12 *μ*M solid line. The high tension voltage of the instrument remained below 600 V (usually considered acceptable) throughout the experiment.[64] See §2.6 for discussions of stray and scattered light.

Some molecules are, however, invisible to *CD* or *LD* and so can be ignored if they do not absorb too much light. In particular achiral molecules are invisible in a *CD* spectrum and unaligned molecules are invisible to *LD* if they do not interact with the analytes of interest to change their molecular or electronic structure.

A second question about sample quality is whether the sample is in the correct state. For example, proteins may have a tendency either to unfold or to aggregate in solution. In the former case the structure of the molecule has changed and so will the spectrum. In the latter case, the aggregates will scatter more light resulting in, at worst, significantly limited ability to record low wavelength data and at best a distortion in the recorded spectrum (Figure 2.15). A simple filtration step prior to measurement (*e.g.* through a 0.22 μm filter) may be all that is needed to substantially reduce scattering of a sample.

Sample concentration

Samples with a low absorbance often (but not always, as discussed in Chapter 9 for magnetic dipole allowed transitions) have a correspondingly lower *CD* or *LD* signals. In extreme cases this results in no signal being detected above the baseline noise. Conversely, too high a concentration can also cause experimental issues since both *CD* and *LD* rely on the transmission of light through the sample. Samples with too high concentrations reduce the amount of light reaching the detector to such an extent that the detector can no longer reliably record data. In some cases achieving an optimum between these two limits is straight forward. However, in others it is not possible to concentrate or dilute a sample. In such cases varying the path length of the cuvette used may solve the problem. A check-list to determine whether the concentrations of the analyte and other sample components is appropriate is as follows.

- Always run a normal absorption spectrum of both the sample and the solvent or buffer (see below) before collecting *CD* or *LD* data, since the concentration and wavelength ranges are the same for both types of spectrum. Do *not* collect the normal absorption spectra against references of the solvent/buffer since *CD* spectropolarimeters are single beamed instruments.

- Sample concentration should usually be such that the total absorbance of the sample is between ~0.2 and ~1.5. Samples with a total absorbance above 2 often give an unreliable *CD* or *LD* signals; an absorbance of 1.1 units is theoretically optimal.

- If the sample plus solvent/buffer has a large absorbance signal, say greater than 1, try using a shorter path length cell together with a more concentrated sample or use a different solvent/buffer (see below). If the solvent/buffer is strongly absorbing then the accuracy of the *CD* spectrum can be significantly reduced.

- Check that the *CD* signal is following the Beer-Lambert law, *e.g.* by using a shorter-path cell or diluting the sample and checking whether the signals reduce in proportion. If not then the *CD* spectrum is unreliable. Most photomultiplier tubes have a voltage above which the Beer-Lambert law will not be followed and below which it probably will. However, note that light scattering samples can give one a false sense of security as scattered rather than transmitted photons reach the photomultiplier tube as illustrated in Figure 2.15 and discussed in

The voltage on the photomultiplier tube (PMT) increases when the number of photons incident on it decreases. The voltage is thus related to the absorbance of the sample. The PMT can therefore be calibrated to give an approximate normal absorbance spectrum.

§2.6. Inhomogeneous samples also often have good PMT readings, but give rise to the phenomenon known as absorption flattening (*cf.* §2.6).

Buffers and solvents

In most cases, samples analyzed by *CD* and *LD* are measured in liquid form, either as a pure liquid or as a solution in an appropriate solvent. It is important to think carefully about the absorbance and any potential *CD* or *LD* of the solvent to be used in the experiment. With biological samples consideration also needs to be given to maintaining the sample in as close to a physiological state as possible which usually entails some form of pH buffering and the addition of salts to maintain a given ionic strength. Indeed the most common buffer used in biology, Dulbecco's phosphate buffered saline (PBS) is a perfect example of an inappropriate buffer for *CD* and *LD* as it contains 136 mM NaCl, 2.7 mM KCl, 0.6 mM $MgCl_2$, 0.9 mM $CaCl_2$, 10 mM Na_2HPO_4, and 2 mM KH_2PO_4. The Cl^- ions are particularly problematic as they absorb light from about 215 nm downwards. For example, PBS in a 0.1 mm path length cell only transmits light to ~200 nm in a bench-top *CD* machine. A simple resolution to this issue is usually to replace Cl^- with F^-, as fluoride ions do not absorb in the same region of the spectrum. Unfortunately chloride ions are not the only problem in biological buffer-design; other common additives, such as dithiothreotol, ethylenediaminetetra acetic acid (EDTA), guanidinium hydrochloride and to a lesser extent urea, all absorb in the far UV region of the spectrum. It is therefore important that the concentrations of these compounds are minimized in buffers used for *CD* and *LD*. As a general rule, if a new buffer is being used for an experiment it is helpful to record a spectrum (in effect a baseline), without any analyte present, in the *CD* instrument. The wavelength cut off from this experiment provides the lower wavelength limit for any experiment that includes this buffer.

2.6   Spectral artifacts

Solid samples

When measuring *CD* in a KBr tablet, of the kind often used for IR spectroscopy, great care has to be taken to avoid artifacts both due to light scattering (*cf.* §2.6) and to birefringence (the tablet is made at high pressure which may cause strain in the matrix). A test of the polarization quality of the light may be performed by putting a solution, with a known high intensity *CD* spectrum where the KBr disc sample does not absorb, in the light path succeeding the solid sample. Any decrease in *CD* intensity due to the KBr disc sample directly reflects the effect of birefringence. For example if the *CD* measured with the KBr disc is 90% that without it, then that means the KBr is causing a 10 % depolarization of the light. If the *CD* = 0 then the KBr disc is adding a quarter wave retardation, similarly if the *CD* has the same magnitude but inverted sign, the KBr disc sample is providing a half-wave retardation. It is also important to test the preparation methodology by comparing the resulting spectrum with a test sample that has a *CD* similar to

Note that even if the circularly polarized light is only partially depolarized by scattering and birefringence, the resulting perturbations to the baseline can be significant.

that of the solid sample, in which case additional baseline problems can be revealed.

Low-temperature and high-pressure measurements

CD experiments run at low temperature often display spurious birefringence artifacts. The test described above, of putting a sample with known *CD* after the cryostat, is an effective way to find out if the optics meet the requirements for trustworthy measurement or if artifacts may be involved. There are some cases in the literature where high-pressure effects reported for chiral substances are almost certainly artifacts due to strain birefringence of the anvil windows.

Light scattering samples

Many of the samples for which we wish to measure *LD* and *CD* data are of comparable size to the wavelength of light. As a result they often scatter light significantly. Furthermore, as well as differentially absorbing the two incident polarizations of light (the basis of the *LD* and *CD* signal), the samples may also differentially scatter the light. Flow *LD* may suffer additional scattering artifacts if the flow is not laminar. The theory of light scattering is complicated, depending not only on size regime of the particles but also on their shape. Thus, if at all possible, one should avoid it occurring rather than trying to correct for it. In general, scattering can be reduced either by reducing the size of the particles or collecting a high percentage of the scattered photons. However, these options may either be insufficient or unavailable.

In general we may write for *LD* (and equivalently for *CD*)

$$LD^{Observed} = LD^{Absorbance} + LD^{Scattering} \qquad (2.4)$$

where the scattering contribution is usually apparent as a sloping baseline outside absorbance bands. We generally only want the absorbance contribution. Nordh *et al.*[65] showed that a simple empirical correction can often be subtracted from the observed *LD* spectrum to remove the sloping baseline [66] as illustrated in Figure 2.16.

$$LD^{Scattering} = a\lambda^{-k} \qquad (2.5)$$

Three methods that may enable scattered photons to be collected (and thus to avoid the problem of scattering) are to (i) have a wide angle photomultiplier tube (PMT) or (ii) place the sample very close to the photomultiplier tube or (iii) have a collecting lens close to the sample to refocus the scattered light onto the PMT.

Stray light

In addition to any light scattering from the sample, so-called 'stray light' from anywhere in the system may cause additional artifacts especially at lower wavelengths for xenon lamp sources where the instruments are struggling for light intensity. The main problem emanates from imperfections of the monochromator: long-wavelength stray light bypasses the monochromator (wavelengths of light that the sample may not even absorb). In such a case, a significant percentage of the light that does reach the photomultiplier tube may be stray light, probably not even of the correct wavelength, rather than unabsorbed photons that have passed through the sample.

An indication that such a problem is occurring is that the high tension voltage trace on the PMT gradually approaches the maximum rather than showing a sharp clean cut-off.

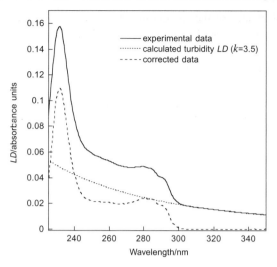

Figure 2.16 A method of light scattering[65] correction applied to an *LD* spectrum of polymerized tubulin (Figure 3.26): the experimental data (——); the calculated turbidity (or scattering) *LD*, using a *k* value of 3.5, with *a* determined by rescaling the curve at 320 nm where there is no intrinsic absorbance (----); and the corrected data (– – –).[67]

Another source of stray light is from room light let in through holes in the spectrometer casing.

All photons get counted as unabsorbed photons of the monochromator's wavelength, causing the photomultiplier to register less absorption, *CD*, or *LD* at the monochromator wavelength than there really is. If the spectrum is 'true' it will follow the Beer-Lambert Law (equation (2.3)). A sharp drop off in intensity on the low wavelength side of a band (Figure 2.15) is often indicative of this problem. However, sometimes, as illustrated in Figure 2.17, the *LD* (or *CD*) spectrum gives no indication that there is a problem until one overlays it with data from other instruments or lower concentration/path length samples, whereupon it becomes apparent that the signal has been seriously attenuated.

Absorbance flattening

The phenomenon of absorption flattening is a suppression of the absorbance signal in regions of high absorbance in non-homogeneous samples causing the Beer-Lambert law to break down. The same issues apply to *LD* and *CD* but the situation is worse as discussed below. [68, 69]

To understand this phenomenon, consider the extreme case where half the light beam passes through pure solvent and half passes through a more dense solution of analytes. Half of the incident photons ($I_o/2$) will thus reach the PMT directly, while the other half will pass through a solution of higher absorbance. The percentage transmittance of the sample is then

$$T = 100\% \times \left(\frac{I_o + I}{2I_o}\right) \tag{2.6}$$

where $A = \log_{10}(I_o/I)$ is the absorbance of the absorbing half of the sample and $\log_{10}(I_o/I_o) = 0$ is the absorbance of the transparent half of the sample. The measured absorbance of the sample may be significantly reduced compared to the true absorbance had the sample been homogeneous:

$$A_{measured} = \log_{10}\left(\frac{2I_o}{I_o+I}\right) = -\log_{10}\left(\frac{1+I/I_o}{2}\right) \qquad (2.7)$$

Thus, for absorbance in the denser part of the sample of 3 or greater, $A_{measured}$ is equal to $\log_{10}(2) = 0.301$. For intermediate and small absorbances the effect of light leakage is significant, as illustrated in Table 2.2.

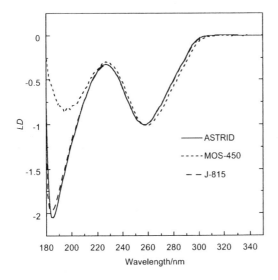

Figure 2.17 *LD* of DNA from *M. luteus* (200 μM) in water measured on three spectrometers: a Biologic MOS-450 spectrometer equipped with degraded mirrors, representing previous generations of bench-top spectrometers; a Jasco J-815, one of the latest spectrometers available (optimized for far UV performance); and the ASTRID UV1 synchrotron beamline.[70]

Table 2.2 Absorbance flattening effects for samples where the light passes through a sample of which half has no absorbance and half has the indicated absorbance.

A of clear part of the solution	A of dense part of solution	$A_{apparent}$ of inhomogeneous solution	A for the analogous homogeneous solution	Error
0	3	0.30	1.5	80%
0	2	0.29	1.0	71%
0	1	0.26	0.5	48%
0	0.5	0.19	0.25	24%

Correcting absorbance flattening

Ideally absorbance flattening can be avoided (see below), however, this is not always possible. This artifact is also known as Duysens flattening in recognition of his work on methods for correcting for it.[71] Duysens considered a suspension containing cubic particles of size d, each particle having an average transmittance

$$T_{particle} = e^{-\alpha d} \tag{2.8}$$

(here expressed using base e, rather than the more common base 10). Duysens divided the solution volume into a cubic network of boxes, each box being either empty (containing only solvent) or containing one cubic particle. If the length of the light path is ℓ, the number of boxes that the light passes is $m = \ell/d$. Due to the possibility that light might sneak through the solution without passing a statistically constant number of particles, the transmission of the inhomogeneous suspension, $T_{suspension}$, can be shown to differ from that expected for a true solution according to:

$$T_{suspension} = \left[1 - q(1 - e^{-\alpha d})\right]^m \tag{2.9}$$

where q is the probability that a box contains a particle and $(1 - q)$ is the probability that it will contain only solvent. It is generally realistic to assume that $q \ll 1$ and that

$$A_{suspension} \approx qm(1 - T_{particle}) \tag{2.10}$$

Then

$$A_{suspension} = \alpha q \ell Q = A_{homogeneous}Q \tag{2.11}$$

where Q is the reduction factor. When $Q = 1$, we have the normal Beer-Lambert law.

Flattening in LD and CD

If dichroism (*CD* or *LD*) is measured with the standard phase-modulation technique, the Duysens flattening effect becomes even stronger than for normal absorbance. This results from the scaling of the alternating (AC) light intensity by the apparent average (DC) light intensity (*cf.* §10.3) performed by the instrument. Because the problem arises for inhomogeneous samples, often the overall intensity of light collected by the PMT seems satisfactory but a true signal is not being obtained. Membrane samples such as liposomes (Figure 2.7) are particularly problematic since the analytes are often close-packed in regions within a liposome rather than being uniformly distributed even within the liposomes which are themselves a partition of the solution. The issues also frequently arise for fibres where the chromophores are close-packed.

This issue was illustrated in reference[69] for flow-orientation of large lecithin liposomes containing high concentrations of membrane-bound β-carotene: the strongly positive *LD* of the 450 nm long-axis polarized band becomes strongly suppressed compared to the normal absorption band. An alternative approach is illustrated in reference [72].

Reference [69] advises a method for correcting the measured *LD* by computing the ratio of correction factors $Q(LD)/Q(A)$ between *LD* and absorbance spectra. The ratio should be determined from a low absorbing part of a band in the spectrum belonging to a single transition polarization (so the *LD/A* ratio is constant). With large liposomes having $d = 10^{-4}$ cm and

$\alpha = 10^5$ cm^{-1}, the correction factors for absorbance and linear dichroism for the β-carotene case are as given in Table 2.3.

Table 2.3 Computed absorbance flattening effect on the *LD* signal.[69] *LD* is measured as the ratio of AC and DC currents from a PMT kept at constant voltage. $Q = 1$ in the absence of any flattening.

$A_{homogeneous}$	αd	$Q(LD)$	$Q(A)$	$Q(LD)/Q(A)$
3.3	4.00	0.018	0.245	0.074
3.0	3.64	0.026	0.267	0.098
2.5	3.03	0.048	0.314	0.154
2.0	2.42	0.089	0.376	0.236
1.5	1.82	0.162	0.460	0.352
1.0	1.21	0.298	0.580	0.514
0.5	0.61	0.543	0.749	0.726
0.25	0.30	0.741	0.864	0.858
0.10	0.12	0.886	0.951	0.930

For membrane and fibre systems, one somehow has to remove the local high concentrations. Mao and Wallace have outlined how this may be done by reducing membrane particle size.[74] Unfortunately, the problem cannot always be avoided. In which case methods such as those of references [68, 69] can be used to correct measured data.

Artifacts in *CD* data due to imperfect optics: *LD* artifacts in *CD* spectra

In a perfect *CD* spectropolarimeter pure left or right handed circularly polarized light is incident on the sample. However, the production of these polarizations is seldom if ever perfect. In most commonly available *CD* instruments, as discussed further in §10.2, the circularly polarized light is produced by the sequential action of two optical elements, a plane polarizer and a photoelastic modulator (PEM, Figure 10.3). In an ideal optical system the light after the polarizer would be perfectly polarized. The PEM then acts to modify the polarized light in a cyclic fashion (governed by an applied alternating voltage) in order to produce the required linear or circularly polarized light. Unfortunately in reality the linear polarization is seldom perfect which leads to 'contamination' of the alternating polarizations produced by the PEM. The oscillation of the PEM adds to the problem, with the net result that when measuring *CD* we are not measuring the difference between pure left and right circularly polarized light but the difference between left (with a small percentage of linearly polarized light) and right (with a small percentage of linearly polarized light).

One of the first detailed analyzes of how non-ideal optical parameters including static birefringence affect the function of a *CD* spectrometer was performed using Mueller-Stokes matrix algebra.[75] The analysis shows how any spurious birefringence makes *CD* measurement extremely sensitive to presence of *LD* in the sample. It also shows what minimal corrections have to be made when measuring *CD* on an anisotropic sample. The combined static birefringence of both the sample and the modulator system is the crucial parameter needed in order to correctly identify the true *CD* in an oriented *CD* experiment (*cf.* §4.6).[76] The birefringence has to be characterized with respect to both its amplitude and direction—which is a challenge for samples in which the *CD* is small compared to any *LD* that is present because the axis of birefringence becomes increasingly difficult to assess in the limit of vanishing birefringence. This leads to an 'uncertainty principle' for measuring oriented *CD*.[76]

Experimental considerations to avoid LD artifacts in CD spectra

Birefringence in the spectropolarimeter optics is generally not an issue with a solution of isotropically (randomly) distributed molecules. However, if some molecular alignment exists in the sample, the *CD* spectra that are collected may contain significant elements of *LD* because *LD* signals are typically several orders of magnitude larger than the *CD* signals. Therefore if the contamination is only 1%, but the *LD* signal is 100 times the magnitude of the *CD* signal, then the *LD* magnitude will be equal to that for the *CD*.

Concerns about aligned samples clearly arise for oriented *CD* experiments (*cf.* §4.6). However, they may also be relevant for *CD* spectra collected using very low path length cuvettes (typically 0.01–0.1 mm). The process of forcing the sample into the narrow space, whether this is by pushing sample in or assembling a demountable cell, can lead to alignment of *e.g.* protein fibres or hydrogels.

Given that considerable inadvertent alignment can occur for fibrous molecules, it becomes important that the presence of such alignment is monitored. Detecting the presence of particles that are anisotropically distributed in a sample turns out to be quite simple. If a sample is truly isotopically distributed then the *CD* signal should be independent of cuvette orientation. So, if the sample is rotated about the light beam axis then the *CD* spectrum should remain unchanged. Therefore the easiest way of detecting anisotropy-induced artifacts is to take two spectra of the sample with the sample oriented orthogonally with respect to the light beam. If these spectra are not identical then there is an artifact in the *CD* spectrum. This issue is considered further in §4.6.

Micropipettes, and indeed glass pipettes, are calibrated for water. To have any hope of measuring accurate volumes with viscous solutions one must pipette very slowly. Cutting the tip may also affect the calibration. To investigate this, weigh samples of known density, or alternatively use a positive displacement pipette.

If a sample is prone to alignment, there are a number of methods that may be sufficient to disrupt or prevent the alignment, these include the following.

- Use of a cuvette with longer path length, typically greater than 1 mm. This reduces shear induced alignment during loading and hence produces a more isotropic sample.
- Reduce the sample concentration since many fibrous systems are inherently viscous and high viscosity systems (as discussed above) are more prone to shear induced alignment than those with low viscosity.
- Collect data as a function of time since some forms of alignment are not stable and decay over a period of minutes or hours.
- In some instances the very act of pipetting the sample into the cuvette can induce alignment, this effect is due to the combination of the pipette tip shape and the size of the tip orifice. In many cases this effect can be negated simply by cutting the tip so that the orifice increases in size.

2.7   Instrumentation considerations

Absolute calibration of *LD* and *CD* spectrometers

Whereas *CD* is calibrated most often using a chemical (chiral) substance (*cf.* §2.4) with established molar ellipticity, *LD* is usually calibrated using a physical device, such as a polarizer or a slab of (isotropic) fused silica tilted 10–45° from normal incidence. From knowledge of the refractive index of silica, the reflection of the tilted plate of silica may be calculated accurately for two orthogonal light beam polarizations, including the effects of multiple reflections, as described in [30]. The calibration is needed for two reasons.

(1)   At low *LD* amplitudes, calibration provides the accurate numerical correlation between instrument signal (photomultiplier current ratio <AC>/<DC>) and the true *LD* of the sample.

(2)   At high *LD* amplitudes (*LD* > 0.1), calibration provides the correction that needs to be applied to *LD* signals when the deflection of <AC>/<DC> as a function of *LD* no longer follows a linear relationship (*cf.* equation (2.12) which is linear in *LD* for small values). For a strongly oriented sample with moderate absorption, *LD* easily exceeds the 0.1 limit, so for accurate quantitative analysis an appropriate careful calibration is required.

In contrast to normal *CD* calibration based on empirical use of a chemical substance to set the scale of the instrument, the *LD* calibration outlined below is absolute. If the *LD* instrument is created from a *CD* instrument plus quarter wave plate (*cf.* §2.2), the *LD* calibration automatically provides a calibrated *CD* instrument in a non-empirical way once the polarization of the instrument is changed from linear polarization to circular polarization. Indeed, as demonstrated in reference [77] a stretched film of PVA containing acridine orange combined with a quarter-wave film at 45° may be used for such a non-empirical calibration of *CD*. The *LD* calibration methodology has also been used to check the enantiomeric purity of two commercial *CD* standards, androsterone and $[Co(en)_3]^{3+}$ which were both found to be only 99.5% enantiomerically pure.[30]

> To calibrate *CD* and *LD* for an instrument where the transition between the two *modus operandi* is achieved by changing the voltage on the photoelastic modulator (PEM), first use a quarter wave plate to change between *CD* and *LD* and calibrate as indicated here. Then check the *LD* measured with a quarter wave plate is the same as *LD* with double the PEM voltage.

High amplitude CD and LD

Both *CD* and *LD*, as measured by the phase-modulation technique, are only linear functions of the true sample dichroisms at small amplitudes. With normal *CD* samples, whose ΔA is generally much less than 0.1, this is never a practical problem; however, for *LD* applications the deviation from linearity, which becomes significant above $\Delta A = 0.1$ absorbance units, generally has to be corrected.

As was first shown in references [29, 30], the instrument signal, LD_i, recorded on the spectropolarimeter, is related to the true *LD* according to the following relation:

$$LD_i(\lambda) = K_1(\lambda) \frac{\tanh\left(\ln 10 \frac{LD(\lambda)}{2}\right)}{\left[1 + K_2(\lambda)\tanh\left(\ln 10 \frac{LD(\lambda)}{2}\right)\right]} \tag{2.12}$$

Its remarkably simple inverse is the dichrometer correction formula:

$$LD(\lambda) = \log\left[\frac{\left(K_1(\lambda) - K_2(\lambda)LD_i(\lambda) + LD_i(\lambda)\right)}{\left(K_1(\lambda) - K_2(\lambda)LD_i(\lambda) - LD_i(\lambda)\right)}\right] \tag{2.13}$$

Here the instrument parameters $K_1(\lambda)$ and $K_2(\lambda)$ in principle depend on wavelength and take care of both the non-linearity of the instrument response as well as any non-ideal polarization due to non-achromacy of quarter wave device *etc*. They are mathematically defined by the inherent effective birefringence of the photo-elastic modulator (including any quarter wave device), according to relations given in references [10, 29, 32] but may be easily calibrated experimentally. Typical values determined for a Jasco spectropolarimeter having a photoelastic modulator supplemented with a high-quality Oxley quarter-wave prism were found to be: $K_1 = 0.879 \pm 0.01$ and $K_2 = -0.103$ between 200 nm and 700 nm, thus, practically wavelength independent.

At small *LD* values, the K_2 terms in equation (2.13) may be neglected. At intermediate *LD* values K_2 is most easily estimated from the asymmetry recorded on the instrument when turning an *LD* standard from parallel to perpendicular orientation (from LD_i^+ to LD_i^-):

$$K_2 = \left[\tanh\left(\ln 10 \frac{LD(\lambda)}{2}\right)\right]^{-1}\left(\frac{1+R}{1-R}\right) \tag{2.14}$$

where $R = LD_i^+/LD_i^-$ is the ratio of recorded positive and negative *LD* values and *LD* is the true dichroism.

This asymmetry becomes negligible if *LD* is measured by driving the photoelastic modulator to peak at half-wave retardation and the *LD* is measured at twice the modulation frequency, provided there is no static birefringence in the modulator.

Note that not only does equation (2.13) account for non-linearity of the *LD* response, it also includes the asymmetry arising from the fact that the modulator (photoelastic or Pockels) and any static quarter wave device (or stray birefringence) generally define a chiral system, thereby making the *LD* response asymmetric: positive and negative *LD* signals are non-linear to different degrees.

For practical purposes, and small *LD* signals, the simplest way to calibrate *LD* is to measure the *LD* of a fused silica plate tilted at 45°. As computed in reference [30] the *LD* should then be as summarized in Table 2.4.

Table 2.4 *LD* of a fused silica plate tilted at 45°.

λ/nm	200	250	300	350	400	500	550	600	650	700
LD	0.0812	0.0740	0.0708	0.0689	0.0678	0.0664	0.0659	0.0656	0.0653	0.0650

3 Linear dichroism of biological macromolecules

DNA is a long fairly rigid polymer that is easy to align by flow, so LD can be used to probe its structure and binding with small and large ligands. This is the subject of the first part of this chapter. Applications for fibrous proteins and membrane peptides and proteins are described. In this chapter, the LD of oriented of membrane proteins attached to model membrane systems (liposomes) that can be distorted and aligned by flow is also described.

Linear Dichroism and Circular Dichroism: A Textbook on Polarized-Light Spectroscopy
By Bengt Nordén, Alison Rodger and Timothy Dafforn
© B. Nordén, A. Rodger and T. Dafforn, 2010
Published by the Royal Society of Chemistry, www.rsc.org

3.1   Introduction

This chapter is organized around a selection of relatively straightforward examples to show how *LD* can be used to study macromolecules of biological interest. DNA (Figure 3.1) and DNA-ligand systems illustrate how *LD* may be used to probe the structure of a biomacromolecule, its conformational changes and the binding of ligands. *LD* is also the ideal technique for studying membrane peptides inserting into membranes. The use of *LD* to understand the structures of a selection of protein fibres concludes the chapter. Chapter 5 contains further examples where the analysis requires more advanced understanding of *CD* and *LD* spectroscopy.

(a)   (b)   (c)

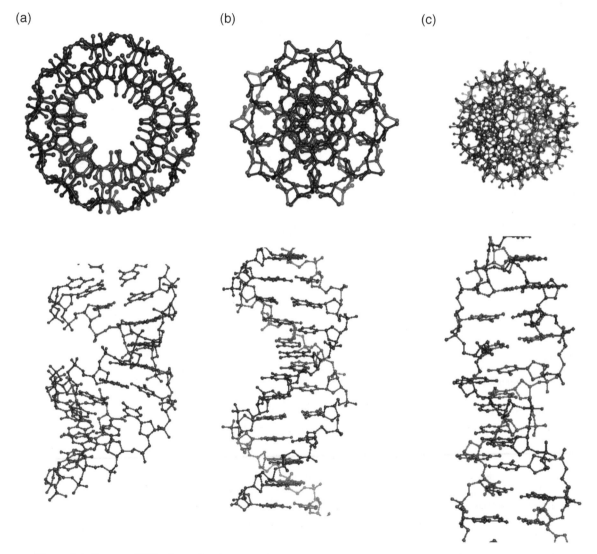

Figure 3.1 Common DNA polymorphs viewed from above and from the side. (a) A-DNA; (b) B-DNA; and (c) Z-DNA.

3.2   Qualitative analysis of *LD*

LD is a conceptually simple technique that gives an indication of relative orientations of chromophores in oriented samples. With biomacromolecules the interpretation is frequently at the level of using the sign of the *LD* to give a qualitative answer as discussed in §1.4. For example, a casual glance at the DNA *LD* spectrum of Figure 3.2 indicates that the π-π* transitions of the DNA bases at 260 nm (§1.3) are more perpendicular than parallel to the DNA helix axis. The transitions of the intercalator ethidium bromide give extra negative *LD* intensity below 300 nm and a new negative *LD* signal above 300 nm (Figure 3.2a) indicating that its transitions are approximately parallel to the DNA bases. By way of contrast, the long axis polarized transition of the minor groove binder Hoechst 33258 gives a positive signal at 350 nm (Figure 3.2b) which means it cannot be intercalated between the base pairs. These examples are discussed further below.

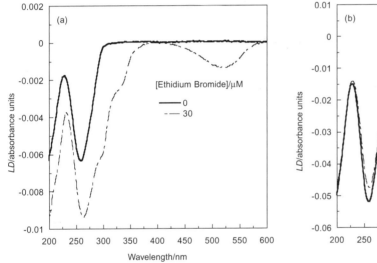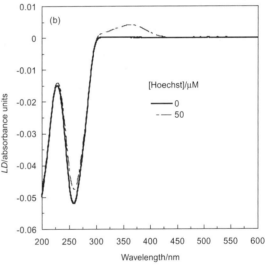

Figure 3.2 Linear dichroism in a 1 mm path length Couette flow cell of (a) calf thymus DNA (200 µM, 20 mM NaCl, pH = 7) and ethidium bromide (concentrations shown on the figure), and (b) calf thymus DNA (1000 µM, 20 mM NaCl, pH = 7) and Hoechst 33258 (concentrations shown on the figure).

A formal mathematical framework for *LD* is presented in Chapter 6 where the focus is on small molecules with single well-defined chromophores. The relevant results are summarized here. *LD* is fairly easy to understand if one considers it in two parts. The first part is to imagine that the molecules are perfectly oriented (as in *e.g.* a crystal), in which case the key parameter is what angle the transition moments of interest make with the local molecular orientation axis (usually the long axis of the molecule). The second part is to realize that the local orientation axes are distributed about the laboratory defined (//, *Z*) orientation axis (*e.g.* the stretch or flow direction). For many situations, we can make the not entirely correct assumption that the orientations of the molecules are uniformly distributed about the orientation direction (so-called uniaxial orientation) with the orientation described by a

The magnitude of an *LD* signal depends on how good the orientation is. Even though Couette flow cells do not provide macroscopically uniaxial orientation (see below), the helical symmetry of DNA and many other samples about their long axis make them effectively uniaxially oriented about their long axis. Equation (3.3) can then be applied.

simple parameter S. Then, as illustrated in Figure 1.9, if a transition is polarized parallel to the local molecular orientation axis

$$LD_{//} = 3A_{iso}S > 0 \tag{3.1}$$

where A_{iso} is the absorbance of an unoriented sample. The factor of three arises because of the three dimensional average of the unoriented sample's isotropic absorbance. If the transition is perpendicular to the orientation direction then

$$LD_{\perp} = -\frac{3}{2}A_{iso}S < 0 \tag{3.2}$$

More generally, for a transition polarized at an angle α to the molecular orientation direction (usually its long axis)

$$LD = \frac{3}{2}A_{iso}S\left(3\cos^2\alpha - 1\right) \tag{3.3}$$

$3\cos^2\alpha = 1$ when $\alpha = 54.7°$, the so-called magic angle.

S is the orientation parameter for the local molecular orientation axis around which the orientation of the transition moment is locally uniaxial. $S = 1$ for a perfectly oriented sample and 0 for an unoriented one. Equation (3.3) is derived in §6.2.

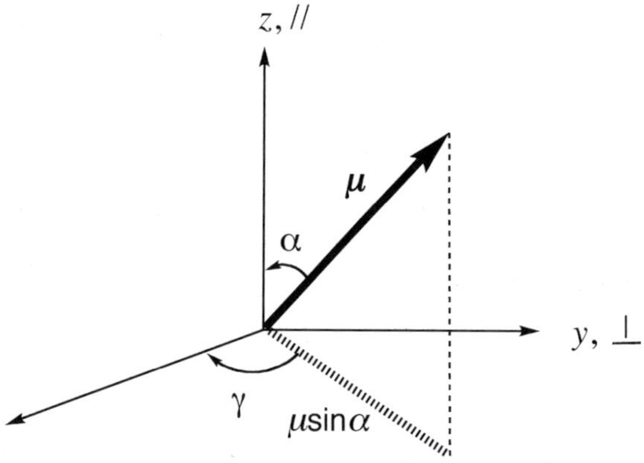

Figure 3.3 The orientations of an electric dipole transition moment, μ, (whose length/magnitude is μ) within a perfectly oriented molecular system where $Z = z$. The magnitude of the projection of μ onto the Z axis is $\mu\cos\alpha$ and onto the X/Y plane is $\mu\sin\alpha$. The Y component of this projection is $\mu\sin\alpha\sin\gamma$.

3.3   Uniaxial orientation

Uniaxial orientation requires an oriented system in which the transition moment lies at an angle α relative to a local reference (or orientation) axis in such a way that all orientations on a cone around this reference axis are equally probable (Figure 3.4). This is the case, for example, for molecules in most stretched films and liquid crystals, and for helical structures such as DNA in flow in which chromophoric units are wound around a central helix axis when one passes from one unit to the other in the structure.[12] The condition of uniform distribution around the reference axis may also be fulfilled for dynamic reasons, such as the free rotation about the backbone of a linear polymer molecule, or due to the symmetry of the orientation methodology about the orientation axis.

Uniaxial orientation

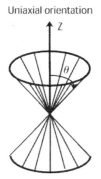

Figure 3.4 Uniaxial orientation.

In cases where the sample fulfils the requirement of being macroscopically uniaxial, such as a polymer film drawn in one direction, or polar molecules in an electric field, or a sample flowing through a cylindrical tube, there is a simple relation beween $A_{//}$, $A_\perp$, and A_{iso} which makes it unnecessary to measure all three quantities:

$$A_{iso} = \frac{1}{3}\left(A_{//} + 2A_\perp\right) \tag{3.4}$$

In such cases, S may be expressed as an average over the orientation distribution of the angle θ between the local orientation axis and the macroscopic orientation direction:

$$S = \frac{1}{2}\left|3\left\langle\cos^2\theta\right\rangle - 1\right| \tag{3.5}$$

where θ is as illustrated in Figure 3.4.

Equation (3.3) also holds for orientation cases of lower macroscopic symmetry than uniaxial, such as in a Couette flow field. In this case S is a more complicated average (over two angles for Couette flow orientation) but still a scalar parameter $0 \le S \le 1$, which may be determined for example from experiment if LD of an absorption band with known α can be measured.

3.4 Reduced *LD*

We may also rewrite equation (3.3) in terms of a 'reduced *LD*'

$$LD^r = \frac{LD}{A_{iso}} = \frac{A_{||} - A_\perp}{A_{iso}} = \frac{\Delta\varepsilon}{\varepsilon} = \frac{3}{2}S\left(3\cos^2\alpha - 1\right) \tag{3.6}$$

LD^r has the advantage of being independent of concentration and path length. Transitions of the same polarization will have the same magnitude LD^r signals. For the example of Hoechst in Figure 3.2, this means that the Hoechst long axis (the polarization direction of its lowest energy transition)

'Environment wiggles' in electronic LD^r are most pronounced for sharp vibrational structure such as the vibronic band sequence at 310–360 nm of pyrene. The L_b fine structure of benzene in some liquid crystals also exhibits remarkable variations, with sharp positive and negative LD peaks evidencing the occurrence of almost orthogonal orientations.[78]

is oriented at less than 54.7° from the helix axis. The 45° angle of a DNA minor groove binding site is consistent with this.

For flexible molecules such as DNA, the angle between a transition moment of interest and the molecular orientation axis may not be the same for all molecules at all times. In that case, $\cos^2\alpha$ above should be replaced by its average, $<\cos^2\alpha>$. We often assume (not necessarily correctly) that $<\cos^2\alpha> = \cos^2<\alpha>$.

If the variance of α is large and it is not near 0 or 90°, $<\cos^2\alpha>$ will differ substantially from $\cos^2<\alpha>$. The 'safe' zone is approximately where we may write $\cos^2\alpha\sim1-\alpha^2$ (for α in radians). Thus we are fairly safe for α within 30° of 0 or 90°.[79]

3.5 Nucleic acids

DNA geometry and spectroscopy

Figure 3.2 illustrates how *LD* can be used to probe DNA structure and ligand binding. In this section more detail is given. To a first approximation, double helical DNA (or RNA) can be viewed as a more-or-less vertical spiral staircase where the steps are made of pairs of purine and pyrimidine nitrogenous bases that are hydrogen bonded together. The support is the backbone of alternating phosphate groups and ribose sugars that adopt a helical structure (Figure 3.1).

Table 3.1 Typical geometric parameters (*cf*. Figure 3.6 and Figure 3.5 for definitions and Figure 3.1 for structures) for standard A-, B-, and Z-forms of DNA.[20, 80, 81] The base pairs in Z-DNA alternate between the values given in column 4 giving it a zig-zag appearance.

A nucleotide is a base plus sugar plus phosphate; a nucleoside is a base plus sugar.

Parameter	A-DNA	B-DNA	Z-DNA
α	−85°	−47°	60°/160°
β	−152°	−146°	−175°/−135°
γ	46°	36°	178°/57°
δ	83°	156°	140°/95°
ϵ	178°	155°	−95°/−110°
ζ	−46°	−95°	−35°/−85°
sugar conformation	$C_{3'}$-endo	$C_{2'}$-endo	$C_{3'}$-endo/$C_{2'}$-endo
glycosidic bond	*anti*	*anti*	*anti* (C), *syn* (G)
base roll (twist)	12°	0°	1°
base tilt	20°	5°	9°
base helical twist	32°	36°	11°/50°
base slide	0.15 nm	0 nm	0.2 nm
helix diameter	2.55 nm	2.37 nm	1.84 nm
bases/turn of helix	11	10	12
base rise/base pair	0.23 nm	0.33 nm	0.38 nm
major groove	narrow, deep	wide, deep	flat
minor groove	broad, shallow	narrow, deep	narrow, deep

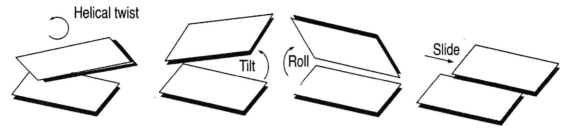

Figure 3.5 Helical twist, tilt, and roll (alternatively called twist), and slide.

Geometric parameters for different polymorphs of DNA are given in Table 3.1. RNA differs from DNA in that every thymine (T) is replaced by uracil (U) and the backbone sugar is ribose instead of deoxyribose. When RNA bases pair up to form a double helix, they adopt the A-form geometry. The repeat unit of Z-DNA is a dinucleotide, necessitating two values of each angular parameter. ζ adopts a range of values in Z-DNA.

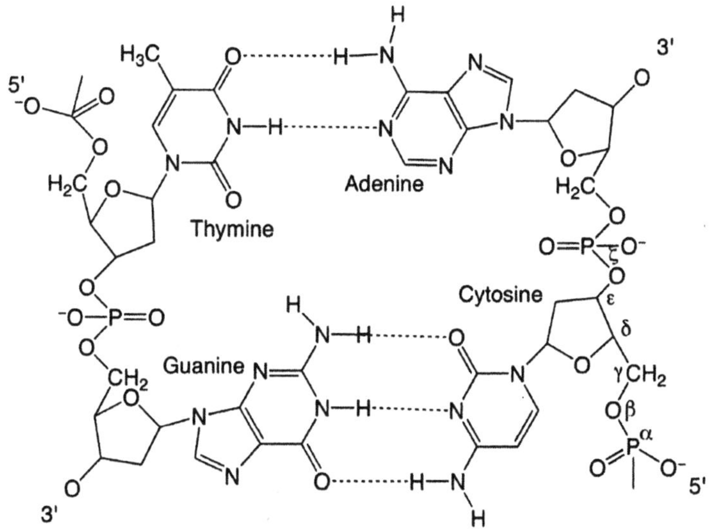

Figure 3.6 DNA nucleotide bases indicating backbone parameter labels used to describe the orientation of DNA bases with respect to one another.

Figure 3.7 shows the absorbance, *LD* and *LD^r* spectra of calf thymus DNA in a Couette flow cell (Figure 2.6). The transitions are all in the plane of the DNA bases so approximately perpendicular to the helix axis. The *LD* is therefore approximately a negative version of the absorbance spectrum.

The fact that the B-DNA *LD* spectrum is similar but not exactly the same shape as −*A* results in the *LD^r* spectrum of Figure 3.7 being fairly constant across the 260 nm band, but not completely so. This tells us that, in solution, the bases are not rigidly perpendicular to the helix axis as implied by Table 3.1. From a careful analysis of solution *LD* studies it has become apparent that the bases of B-DNA in solution lie at an average angle of ~80° (or even

There may be some $n \rightarrow \pi^*$ transitions present in the DNA base spectra whose intensity is so small that they have not been definitively identified.

less) from the helix axis[20] (though we typically assume 86° in a calculation [82]). By way of contrast, the almost perpendicular orientation seems to apply to DNA in stretched films.[35] This is discussed further below and in §5.13.

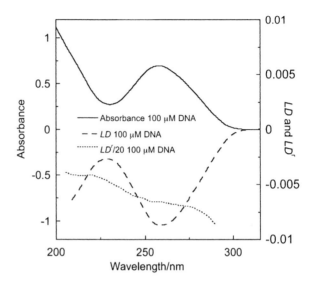

Figure 3.7 Absorbance, *LD* and *LD*[r] spectra of calf thymus DNA (100 μM) in a 1 cm path length cuvette and a 1 mm path length Couette flow cell. *LD*[r] ~ –0.15, so *S* ~ 0.05.

The shortest piece of DNA that can be flow-oriented is about 250 base pairs. DNAs of this length, however, give very weak *LD* signals. To use *LD* quantitatively, lengths should be 800 or more base pairs. Naturally occurring DNA samples such as calf thymus DNA usually end up being tens to hundreds of thousands of base pairs in length which proves to be ideal.

Determining the orientation parameter *S* for DNA

In any analysis of DNA *LD*, the most difficult part is usually determining the degree of orientation of the DNA in the experiment. The theoretical limiting value of *LD*[r] for a transition polarized perpendicular to an orientation axis is –1.5 (equation (3.6)). This value is never reached with DNA in solution. Calf thymus DNA in a cylindrical coaxial flow device has a theoretical maximum *LD*[r] value of –1.48 at 260 nm,[82] however, values of –0.1 in aqueous solution at shear gradients of 1000 s[-1] are typical. In principle, at least, *S* may be calculated. In practice, for macromolecular systems, *S* is usually determined more or less reliably by empirical means either by using known maximum *LD*[r] values (perhaps from a parallel experiment assuming the DNA orientation does not change), or by calibrating with a probe dye whose binding geometry is known and whose spectroscopy is in a different place from that of the DNA and any other ligand of interest.[45, 83] Methylene blue is often used as it has a clear band well-separated from that of the DNA. Its *LD*[r] magnitude is very close to that of the DNA bands.[84] It is really only when α values are close to 0 (parallel to helix axis) or close to 90° (perpendicular to helix axis) that very precise determination of *S* is required. Generally, assuming $\alpha = 86°$ for the intrinsic DNA *LD* of the absorption at 260 nm gives a sufficiently good estimate of *S* (equation (3.6)) to allow the application of the same equation to determine α values for transitions in bound dyes.

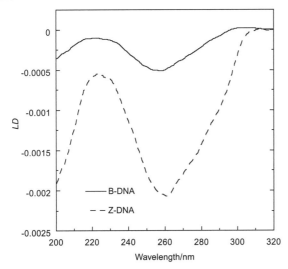

Figure 3.8 *LD* spectra of B-DNA and Z-DNA poly[d(G-C)]$_2$ (50 μM, 5 mM NaCl, 1 mM cacodylate). Z-form is induced by 50 μM [Co(NH$_3$)]$^{3+}$.

LD to determine DNA bending or stiffening

The magnitude of the orientation parameter, *S*, can be used to provide information about the length, flexibility, or bendability of DNA or about the introduction of kinks into DNA. Thus Figure 3.8 indiates Z-DNA is stiffer than B-DNA. If *S* increases as the result of some change imposed on the system (such as the addition of a DNA-binding ligand), then the *LD* increases indicating the DNA has become more oriented. This generally means it has lengthened, or stiffened (or both), or some bend or kink has been removed. More information about the system is required to distinguish between these effects.

Molecules that intercalate between DNA bases push the base pairs apart by ~3.4 Å and unwind the helix by up to 36° (Table 3.1). Thus intercalators both lengthen and stiffen DNA.

Flexibility is an intrinsic property of the polymer. Bends may be induced by *e.g.* ligands such as cisplatin binding to DNA.

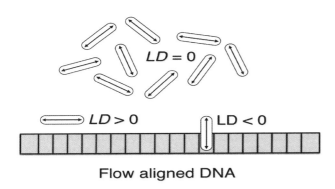

Figure 3.9 Schematic illustration of ligand *LD* when free (*LD* = 0) or bound to flow oriented DNA. Double-headed arrows indicate transition moment polarizations.

Examples of DNA conformational changes easily detected by *LD* are those induced by spermine, $[NH_3(CH_2)_3NH_2(CH_2)_4NH_2(CH_2)_3NH_3]^{4+}$, or $[Co(NH_3)_6]^{3+}$ which when added to DNA decrease the *LD* as the DNA is first bent and then condensed.[85, 86] Similarly the tetra-cationic di-iron triple helicate of Figure 3.10 bends the DNA and intramolecularly coils it.[87, 88] Silver(I) ions[89] and cis-platin[90] are not as highly charged as these ions, but interact specifically with the DNA to bend it significantly.

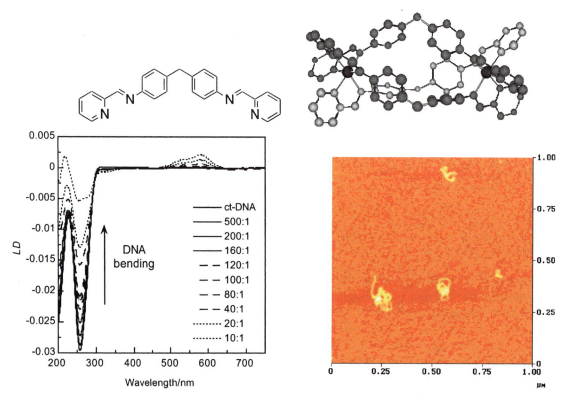

Figure 3.10 The *LD* of a tetracationic di-iron triple helicate $[Fe(LL)_3]^{4+}$ ($LL=C_{25}H_{20}N_4$) illustrated) binding to calf thymus DNA showing the effect on the DNA *LD* as the ligand bends the DNA. ct-DNA (500 mM, 20 mM NaCl, 1 mM sodium cacodylate buffer pH = 7); DNA:ligand ratios are shown on the figure.[91] The AFM picture shows the extreme DNA bending (intramolecular coiling) at high di-iron triple helicate loading with a linearized pBR322 plasmid DNA.

3.6   *LD* of DNA-bound ligands

Once the orientation of DNA itself is understood, a next step is to use *LD* to probe the binding and orientation on DNA of small ligand molecules. A key feature of such *LD* studies is that if there is no specific binding, then there will be no *LD* of the ligand transitions as schematically illustrated in Figure 3.9. Conversely, the presence of an *LD* signal from an absorption band of the putative ligand immediately indicates that it is bound as illustrated by the 550 nm signal in Figure 3.10.

Although *LD* *cannot* tell us *where* a ligand binds to DNA, it can be used to determine the *orientation* of the ligand (if the ligand's transition moments have been assigned). This often provides significant clues as to the binding

In most situations not all the ligands in the solution will be bound to DNA, so the isotropic absorbance is a mixture of that due to bound and free ligands. Thus care must be taken in calculating LD^r to use only the absorbance of bound ligands.

mode. A ligand may bind to DNA in a number of different types of site, as well as with sequence and orientation preferences. Broadly speaking a ligand may bind in the different ways outlined below.

- **Externally** bound to the phosphate backbone: this is usually orientationally fairly non-specific and the induced *LD* (and *CD*) is correspondingly small.
- **Intercalated** between DNA bases: this mode requires planar aromatic molecules (or parts of molecules) which insert between DNA base pairs causing the DNA to unwind (reducing the helical twist, Figure 3.5) and open up a slot between adjacent base pairs so the ligand may be sandwiched between them (Figure 3.12). The *LD* of the intercalated part is then negative, as the transitions we probe are in-plane polarized π→π* transitions.

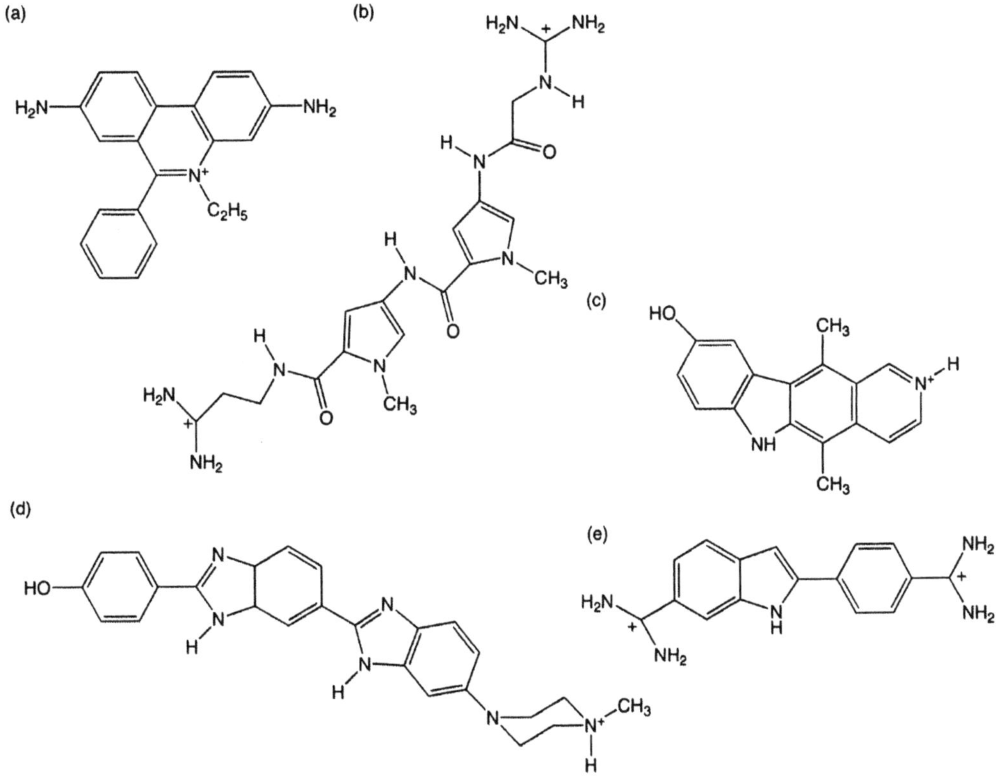

Figure 3.11 DNA binding ligands: (a) ethidium, (b) netropsin, (c) 9-hydroxyellipticene, (d) Hoechst 33258, and (e) 4',6-diamidino-2-phenylindole (DAPI).

- In the **minor groove**: this mode is frequently adopted by aromatic molecules containing internuclear bonds with some rotational freedom. For example, long crescent-shaped molecules such as netropsin, 9-hydroxyellipticene, distamycin, Hoechst 33258, and DAPI (Figure 3.11) can fit snugly into the minor groove with the molecule following the curvature of the groove. The long axis of such

Minor groove binders tend to be selective for A-T rich sequences which have a deeper electrostatic potential than G-C rich sequences. Also G sterically hinders minor groove binding as it has an amine group in the groove (Figure 3.6).

molecules is oriented at about 45° to the helix axis making the *LD* of the long-axis polarized transition positive as illustrated in Figure 3.2b. The short axis polarized transition of minor groove binding molecules are perpendicular to the helix axis so approximately parallel to the base transitions resulting in negative *LD* signals.[10]

Minor groove binding is stabilized by the strong electronegative potential in AT rich DNAs and formation of hydrogen bonding to thymine carbonyl oxygens. In alternating GC the exocyclic amino group of guanine forms a steric barrier to binding in the minor groove (*cf.* Table 3.3).

• In the **major groove**: this type of binding is found for several regulatory proteins and there is enough space to accommodate most smaller ligands including fairly bulky metal complexes, *e.g.* the triple helicate of Figure 3.10, in a variety of orientations.

(a)   (b)

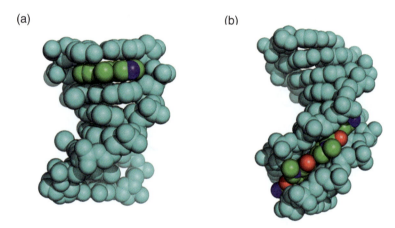

Figure 3.12 (a) An intercalator (9-hydroxyellipticine, Figure 3.11) inserted into 5'-dCdGdAdTdCdG-3' B-DNA.[10] (b) A groove binder (netropsin, Figure 3.11) bound to 5'-dCdCdIdIdCdIdCdI-3'.

DNA intercalating ligands

π-π stacking effects are usually more obvious in the *LD* spectrum than in the absorbance since, in most experiments, the absorbance signal has contributions from free ligand.

The binding energy of an intercalating molecule with DNA is due to a combination of electrostatic interactions, the increase in entropy from displaced waters and sodium ions, and π-π stacking interactions. The π-π interactions also cause the electronic transitions to shift to longer wavelengths, to lose intensity, and to lose fine structure of the spectrum.

Ethidium bromide

That the *LD*^r^ is fairly constant in the 250–290 nm region is consistent with the B conformation of DNA as discussed above. See §5.13 for further discussion.

We first consider the classical intercalator: ethidium bromide (EB, Figure 3.11). Ethidium is a monovalent cation known to bind by intercalation (with the long axis of the phenanthridine parallel to the long axis of the base pair pocket)[92] to various sequences of double stranded B-DNA. Its *LD* spectrum is shown in Figure 3.2. It has a transition occurring from 250–350 nm with a polarization approximately parallel to the long axis of the phenanthridinium moiety and one at 450–550 nm with a polarization at an angle of 60° from the long axis. The LD^r values for some DNA/EB transitions are summarized in Table 3.2.

A first qualitative conclusion from the data in Table 3.2 is that the strongly negative LD^r values for both of the in-plane polarized transitions of EB are consistent with the plane of the latter being approximately perpendicular to the helix axis of DNA. Secondly, the similar LD^r magnitude for the nucleobase absorption region and the EB supports the conclusion that the bases and EB are coplanar. The fact that the LD^r signals for the 250–290 nm DNA and the 330 nm EB transitions are closest suggests that the EB long axis lies along the average polarization direction for the base transitions, whereas the larger value for the 60° polarized EB transition suggests it lies slightly more perpendicular to the average DNA helix axis. A larger magnitude ligand LD^r than DNA-base LD^r is frequently observed for intercalators as the DNA helix is locally stiffer about the intercalation site so orients better than the average DNA helix. The difference between the two EB transitions is discussed further below.

Table 3.2 LD^r data for some ethidium bromide (EB) and DNA transitions.

Wavelength/nm	250–290	330	500
Transition identity	DNA	EB, long axis	EB, 60° polarization
LD^r	−0.023	−0.022	−0.031

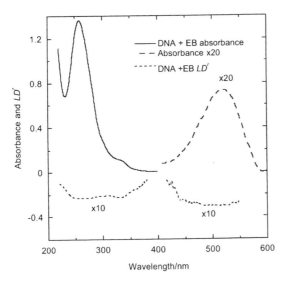

Figure 3.13 Absorbance and LD^r spectra of ethidium bromide in 5 mM phosphate buffer, pH = 6.9, bound to calf thymus DNA (200 μM), DNA base:ligand ratio is 20:1.[93]

Further analysis of the EB/DNA data can provide more geometric information: assuming that the most negative LD^r value of the 500 nm transition represents a transition that is oriented perfectly perpendicular to the helix axis enables us to determine a lower bound on the orientation factor S. Insertion of $\alpha = 90°$ and $LD^r = -0.031$ into equation (3.6) gives $S = 0.021$

Inserting $LD^r = -0.022$, $S = 0.021$ into equations (3.3) and (3.6) gives $\langle \alpha \rangle = 71°$ upon assuming that $\langle \cos^2 \alpha \rangle \sim \cos^2 \langle \alpha \rangle$.

(which shows the low level of alignment commonly observed in flow aligned *LD* experiments). From this it follows that the long axis of EB (as represented by the 250–290 nm transition), as well as the nucleobase planes, must be inclined on average at about 20° from perpendicular to the helix axis (*cf.* Figure 3.7 and related discussion).

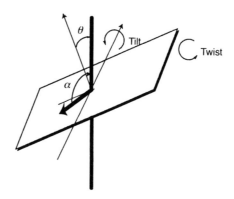

Figure 3.14 Orientation of DNA bases with respect to the helix axis.

Intercalating dyes and DNA base tilt

Consistent with the EB experiment's implications for DNA base tilting, it has been observed for other intercalating dyes, such as 9-aminoacridine and quinacrine, that their long axes (which are believed to be oriented parallel to the longest dimension of the base pair pocket) are inclined some 20° from perpendicularity, whereas their short axis is oriented more perpendicular to the helix axis. It may thus be inferred that the long axis of the base pairs is rotated by 20° about the short axis as schematically illustrated in Figure 3.14.

Intercalating metal complex

The absorption spectra of $[Ru(tpyanth)_2]^{2+}$ in absence and presence of DNA are illustrated in Figure 3.15. As both calf thymus DNA and the ruthenium complexes have overlapping transitions below 300 nm, analysis of the metal complex spectral changes can be most easily interpreted above this wavelength where calf thymus DNA has no absorbance.

We see that the absorption maxima shift to longer wavelengths and the intensity is lower than the spectrum made from the sum of the component parts. This is consistent with enhanced π-stacking upon addition of the DNA which could arise either from intercalation of part of the ligand—probably the anthryl group—between DNA base pairs or metal complex stacking mediated by the DNA template.

When the metal complex of Figure 3.15 is added to the DNA, both the DNA region and the metal complex region of the spectra show *LD* signals, which means that $[Ru(tpyanth)_2]^{2+}$ binds in a specific orientation rather than randomly along the backbone. The orientation of the DNA bases can not be determined directly in the presence of the metal complexes as the metal complex absorbances overlap with those of the bases. This was a case where methylene blue was used.[83] The methylene blue *LD* (*cf.* §3.5) increased when the metal complex was added, suggesting that the complex either lengthens (by unwinding due to intercalation) and/or stiffens the DNA. The data summary for the $[Ru(tpyanth)_2]^{2+}$ complex is consistent with the anthracene group intercalating—negative LD^r for the anthracene short axis transition from 350–400 nm. The two terpyridine groups lie perpendicular to one another and to the z-axis (Figure 3.15), they are therefore fitting into the grooves resulting in the observed positive terpy *LD* signal.

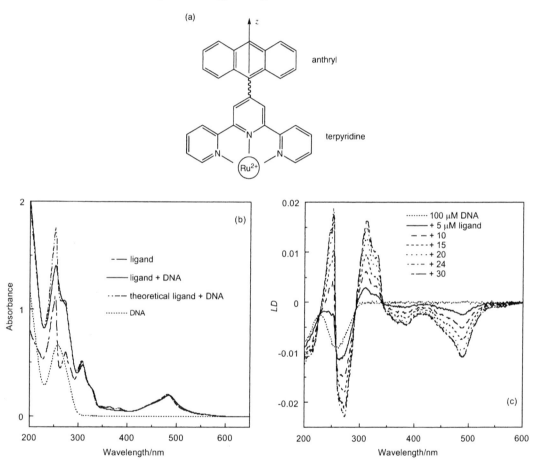

Figure 3.15 (a) $[Ru(tpyanth)_2]Cl_2$, z indicates the approximate position of the metal complex long axis; a second anthrylterpyridine ligand is oriented perpendicular to the one illustrated making an approximately octahedral coordination geometry. (b) UV–visible absorption spectrum in water of $[Ru(tpyanth)_2]^{2+}$ (10 μM); $[Ru(tpyanth)_2]^{2+}$ (10 μM) plus calf thymus DNA (100 μM); and calf thymus DNA (100 μM) together with the 'theoretical' sum spectrum. (c) *LD* spectra from a titration series of $[Ru(tpyanth)_2]^{2+}$ with ct-DNA (100 μM). Metal complex concentrations are indicated in the figure.[83]

DNA groove binding ligands

Methyl green

In the case of pure (non-overlapping) transitions, equation (3.6) can be applied to each transition. Consider the simple case of the DNA groove binder methyl green (Figure 3.16). It has two transitions at 425 nm and 630 nm polarized, respectively, along the pseudo-twofold axis (through the bulky trimethyl amine moiety) and perpendicular to this direction as indicated in Figure 3.16. The *LD* spectrum of methyl green bound to DNA has a negative $LD^r = -0.21$ at 425 nm (compared to DNA's -0.18 at 260 nm) and a positive $LD^r = +0.07$ at 630 nm. From this we may deduce that $\alpha \approx 90°$ for the 425 nm transition and $\alpha \approx 48 \pm 5°$ for the one at 650 nm. This is suggestive of the methyl green binding along either the minor or the major groove with the less bulky axis (630 nm) pointing along the groove. The orientation at 90° for the 425 nm bulky group axis could be consistent with intercalation; however, independent evidence, including the charge and bulkiness of this group, instead suggests that it is pointing away from the DNA centre and that methyl green binds in the major groove.[94]

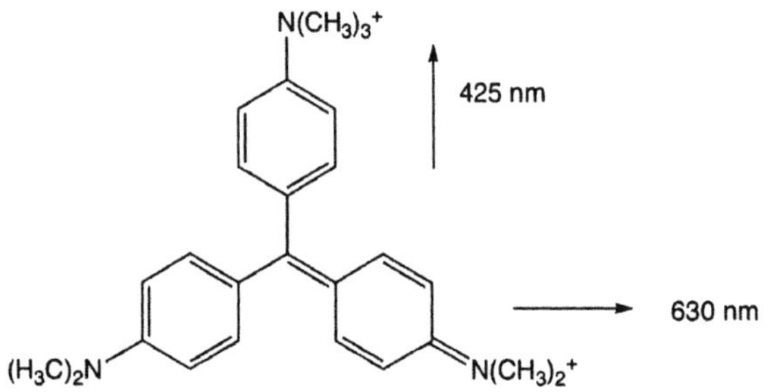

Figure 3.16 Methyl green.

Diamidinophenylindole

Unfortunately most DNA binding dyes do not have conveniently spaced transitions of orthogonal polarization. Our next example refers to a monocationic dye molecule diamidinophenylindole (DAPI) (Figure 3.11) which is known to prefer AT sequences of duplex DNA to which it binds in the minor groove. It has a strong long-axis polarized absorption band near 360 nm. Recording the flow linear dichroism in the DNA region as well as for the dye gives the LD^r values of Table 3.3.

Table 3.3 LD^r for DAPI mixed with calf thymus DNA and the alternating duplexes poly[d(A-T)]$_2$ and poly[d(G-C)]$_2$ at base:dye ratios of 50:1.

DNA	calf thymus DNA	poly[d(A-T)]$_2$	poly[d(G-C)]$_2$
LD^r at 260 nm	−0.18	−0.028	−0.033
LD^r at 360 nm	+0.09	+0.012	−0.030
S	0.12	0.019	0.022
α	45°	46°	80°

The markedly lower degree of alignment of the synthetic polynucleotides is a result of their significantly smaller contour lengths (<1,000 base pairs *versus* >10,000).

A number of conclusions may be drawn from the data of Table 3.3 that illustrate the various pieces of information that may be provided by flow *LD* studies of drug–DNA systems.[10]

- Assuming that the DNA bases are perpendicular to the helix axis, *i.e.* $\alpha = 90°$, and assuming that DAPI does not have a significant *LD* signal at 260 nm allows us to estimate the orientation parameter *S* in each of these experiments as summarized in Table 3.3.

- Using these calculated *S* values, the binding geometries of the DNA-drug complexes may be characterized in terms of the angle between the long axis of the drug molecule and the DNA helix axis. As mentioned above, a 45° orientation for the long axis of a drug is expected for a minor-groove binding geometry, so we conclude that this is indeed the binding mode of DAPI in alternating poly[d(A–T)]$_2$ as well as in mixed sequence DNA at these low binding ratios. In alternating poly[d(G–C)]$_2$, however, as a consequence of the exocyclic amino group of guanine protruding into the minor groove, a different binding mode is found. DAPI with poly[d(G–C)]$_2$ is oriented almost perpendicular to the helix axis and also shows a larger hypochromicity and stronger red-shift of the ligand transition than with the other two DNAs. These observations are suggestive of an (at least partial) intercalative binding mode with GC, despite the size of the DAPI molecule. This indicates that the preferred binding mode in mixed sequence DNAs is along AT-rich stretches of the minor groove.

> Using a more accurate value of $\alpha \sim 75–86°$ does not make a large difference to the final geometrical conclusions.

This example thus illustrates the potency of *LD* to discriminate between groove-binding and intercalation and, in addition, demonstrates that a given DNA ligand may bind in several different ways depending on nucleobase sequence and binding ratio. *LD*, in common with *CD*, may also be used to directly determine equilibrium binding constants for DNA-drug systems. Some methods to do this are outlined in §11.2.

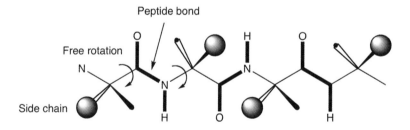

Figure 3.17 L-Amino acids joined *via* rigid peptide bonds, indicated by the bold lines.

3.7   Proteins

Attempts to use *LD* to study protein systems usually encounter one of two problems: either (i) the system is difficult to orient (as is certainly the case for globular proteins); or (ii) once oriented, the environment that is required to create orientation and surrounds the oriented protein chromophores (such as a cell membrane or a filamentous aggregate) may give rise to artifacts caused by light scattering (*cf.* §2.6). These problems are, however, generally surmountable and *LD* has been used to good effect to determine orientations

> Proteins are (usually large) naturally occurring sequences of amino acids. Peptides (or polypeptides) are smaller chains of amino acids (typically < fifty amino acid residues). In most cases, their spectroscopy goes can usually be treated as equivalent.

of chromophores, mobility of particular centres, and for studying ligand binding, DNA-protein interactions, fibre assembly, and protein insertion into membranes. Some of the more straight forward examples are given below. Other applications are considered in Chapter 5, including the site specific linear dichroism (*SSLD*) methodology for systematic determination of protein structure.

Protein geometry

Proteins are linear polymers of well-defined sequences of amino acids. Each amino acid is usually one of the twenty commonly occurring L-amino acids illustrated in Figure 1.5; they are joined *via* what is known as the peptide bond (Figure 3.17) which is formed between the acid group of one amino acid and the amine group of another by the elimination of a water molecule.

Protein primary structure refers to which is bonded to which other amino acid via a peptide bond. It results in one end of the protein molecule being the amino end (by convention this is the beginning) and the other end being the carboxylate end; these are usually referred to as the N terminus and C terminus respectively. The molecular identity of the protein is specified by listing its unique sequence of amino acids.

(a)   (b)   (c)

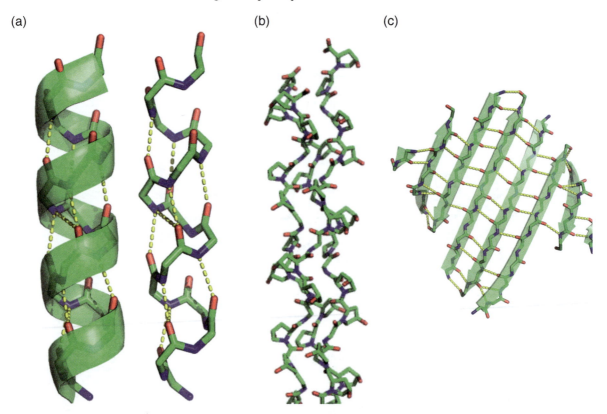

Figure 3.18 (a) α-helix, (b) polyproline II helix, and (c) anti-parallel β-sheet protein secondary structures. Dotted yellow lines indicate H-bonds.[95]

The overall shape of a protein molecule is crucial for its biological activity. This is summarized in its secondary structure, which is the arrangement in space of neighbouring amino acids, and its tertiary structure which is how the local motifs containing a number of aminoacids are arranged with respect to one another. Quaternary structure is the next layer of organisation and describes the arrangement of individual polypeptide chains with respect to one another. In nature, proteins usually adopt unique folded conformations that optimize their activity. Thus, any techniques that can provide information about protein structure, particularly about proteins in solution, are potentially valuable to scientists endeavouring to understand how proteins function.

Proteins form regular secondary structural units because the peptide unit O=C–N– is planar and rigid yet it has a large degree of rotational freedom about its links to the rest of the protein chain (Figure 3.17). This both constrains the possible relative orientations of neighbouring residues and allows a variety of possible intramolecular hydrogen bonding arrangements between the C=O of one peptide unit and the N–H of another unit.

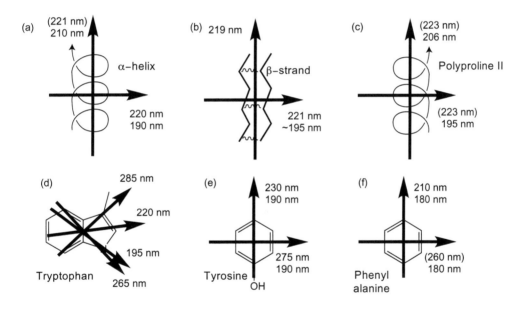

Figure 3.19 UV transition polarizations for (a) α-helix, (b) β-sheet, (c) polyproline II helix,[42] (d) tryptophan,[96] (e) tyrosine,[96,97] and (f) phenyl alanine. Transitions indicated in parentheses are weak.

The twenty common naturally occurring amino acids in most biological systems all have the same handedness (except for glycine which has a hydrogen atom as its side chain (Figure 1.5) and is therefore achiral). As a result of the chirality of the amino acid residues, the secondary structure features they form are also chiral. The most common structural motifs that are found in nearly all proteins are: the α-helix, parallel and anti-parallel β-sheets, and the turns that join them (Figure 3.18).

The right-handed (for L-amino acids) α-helix results when the nth peptide

The 3_{10} helix is similar in structure to the α-helix but has 3.2 amino acids per turn of its right handed helix and has hydrogen bonds between the nth amino acid and tha in the (n+3) position, thus making a 10-bond ring.

unit forms hydrogen bonds between its C=O and the N–H of the (*n*+4)th peptide and between its N–H and the (*n*–4)th C=O; there is a 1.5 Å translation and 100° turn between two consecutive peptide units, giving 3.6 amino acid residues per turn.

An alternative efficient formation of hydrogen bonds occurs between a sheet of parallel or antiparallel runs of amino acids; these are known as a β-sheets (Figure 3.18). Typically the strands of an anti-parallel β-sheet are linked by β-turns where the *n*th peptide unit forms hydrogen bonds with the (*n*+3)rd peptide unit. If a β-sheet extends over more than two strands, then the relative arrangements of the strands in space must be considered—this often forms a chiral 'superstructure'.

Other structural forms are known but, being less well defined, have less well-defined spectral signatures. The exceptions to this are the left-handed polyproline type II helix and the random coil which have very similar spectral features as discussed in §4.5.

Absorbance and *LD* of polypeptides

Typical absorption spectra of oligo- or poly-peptides in the 180–240 nm 'backbone' region show features due mainly to two electronic transitions on each residue: a stronger $\pi \rightarrow \pi^*$ amide transition at ~195 nm and a weaker $n \rightarrow \pi^*$ transition at about 220 nm. For the α-helix (and also some other helices) the $\pi \rightarrow \pi^*$ transitions on neighbouring residues couple together resulting in two transitions (*cf.* Chapter 8) on either side of 200 nm. These two component transitions are polarized at right angles to each other: the α-helix transition at ~208 nm is polarized along the helix axis and the other at ~192 nm is polarized perpendicular to the helix axis as illustrated in Figure 3.19.

If the $n \rightarrow \pi^*$ transition belonged within a pure carbonyl chromophore (*cf.* Chapter 9), it would be electric dipole forbidden and so have very weak absorbance. However, for peptides the lower symmetry of the amide chromophore enables coupling with the π system. The n-π^* transition thus acquires some electric character and an electric transition moment on each amide approximately as indicated in Figure 3.20. It has a net perpendicular polarization as illustrated in Figure 3.19.

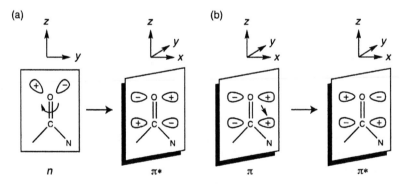

Figure 3.20 Schematic illustrations of (a) $n \rightarrow \pi^*$ and (b) $\pi \rightarrow \pi^*$ transitions in peptides.

To illustrate this discussion, Table 3.4 gives data for the polarized spectra, $A_{//}$ and $A_{\perp}$, of a thin humid film of α-helical poly-L-glutamic acid oriented by stroking on a quartz plate. The parallel of $A_{//}$ means that the light is polarized parallel to the orientation direction of the film, which should also be the direction of preferred orientation (helix axis) of the α-helical polypeptide filaments. The LD can be determined from equation (1.1) and the isotropic absorbance from equation (3.4); the concentration and path length independent reduced LD, LD^r, follows from equation (3.6).

The negative LD signals at 190 nm and 230 nm and the positive signal at 210 nm in Table 3.4 are consistent with absorbances that are polarized, respectively, perpendicular and parallel to the polymer axis as illustrated in Figure 3.19. If we assume that the absorbance at 190 nm is perpendicularly polarized with $\alpha = 90°$, we may conclude from equation (3.6) that the orientation factor S of the system is 0.21. Deviation from perfect perpendicular polarization (*i.e.* $\alpha < 90°$) for the 190 nm $\pi \rightarrow \pi^*$ transition would result in our calculated value of S being too low.

Table 3.4 Typical polarized and unpolarized absorbance data for poly-L-glutamic acid in an oriented film (created by stroking a concentrated solution onto a quartz plate). *LD* and *LD^r* are determined using the absorption data.

Wavelength/nm	190	200	210	220	230	240
$A_{//}$	0.45	0.55	0.56	0.38	0.22	0.18
$A_{\perp}$	0.63	0.55	0.47	0.38	0.27	0.18
A	0.57	0.55	0.50	0.38	0.25	0.18
LD	−0.18	0.00	0.09	0.00	−0.05	0.00
LD^r	−0.32	0.00	0.18	0.00	−0.20	0.00

Overlap of transitions of different polarizations also leads to calculated values of S being too low. For example, let us assume that the intensity at 210 nm is due to a parallel polarized transition with no overlap by transitions of other polarizations, so $\alpha = 0°$. Substituting $LD^r = 0.18$ (Table 3.4) into equation (3.6) gives $S = 0.09$, which is less than half of the lower bound value for S determined from the 190 nm signal. This indicates that there is substantial overlap from the surrounding oppositely polarized transitions. The 210 nm transition is, in fact, only seen as a shoulder in the absorption spectrum.

In solution, the 210 nm *LD* is usually dominated by the neighbouring transitions so appears as a minimum in the spectrum rather than a band of opposite sign from those at 220 and 195 nm.

3.8   Membrane proteins

Membrane proteins and peptides may be peripherally associated with or more deeply embedded within the membranes of a cell. They play key roles in the function of cells in health and disease. Some membrane proteins act as channels to transport species into and out of cells or organelles, others may disrupt the membrane to which they bind, yet others transmit signals by changing their structure upon ligand binding. Membrane proteins are therefore important drug targets and also drug candidates.

Despite their importance (approximately one third of proteins coded by the human genome are membrane proteins), we know comparatively little about membrane proteins and membrane peptides because they are very challenging to study. The normal environment of a membrane protein includes a lipid bilayer and its associated surface and integral small molecules and proteins, all solvated by an aqueous solution that also contains a wide range of molecules. The lipids within the bilayer also vary significantly in structure depending on what cell or organelle one is considering. Many techniques have difficulty with lipid environments, but the exchange of the normal environment of the membrane protein for an aqueous environment usually leads to the protein changing structure or becoming insoluble. However, both *CD* and *LD* are able to be used with a wide range of buffers or solvents including lipids and detergents. They are therefore almost as useful for membrane proteins as for the better studied globular proteins.

Membrane proteins are naturally oriented systems, so *LD* is in principle an ideal technique to study them. However, it is something of a challenge to orient them in a spectroscopically and biologically friendly manner. Squeezed gels is one option as discussed in §2.3. Griebenow *et al.* have used this method to deduce the relative orientation of bacteriochlorophyll transition moments relative to the long axis of the chlorosomes.[98] However, gels are not suitable for *LD* studies of the backbone region of proteins as they are not sufficiently transparent. A further issue with gels as an orientation matrix is that in order to analyze experimental *LD* data quantitatively, it is important to determine *S* (equation (3.6)). With gels, *S* has often been assumed to be 1, which leads to the final conclusions being suspect, especially if *S* is actually small. This problem is not intrinsic to gels, simply to the assumptions typically used in data analysis. An attempt to determine the true *S* for a gel oriented DNA-protein complex is given in reference [48].

Single crystal orientation of photochemical reaction centres have also been used to determine orientations of chromophores[48] and Wan *et al.* have used time-dependent *LD* studies to probe the conformational motion in bacteriorhodopsin over the first 10 μs after excitation.[99]

Membrane proteins can also be studied directly in membrane films dried onto surfaces of the kind used for oriented *CD* (*cf.* §4.6). Experimental difficulties arise because the sample is a thin layer through which the light beam is propagated, so the optical axis of the sample aligns with the propagation direction of the light. Nordén *et al.* have developed a method to study such uniaxial samples by inclining the optical axis to the direction of the incident radiation as discussed in §2.3 and §6.6. A further problem with dried membrane films is that the removal of the aqueous boundaries either side of the membrane bilayer can peturb the structure of parts of the membrane protein located there, *e.g.* loops that are not completely buried in the membrane.

Membrane peptides in liposomes

Nordén et al.[41] opened up the possibilities for solution phase lipid membrane *LD* as discussed in §2.3 when they found that unilamellar liposomes (model membrane systems with a single bilayer of lipid enclosing a central space, Figure 2.7) could be simultaneously distorted and oriented in shear flow. Thus, small molecules such as pyrene which were bound to the liposome could also be flow oriented in liposomes. Rodger *et al.*[42] showed this could be extended to proteins and peptides. Thus anything bound on or in the liposome lipid bilayer is oriented and expected to give an *LD* signal in its absorption bands. Conversely, any species not bound is not oriented so invisible to the *LD* experiment.

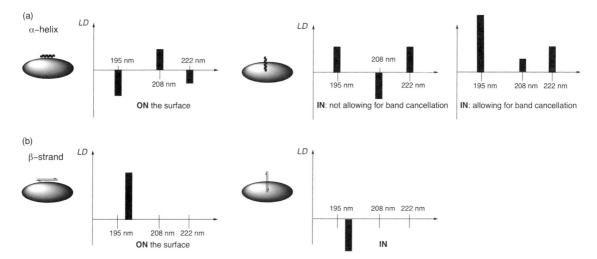

Figure 3.21 Schematic illustration of expected *LD* signals for (a) an α-helix on the surface of or inserted into a membrane, and (b) a β-sheet lying flat on the surface of or inserted into a membrane.

LD is thus an extremely attractive technique for probing whether or not a species such as a proposed trans-membrane antibiotic peptide binds to a given membrane. When the secondary structure of the peptide is known (*e.g.* α-helical or β-sheet, *cf.* Chapter 4) from *CD*, then its orientation can be approximately determined from the sign of the *LD* as summarized in Figure 3.21. If *S* is known then the angle can be determined more accurately. *LD* is currently the only technique that can provide this information in solution phase. It is also ideally suited for following the kinetics of insertion of peptides or proteins.

The possibilities are illustrated in Figure 3.22 for a simple α-helical peptide and a β-strand peptide. Further examples are given in Chapter 5. In interpreting membrane *LD* data one must note that the geometry of the liposome experiment is different from that of long polymers since the flow direction creates the long axis of the liposome (Figure 2.7) but the membrane normal (which is perpendicular to the long axis of the liposome) is the molecular orientation direction. The equation analogous to equation (3.6) for liposome (and geometrically related) systems is therefore:[41,42]

Note the baseline slope in Figure 3.22 is due to the scattering of light by the 100 μm diameter liposomes (§2.6).

In flow *LD* experiments with proteins in liposomes, *S*, is much smaller than discussed above for polyglutamic acid in a film.

$$LD^r = \frac{LD}{A_{iso}} = \frac{3S}{4}\left(1 - 3\cos^2\beta\right) \qquad (3.7)$$

where β is the angle between the lipid normal and the transition moment polarization as illustrated in Figure 2.7. Its derivation can be found in §7.6.

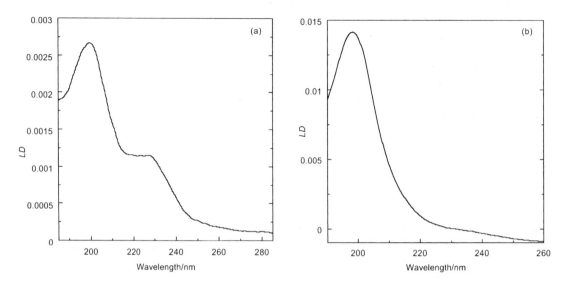

Figure 3.22 (a) *LD* of the α-helical frog peptide aurein 2.5 with sequence: GLFDIVKKVVGAFGSL-NH$_2$ inserted into a lipid bilayer liposome composed of 50:50 DPPC (dipalmitoylphosphatidylcholine) and DPPG (dipalmitoylphosphatidylglycerol) at 50 °C (*i.e.* in the liquid phase).[100] (b) *LD* of the β-sheet (by *CD cf.* §4.5) core peptide fragment GLRILLLKV (derived from the T-cell antigen receptor with liposomes made from posphatidyl choline extracted from soya beans, showing that it sits on the membrane surface.[101]

If the transition of a bound molecule is oriented at 54.7° the *LD* will also be zero.

Two special cases of equation (3.7) are: (i) when a transition moment is parallel to the cylinder normal (the lipids), for which $\beta = 0$ and $LD^r/3S = -1/2$; and (ii) when it is perpendicular to the cylinder normal for which $\beta = 90°$ and $LD^r/3S = +1/4$. The 210 nm transition of an α-helix inserted into the membrane is an example of case (i)—but it usually only appears as a small dip between two larger positive signals. The same helix has $\beta = 90°$ for its 222 nm and lower wavelength π-π* transitions. Conversely it is the 210 nm transition of a helix lying on the surface that satisfies case (ii) as illustrated in Figure 3.21. The 220 nm and 190 nm short-axis polarized bands of an α-helix lying on the surface are a combination of these two extremes (because of the screw symmetry of the helix), resulting in net negative *LD* signals.

The *LD* shown in Figure 3.22 for an 18-residue α-helical peptide is consistent with the schematic of an α-helical peptide inserted into membrane when one allows for band cancellation so that the 208 nm band looks like a positive minimum rather than a negative maximum (Figure 3.21(a)). By way of contrast, the nonapeptide GLRILLLKV derived from the T cell antigen receptor,[102] which adopts a β-sheet structure (as determined from the *CD* spectrum of Figure 4.14, *cf.* §4.5) with PC soya bean liposomes, lies flat on the PC soya bean liposome surface only showing significant *LD* at ~200 nm since the *LD* of its longer wavelength transitions cancel.

3.9 Fibrous proteins: cytoskeletal proteins

The cytoskeleton in both prokaryotic and eukaryotic cells is dependent on the rapid and controlled assembly and disassembly of polymers (fibres) whose monomeric units are themselves folded proteins. The structures of the monomers change little, if at all, when they form the fibres. In some cases a single linear polymer forms, in others these protofilaments assemble to form bundles of some kind. Others, such as tubulin (Figure 3.26), assemble to form more complicated structures. The monomeric or low order oligomeric protein units of fibres such as FtsZ and tubulin have no flow *LD* signal, thus they provide no background signal to interfere with attempts to follow kinetics of fibre assembly, disassembly and reorganisation using *LD*. The possibilities of *LD* spectroscopy can be illustrated by looking at the bacterial homologue of tubulin, FtsZ, and tubulin itself.

FtsZ polymerization and bundling

FtsZ is a bacterial protein that polymerizes to form the so-called Z-ring. The role of the Z-ring is to contract and pull in the cell membrane to enable one cell to divide into two daughter cells. *In vitro* (and presumably *in vivo*), FtsZ polymerizes to form linear polymers or protofilaments on the seconds timescale when GTP (guanine triphosphate) and Mg^{2+} are present. The protofilament *LD* spectrum for *E. coli* FtsZ is the 0 mM Ca^{2+} spectrum of Figure 3.23.

E. coli FtsZ has no tryptophans to interfere with the 280 nm region of the spectrum, so the *LD* in that region is dominated by GTP.

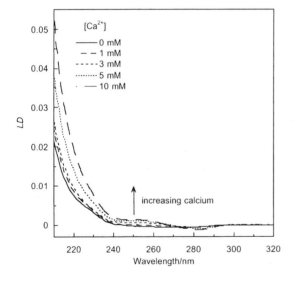
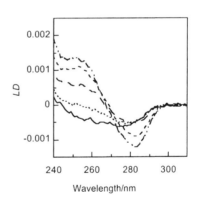

Figure 3.23 FstZ (11 μM) polymerization in the presence of $MgCl_2$ (10 mM) and varying amounts of Ca^{2+} (MES buffer (50 mM), pH 6.5, KCl (50 mM), EDTA (0.1 mM) and GTP (0.2 mM)). The GTP region is expanded on the right.

Upon addition of Ca^{2+} the FtsZ protofilaments bundle together.[103] The bundling process results in the guanine chromophores of the GTPs (which are between each pair of monomer units) tilting as shown by the positive peak that appears at 260 nm when the Ca^{2+} concentration is increased.

FtsZ achieves its objective in cells by forming a well-orchestrated complex with more than 10 other proteins, one of which is YgfE.

By considering the polarizations of the guanine transitions (Figure 1.4), one can deduce that the guanine in the protofilaments is more or less perpendicular to the fibre axis, whereas in the bundles the guanines have been tilted as suggested in the schematic diagram of Figure 3.24. The protein YgfE performs the same function as the calcium ions illustrated in Figure 3.23, but at biologically realistic concentrations.[103] *LD* is currently the only technique that can follow such a reorganisation of the components of this complicated molecular system in real time.

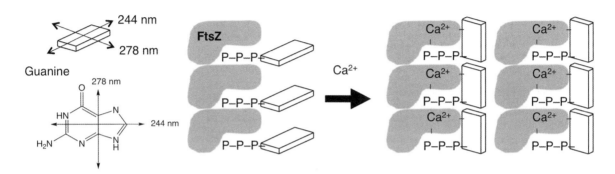

Figure 3.24 Schematic of guanine reorientation upon calcium-induced bundling of FtsZ protofilaments into fibres.

Tubulin polymerization and depolymerization

The eukaryotic protein tubulin is also a polymer of protein monomers with a GTP between each monomer unit. As its name implies, tubulin forms tubules. The proteins α-tubulin and β-tubulin first dimerize, then the dimers assemble into linear protofilaments, which in turn assemble into hollow cylindrical filaments as illustrated in Figure 3.25.

Figure 3.25 Tubulin assembled into a microtubule.[104]

Rather than forming a Z-ring, tubulin polymers radiate from the centrosome of a cell to attachment sites just under the cell membrane and from mitotic spindles to chromosomes undergoing separation during cell division. Microtubules also play a role in moving cells and organelles and interact with motor proteins. They are thus very attractive dynamic drug targets. *LD* is the ideal (and perhaps only) technique to follow the kinetics of processes such as the fibre assembly and disassembly as illustrated in Figure 3.26. Microtubules constantly polymerize at one end and depolymerize at the other as long as a supply of GTP is present.

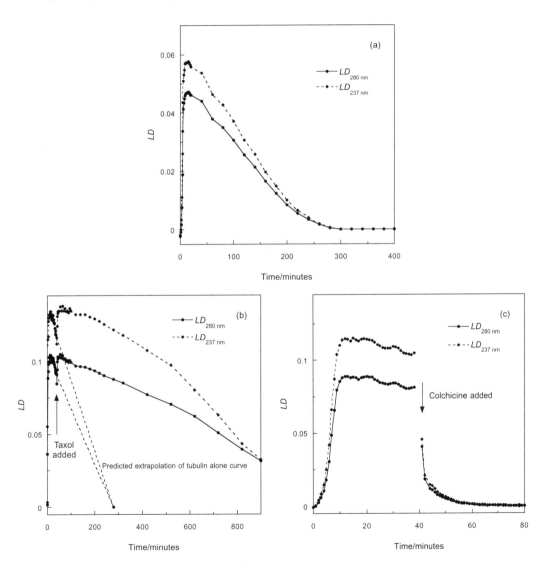

Figure 3.26 Capillary *LD* at two absorbance wavelengths of tubulin polymers (28 μM monomer) at 37 °C: (a) tubulin only; (b) tubulin before and after addition of taxol (final concentrations 25.5 μM and 18.2 μM respectively); (c) tubulin before and after addition of colchicine (final concentrations 25.5 μM and 18.2 μM respectively).[67]

Figure 3.26 shows the kinetics of polymerization and depolymerization of tubulin.[67] The rates are the same in the aromatic region (280 nm) and backbone region (237 nm) which means that the backbone and side chains adopt their final orientation simultaneously. The high concentrations of tubulin needed to initiate polymerization mean that the *LD* data are only reliable down to about 235 nm as the sample absorbance is too high at lower wavelengths. The polymerization in the absence (Figure 3.26a) and presence (Figure 3.26b) of the taxol illustrates the stabilisation effect of taxol on the microtubules. Colchicine (which is used in the treatment of gout), by way of contrast, inhibits microtubule formation by binding to tubulin monomers. This has the consequence that the polymerization reaction does not occur but the depolymerization does, resulting in a net tubulin depolymerization (Figure 3.26c).

4 Circular dichroism of biological macromolecules

CD depends on the underlying helicity of the electronic structures of light-absorbing molecules. Therefore, it is ideal for probing asymmetric aspects of nucleic acids and proteins and their interactions with their environments and other molecules. How to interpet the data empirically forms the basis of this chapter.

Linear Dichroism and Circular Dichroism: A Textbook on Polarized-Light Spectroscopy
By Bengt Nordén, Alison Rodger and Timothy Dafforn
© B. Nordén, A. Rodger and T. Dafforn, 2010
Published by the Royal Society of Chemistry, www.rsc.org

4.1   Introduction

CD is now more widely used to study biological molecules than non-biological systems. This is despite the fact that the molecules concerned are usually large and therefore detailed structural analysis of their *CD* data is seldom possible. In this chapter the focus is on the *CD* of nucleic acids, nucleic acid-ligand systems, and proteins. The content is designed to be illustrative rather than comprehensive. It should enable one to read the literature with discernment.

The *CD* of biological macromolecules is mostly measured to investigate one of four things:

(1) the conformation of the macromolecule itself,
(2) changes in the conformation of the macromolecule,
(3) interactions of the macromolecule with small molecules, especially achiral ones whose induced *CD* is due solely to their interaction with the macromolecule,
(4) the interactions between the macromolecule and other chiral molecules, including DNAs and proteins.

The data analysis that is performed for such systems is usually qualitative or empirical in nature, however, the attractions of *CD* include the following.

- The experiments are simple and quick to perform.
- *CD* is uniquely sensitive to the *asymmetry* of a system.
- *CD* experiments deal with the solution phase.
- By changing the path length from 10 μm even up to 10 cm a wide range of concentrations can be used effectively.
- *CD* can use much lower concentrations than are required for other structural techniques such as nuclear magnetic resonance spectroscopy (NMR), thereby also avoiding solubility problems.
- The intrinsic timescale of *CD* measurements (10^{-15}s for absorption of a UV photon) is much shorter than that for NMR (10^{-9} s) which uses radio waves.
- A *CD* spectrum may be collected in a few minutes and single wavelength measurements require only a few milliseconds.
- *CD* data can be collected on any sized molecule. It has neither an upper nor lower limit in contrast to NMR.
- *CD* data can be collected on both rigid and flexible systems.

4.2   Qualitative description of the origin of *CD* signals

Chapters 8–10 contain the underlying theory of *CD*. A semi-quantitative outline of where *CD* originates for chiral dimers is given in §4.7. One can use *CD* effectively without that level of understanding, however, it is helpful to have at least a pictorial understanding of the origin of *CD* before attempting to interpret data. This section contains a description of why the handedness of a molecule (also called its absolute configuration) relates to the sign, or sign pattern, of its *CD* spectrum. Since a crucial point in our discussion is the direction of the net electric transition moment, a by-product

Spectroscopic techniques give an average over the chromophores in the sample.

Being able to get structural information on biomacromolecules in solution is important because the crystallization process that underlies the major high resolution structural technique X-ray crystallography can change the structure of a molecule.[105, 106]

The sample mixing process defines the lower-limit timescale for kinetic *CD* data. Rapid mixing systems are commercially available for *CD*, allowing samples to be mixed on the millisecond timescale. Such systems have proved ideal for protein folding studies.

of understanding *CD* is also information that relates molecular structure and *LD* spectra to each other for systems of interacting chromophores.

The key feature that gives rise to a non-zero *CD* spectrum is that the electron redistribution which happens during the transition is helical. This is mathematically expressed by two vectors, one of which describes the linear direction of the electron density change and one of which describes the circling of charge about that direction. After a brief illustration of this concept, the remainder of this chapter describes what we measure and how we can interpret data empirically with only minimal understanding of the underlying mechanisms presented.

Pictorial description of *CD* spectroscopy

An electronic transition occurs because either the electric or magnetic field (or both) of the radiation 'pushes' the electrons to a new stationary state. The effect of the electric field is to cause a linear rearrangement of the electrons; the net linear displacement of charge during any transition is called, as discussed in Chapter 3, the *electric dipole transition moment* of the transition and is denoted by the vector μ. The direction in which μ points is the same as the transition polarization—it is the direction in which the electrons are pushed during the transition. In contrast to the linear effect of the electric field, the magnetic field induces a circular rearrangement of electron density (Figure 4.1). The net circulation of charge is the *magnetic dipole transition moment*, m.

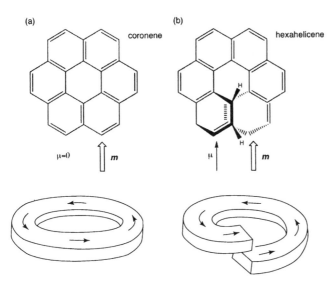

μ is known as a *polar* vector: it has magnitude and direction. m describes rotation about an axis and is known as an *axial* vector. Its direction is such that when viewed from its tip the rotation is anti-clockwise. Alternatively, if one aligns the right-hand thumb along m and holds the fingers in a curve, the circulation of charge follows the fingers (Figure 4.24).

We use bold face type, *e.g.* u, to denote vectors; the same symbol in normal face type, u, is the length or magnitude of the vector; a vector of length 1 unit along the direction of a unit vector is denoted by $\hat{u}$. The components of the vector are denoted by subscripts, so an electric dipole transition moment is
$$\mu = (\mu_x, \mu_y, \mu_z).$$

Figure 4.1 Electron redistribution upon light excitation for a transition of (a) an achiral molecule and (b) a chiral molecule. The electric dipole transition moments are represented by line arrows and the magnetic dipole transition moments by outline arrows. The choice of direction of the 'current' dipoles, *i.e.* → or ←, is arbitrary , but once the direction of μ is assigned, the direction of m is defined (or conversely).

To understand how *CD* can arise, consider the hexahelicene example of Figure 4.1. As illustrated in the bottom part of the figure, if we put arrows for 'local excitations'—charge displacements—across the aromatic rings, the resultant helix of electron redistribution is shown. It is

(1) a net electric displacement directed parallel to the helix axis (a resultant electric dipole transition moment), and

(2) a right or left handed circulation of charge in a plane perpendicular to the direction of the resultant electric dipole transition moment, *i.e.* a magnetic dipole transition moment either parallel or antiparallel to the resultant electric moment. The combination of these two electron motions is a helical electron movement.

In this way a positive or negative *CD* is produced depending on whether the hexahelicene is a right-handed or left-handed helix.

In an achiral molecule the net electron redistribution is always planar. It is usually linear ($\mu \neq 0$, $m = 0$) or circular ($\mu = 0$, $m \neq 0$). In a chiral molecule the electron rearrangement of a transition is always helical as a result of μ and m having parallel (or anti-parallel) components that combine to make a helix of electron redistribution. If the helix of electron motion is right-handed then it is more easily induced by left circularly polarized light (of the appropriate frequency) than by right circularly polarized light (*cf.* §1.2 and §10.3) and we observe a positive *CD* signal.

4.3 *CD* of polynucleotides: DNA and RNA

CD of nucleotide bases

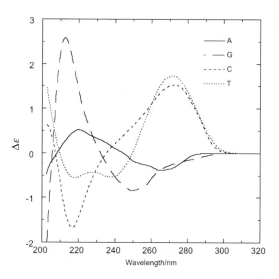

Figure 4.2 *CD* spectra of DNA nucleotides: deoxyadenosine 5'-monophosphate (A), guanosine 5'-monophosphate (G), deoxycytidine 5'-monophosphate (C); and thymidine 5'-monophosphate (T) (that of uracil is similar).

DNA is composed of nucleotides which are planar aromatic bases linked to a sugar and a phosphate (Figure 3.6). Most of the nucleotide UV spectroscopy we measure (*e.g.* Figure 1.4) is due to the π-π^* transitions of the bases (§3.5). These transitions have no intrinsic *CD* signal since the bases are planar and hence they are achiral. However, the bases acquire asymmetry in their electronic transitions by coupling with the chiral ribose sugar units of the backbone. The magnitude of $\Delta\varepsilon_{max}$ for each base is of the order of 2 mol^{-1} dm^3 cm^{-1} at 270 nm; the purine bases have a negative signal whereas the pyrimidine ones have a positive *CD* (Figure 4.2). When the bases are linked by phosphodiester bonds to form DNA, an additional source of chirality is the helical stacking of the bases.

Empirical structural analysis of DNA CD

CD spectra are sufficiently easy to measure that it is often the simplest technique to use to probe DNA conformational changes as a function of ionic strength, solvent, ligand concentration, *etc*. The simplest application of *CD* to DNA (or RNA) structure determination is to identify which polymorph is present in a sample. Since the DNA *CD* from 190–300 nm is mainly due to the skewed orientation of the bases, if the DNA is untwisted or the bases are tilted a change from what is observed for B-DNA would be expected in the *CD* spectrum. Somewhat surprisingly, observed *CD* spectra often vary more with changes in base orientation (DNA polymorph) than as a function of the base composition of the DNA, though they are of course a sensitive function of DNA sequence as well.

The couplings are short range, so only neighbouring bases have a significant effect on each other.

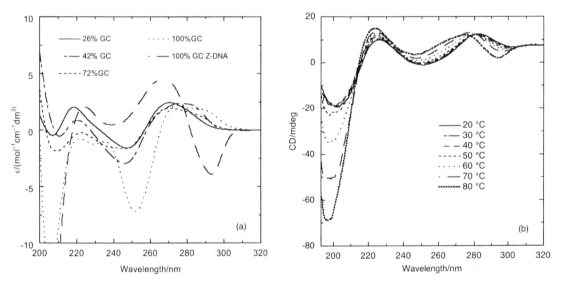

Figure 4.3 (a) *CD* spectrum of DNAs as a function of GC content. All samples in pH = 6.8, sodium cacodylate (1 mM), NaCl (5 mM). *Clostridium* perfringens (26% GC), calf thymus DNA (42% GC), *Micrococcus luteus* (72% GC), poly[d(G-C)]$_2$ (100% GC). Also shown is Z-form poly[d(G-C)]$_2$ induced by 50 μM [Co(NH$_3$)$_6$]$^{3+}$. **(b)** Poly[d(G-C)]$_2$ (50 μM) and spermine (50 μM) in tris buffer (5 mM, pH = 7) and NaF (20 mM) as a function of temperature. The spectra represent the temperature dependent equilibrium between B- and Z-DNA. Increased temperature facilitates the B to Z transition in the presence of spermine.

The *CD* signature of B-form DNA (Figure 4.3), as read from longer to shorter wavelength, is a positive band centred at ~275 nm, a negative band at ~240 nm, with the zero crossing around 258 nm. These two bands are the net result of all the couplings of the transitions present in all the bases. At

220 nm the *CD* signal either becomes positive or less negative (depending on sequence and ionic strength), a small negative peak lies between this and a large positive peak from 180–190 nm.

If B-DNA is compacted and the bases tilted and radially displaced from the centre of the helix (thus creating a hole when one looks down the helix axis) (Figure 3.1a and Table 3.1) then A-form DNA results. A-DNA is characterized by a positive *CD* band centred at 260 nm that is larger than the corresponding B-DNA band, a fairly intense negative band at 210 nm and a very intense positive band at 190 nm. The 230–250 nm region is fairly flat (though not necessarily zero). Naturally occurring RNAs adopt the A-form if they are duplex. A typical natural A-form RNA spectrum is shown in Figure 4.4.

RNAs are stabilized by hydrogen bonding, thus they base pair as much as they can, but are not necessarily restricted to standard Watson-Crick base pairs. siRNA and tRNA are well-defined double-stranded structures.[107]

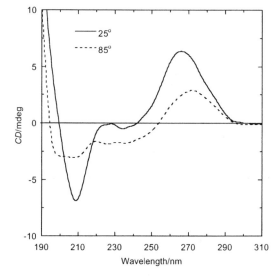

Figure 4.4 *CD* of a long (372 nucleotide) naturally occurring mRNA. At 25 °C the RNA is largely in the A-form; at 85 °C the bases are not rigidly stacked in a helical geometry and the 210 nm band disappears. $A_{260\ nm} = 1$ at 25 °C for this sample.

When the twist direction is reversed from B-DNA to Z-DNA, an inversion of the *CD* spectrum might be expected. While this is indeed true to some extent, we must not assume that B-DNA (which has parallel bases and a right-handed twist) and Z-DNA (which also has parallel bases and a left-handed twist) have equal and opposite *CD*s. This would only be true if they were true mirror images of one another. In going from B-DNA to Z-DNA the *average* skew angle does (more or less) reversed, but the skew angle of any particular pair of coupling dipoles does not. The intrinsic chirality of the backbone ribose units is of course the reason for this lack of reflection symmetry.

Left-handed Z-form DNA (Figure 3.1c and Table 3.1) does not readily form for all sequences. However, it is formed by poly[d(G-C)]₂ in the presence of highly charged ions as illustrated in Figure 4.3 and Figure 4.5. Z-DNA has a negative *CD* band at 290 nm and a positive band at 260 nm. However, care must be taken in using these signatures to identify Z-form DNA, since the same DNA in the A-form has a negative band at 295 nm and a positive band at 270 nm.[108] A more definitive signal is the large negative *CD* signal in the 195–200 nm region for Z-DNA shown in Figure 4.3 a and b, whereas B-form DNA *CD* is near zero or positive in this region. For Z-DNA the *CD* passes through zero between 180 and 185 nm.

A *CD* spectrometer is an ideal tool for probing condensed DNA structures, although what is measured then is not strictly *CD*, but rather a form of light scattering. The results of intramolecular condensation are particles of size comparable with the wavelength of the light. The particles interact differently with left and right circularly polarized light and, at the wavelengths where the chromophores absorb light, give a large apparent *CD* signal—though it is a differential scattering signal not a differential absorbance signal. The sign of the *CD* indicates the helical handedness of the condensed DNA particle, with a negative signal above 250 nm (Figure 4.6) corresponding to a left-handed helix.[109]

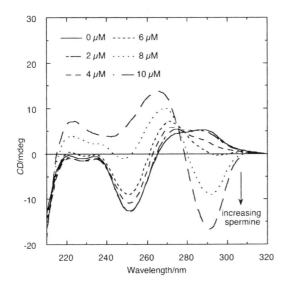

Figure 4.5 Poly[d(G-C)$_2$] as a function of increasing spermine concentration, [NH$_3$(CH$_2$)$_3$NH$_2$(CH$_2$)$_4$NH$_2$(CH$_2$)$_3$NH$_3$]$^{4+}$ showing the B-DNA to Z-DNA transition. Spermine concentrations are indicated in the figure.

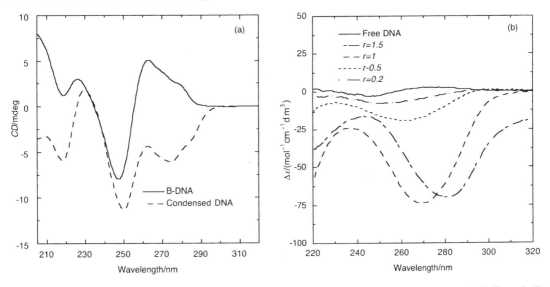

Figure 4.6 (a) Poly[(d(A-T)$_2$] (44 μM, 1 mM phosphate) in B-form (20 mM NaCl) and condensed (1 M NaCl and 45% v/v ethanol). (b) DNA condensed with a polyamidoamine dendrimer as a function of DNA:dendrimer charge ratio, *r*.[110]

CD is also useful for probing structural changes such as those from quadruplex to non-quadruplex structures. However, it should be noted that the accepted spectral signatures for such transitions need to be validated by considering data from other techniques, *e.g.* NMR, from samples at the same concentrations and conditions.

Quantitative structural analysis of DNA CD

With DNA, as discussed above, it is usually possible to qualitatively identify the overall polymorph that is present from its *CD* signature. However, a quantitative empirical analysis, analogous to the secondary structure determination for proteins, has defied extensive efforts. Somewhat perversely, however, theoretical analysis of DNA *CD* performed by calculating the spectrum for a proposed geometry using the so-called matrix method is easier than for proteins. The method used to perform *CD* calculations for DNA and also proteins was developed by Tinoco, Schellman and others, references in [111] and has recently been exploited and extended by Hirst *et al.*[112, 113] The label 'matrix method' was coined because the key computational step involves the diagonalization of an interaction matrix. Instead of using transition dipole vectors as direct input for the exciton calculations as described in Chapter 8, in the matrix method the transitions are expressed as collections of transition monopoles—with the transition monopoles being the changes in charge occurring at various points, usually the atoms, during the transition. The couplings between the transitions are determined from the monopole description of the transition, and the resulting *CD* intensities are then calculated. The way in which *CD* can potentially be used as a detailed probe of DNA (or protein) structure is schematically illustrated in Figure 4.7.

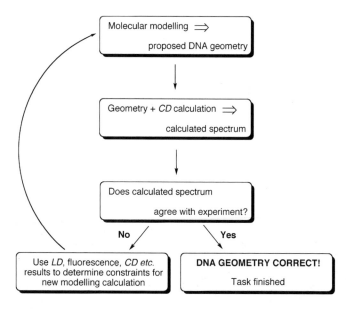

Figure 4.7 Schematic illustration of a self-consistent method for determining DNA structure from *CD* data by coupling calculations with molecular modelling studies of DNA systems.

4.4   DNA/ligand interactions

Many DNA binding molecules (usually referred to as ligands or drugs even when they have little or no therapeutic value) are themselves achiral. However, when they bind to DNA they acquire an induced *CD* (*ICD*) that is characteristic of their interaction. The ligand *ICD* can be used on a number of levels. The simplest application is to note that *ICD* ≠ 0 and therefore to conclude that the molecule does bind to DNA. However, more information can usually be extracted as discussed in reference [114]. Some illustrative examples are given below.

If the ligand transitions of interest occur in the DNA region of the spectrum, we take the *ICD* to be the total *CD* of the system minus that of the DNA at the same concentration. This means any DNA conformational changes that alter the DNA *CD* are included in the *ICD*.

Empirical analyses of ligand *ICD*

It is often useful to measure a series of spectra where some variable such as the ionic strength, or mixing ratio (drug:DNA ratio), or temperature is changed. If the *CD* intensity changes but the shape of the spectrum remains the same during such an experiment, then it may be concluded that the ligand is binding in the same mode (or in a number of sites whose relative proportions are independent of DNA:drug ratio) over the whole concentration range, though the amount of ligand actually bound may have changed (*cf.* Figure 4.8). In such a case, we may write

It is easiest to deduce the *CD* changes per ligand if the ligand concentration is kept constant. A method for doing this is given in §2.4.

$$L_b = \alpha \times ICD \qquad (4.1)$$

where L_b is the concentration of bound ligand, and α is a proportionality constant.

If the ligand causes the DNA to precipitate out of solution, constant spectral shape with decreasing magnitude may be observed; the data must not be used to calculate binding affinities.

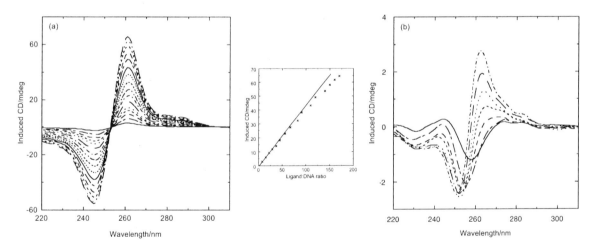

Figure 4.8 Induced *CD* of DNA with an intercalated anthracene chromophore that is coupled to a cationic spermine tail. *ICD* signals are the total *CD* minus the free DNA *CD*. The smallest magnitude *ICD* signal corresponds to the lowest ligand concentration. (a) Poly[d(G-C)]$_2$ (160 μM base). Ligand:DNA ratios are: 0.008:1; 0.015:1; 0.022:1; 0.029:1; 0.037:1; 0.045:1; 0.056:1; 0.067:1; 0.083:1; 0.091:1; 0.11:1; 0.12:1; 0.14:1; 0.15:1; 0.16:1; 0.17:1. The middle graph shows the *CD* at 262 nm as a function of ligand:DNA ratio.[115] (b) Poly[d(A-T)]$_2$ DNA (130 μM base). Ligand:DNA ratios are: 0.05:1; 0.13:1; 0.19:1; 0.23:1; 0.30:1; 0.40:1; 0.56:1.[116] Pathlength equals 1 cm.

For very high DNA concentrations, the concentration of ligand that is bound to the DNA approximately equals the total ligand concentration. Some of the simpler analysis methods that may be used to determine binding constants and site-sizes are outlined in §11.2. The data in Figure 4.8, *e.g.*, lead to the equilibrium association constant of anthracene-9-carbonyl-N^1-spermine with poly[d(G-C)$_2$] being $(2.2 \pm 1.1) \times 10^7$ M^{-1} in a mode that spans six bases (Figure 11.2). Other data, including *LD* data, showed that the binding mode involves the anthracene moiety being intercalated.[115]

If there is a change in the shape of the spectrum as the mixing ratio or other experimental variable is changed (*e.g.* Figure 4.8b), this implies there is a change in the DNA-drug interaction as a function of the experimental variable. The change is usually due to occupancy of more than one binding mode as the drug load on the DNA increases, but may also be due to changes in the DNA conformation or to ligand-ligand interactions. Thus, the high ligand load limiting spectrum for single particular binding mode is seldom available.

If an experiment is conducted at constant ligand concentration and varying DNA concentration, then, unless the ligand binding mode is changing, there should be no change in the observed ligand *ICD* outside the DNA absorbance region, *i.e.* above 300 nm. A particularly dramatic change is observed, however, if as the DNA concentration is decreased (so one would expect fewer ligands to bind and so a decrease in *ICD*), the ligands begin to stack together and there are significant ligand-ligand interactions occurring on the DNA support. Under these circumstances, we generally observe a large excitonic *CD* (Chapter 8) as illustrated in Figure 4.9 and the *ICD* magnitude increases rather than decreases.

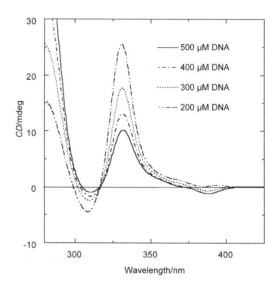

Figure 4.9 *CD* of constant concentrations of 9-hydroxyellipticene (50 μM) in the presence of calf thymus DNA (concentrations indicated in figure, 5 mM NaCl, 1 mM phosphate buffer). Note *ICD* signals increase with decreasing DNA concentration.

To be able to determine a binding constant from *CD* or *LD* data, the plot of intensity *versus* concentration should have a reasonably straight part (where most ligands bind), followed by a curved part (where the equilibrium between bound and free is apparent), and ideally a levelling off where no further change occurs (since all DNA or ligand sites are occupied) (*e.g.* Figure 11.3). Some methods for analyzing such spectral data are outlined in §11.2.

Many DNA binding ligands have sufficiently high binding constants that a plot of induced *CD* signal *versus* ligand:DNA mixing ratio is linear. In such a case all one can deduce is that the binding constant is high. One such example is given in Figure 4.10. In this case a clear change in binding mode occurs at 10:1 DNA base:ligand mixing. Since the binding constant is high (so essentially all available ligand is bound) and the data are for the racemic mixture of the two ligand enantiomers, this suggests that one of each enantiomer is binding in every turn of the B-DNA helix. Alternatively, a linear relation may also be observed in the limit of extremely weak binding. In this case saturation is never observed.

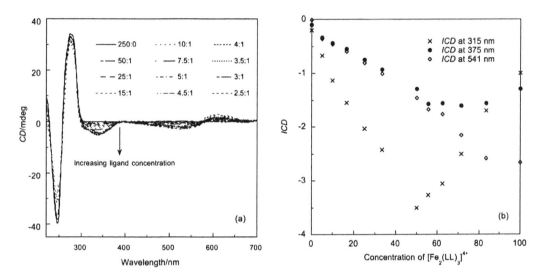

Figure 4.10 (a) *CD* calf thymus DNA (500 μM (in bases) in 20 mM NaCl and 1 mM Na$(CH_3)_2$AsO$_2$.3H$_2$O buffer pH = 6.8) with increasing concentrations of (±)-[Fe$_2$(LL)$_3$]$^{4+}$.[91] 1 cm path length. (b) *ICD* as a function of ligand concentration from (a). LL is illustrated in Figure 3.10.

The origin of ligand *ICD*

The *CD* induced into ligand transitions upon binding to DNA almost always arises as a result of non-degenerate exciton coupling (*cf.* Chapter 8) between the electric dipole transition moment of the ligand and those of the DNA bases. The simple exciton approach of Chapter 8 can be used to derive an equation for the net effect of all the bases in a piece of DNA on a given ligand transition. This is pursued in more detail in §8.8 for the case where the ligand is intercalated between two DNA base pairs.

Based on coupled-oscillator theory (Chapter 8) relations have been derived for the *CD* of DNA-bound ligands[45] which have enabled structural characterisation of both intercalative as well as groove-bound complexes. When combined with *LD* data, which provide orientation angles of the

ICD is obtained by subtracting the free metal complex *CD* spectra from the calf thymus DNA plus metal complex *CD* spectra.

ligands relative to the DNA helix axis, the *CD* adds an extra dimension by probing the 'azimuthal' angles, *i.e.* directions referring to projections on the plane perpendicular to the helix axis.

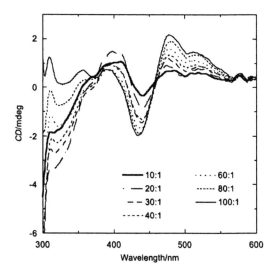

Figure 4.11 *ICD* spectra (50 mM NaCl and 1 mM sodium cacodylate buffer) for constant concentration (P)-[Ru$_2$(phen)$_4$L]$^{4+}$ (L is a bis(pyridylimine) tetradentate ligand that spans the two ruthenium centres, 15 μM) and varying calf thymus DNA concentrations. DNA base to metal complex ratios are indicated on the figure. 1 cm path length.[117]

The *ICD* of intercalated chromophores is generally rather weak, with a sign that depends on how the transition moment is directed within the intercalation pocket. A simple expression applies for the case of an electric-dipole allowed transition in an intercalated chromophore[118, 119] as discussed in §8.8. From this work, the orientation of intercalators within the intercalation pocket can be determined. It predicts a positive *ICD* for transitions polarized poking out into the groove and negative for those polarized along the longest dimension of the intercalation pocket.[84, 120-122] For non-intercalated, groove bound ligands a similar relation was derived to explain the markedly stronger *ICD* for such complexes, predicting strong positive *ICD* for transitions polarized parallel to the grooves.[123]

Using the matrix method and established transition moments of guanine and cytosine, the *CD* of [poly(dG-dC)]$_2$ and the *ICD* of ligand transitions for various geometries of binding to [poly(dG-dC)]$_2$ have been computed.[124] Similar work with the alternating AT DNA has been undertaken.[125] The calculations show, for example, that the *ICD* of a groove-binder is an order of magnitude stronger than that of an intercalated molecule. This finding is in accord with experimental observations, as is the predominantly positive sign of groove bound *ICD* signals.

It should be emphasized that the positive *ICD* is for *transitions* poking out into the groove. This may or may not be the same as the long axis of the molecule poking out into the groove. The assignment of transitions within the intercalator is required to make use of the *CD* sign to determine the orientation of intercalators in the intercalation site.

4.5   *CD* of polypeptides and proteins

Introduction

The *CD* spectrum of a protein contains information about the asymmetric features of the backbone of proteins as well as about the orientations and environments of the side-chains. The challenge is to extract that information. The most common reason for collecting protein *CD* data is to assign secondary structure content by expressing the spectra as a combination of standard spectra which are then deconvoluted to give the percentages of a limited number of well-defined backbone geometries. In the absence of any contributions to the *CD* from side chain transitions, this has proved to be a very successful approach. As discussed in §1.3, protein spectroscopy is usually referred to as either (i) far UV or backbone (meaning the amide transitions, Figure 3.20) with data collected from ~190 to 250 nm or (ii) near UV or aromatic with data collected from 250 to 300 nm. The main reason for the division is that the absorbance magnitudes of the two regions for a typical protein differ by ~2 orders of magnitude.

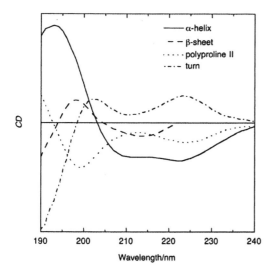

One routinely collects protein backbone *CD* for a 0.1 mg/mL protein solution in a 1 mm path length cuvette. Although it depends on the aromatic chromophores present (which thus affects the absorbance intensity), a good starting point for aromatic region spectroscopy is 1 mg/mL protein in a 1 cm path length cuvette. If the concentrations of the sample of interest differs from these, then the path length can be adjusted to compensate according to the Beer-Lambert Law (equation (2.3)).

Figure 4.12 Approximate shapes of *CD* spectra expected for standard protein motifs, see *e.g.* reference [126].

Distinctive *CD* spectra (Figure 4.12) have been described for pure conformations such as the α-helix, β-sheets (with different spectra sometimes being given for parallel and anti-parallel sheets), β-turns, poly-proline type II helix, and also the 'random' coil conformation. At least in principle, the *CD* spectrum of a native protein is then the sum of the appropriate percentages of each component spectrum. As discussed below, there are a variety of methods for working backwards to determine the percentage α-helical *etc.* content of a given protein. In practice it has proved better to use a basis set of real proteins of known secondary structure content rather than determining idealized spectra secondary structure conformation.[97, 126-128]

Protein backbone *CD*

α-helix CD

The α-helix is the dominant secondary structure in many proteins and on average accounts for about one third of the residues in globular proteins. It is a well-defined motif and is also adopted by many polypeptides, particularly membrane ones.

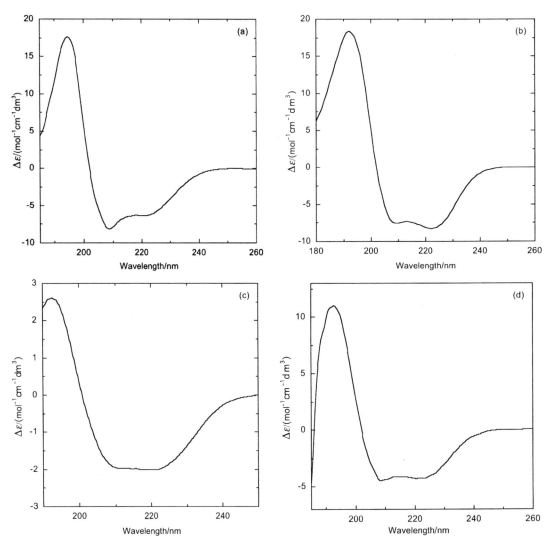

Figure 4.13 (a) *CD* of an α-helical frog peptide aurein 2.5 with sequence: GLFDIVKKVVGAFGSL-NH$_2$ inserted into a lipid bilayer composed of 50:50 DPPC (dipalmitoylphosphatidylcholine) and DPPG (dipalmitoylphosphatidylglycerol). (b) *CD* of the 73% helical protein myoglobin. Δε is per residue. (c) *CD* of a 20% helical protein fragment from the C-terminal domain of TolA (TolA-(296–421)). Δε is per residue. Data were provided by J. Lakey.[129] (d) *CD* of the 29 residue helical peptide melittin from bee venom.

The α-helix *CD* is larger in magnitude than that due to other motifs and has a distinctive spectral form in the backbone region of the spectrum; it is apparent upon the most casual inspection of a spectrum. The *CD* spectra of α-helices have two negative peaks of similar magnitude at 222 nm (the $n{\rightarrow}\pi^*$ transition) and 208 nm (part of the $\pi{\rightarrow}\pi^*$ transition) and a larger positive peak at about 195 nm. For further discussion of the 208 nm band, which only appears for the α-helix, see §8.6. The relative magnitudes of the 222 nm and 208 nm peaks may vary, particularly as a function of the hydrophobicity of the environment which is increased in a membrane or by proteins dimerizing or folding and protecting their surfaces from the solvent (resulting in a relative increase of the 222 nm band).

The α-helical content of a protein may be estimated approximately by assuming that a 100% helical protein has a mean residue signal of $\Delta\varepsilon \sim 12$ mol^{-1} cm^{-1} dm^3.[131, 132] The *CD* of the aurein 2.5 frog peptide in Figure 4.13a (*cf.* Figure 3.22) therefore shows that the peptide is about 64% folded into a helix, whereas the myoglobin *CD* spectrum of Figure 4.13b correctly suggests that myoglobin is approximately 70% α-helical. The various empirical *CD* fitting programs that are available are fairly successful in determining the percentage of α-helical content of a protein—particularly if the concentration is known accurately.

It should, however, be noted that the magnitude of the *CD* signal does vary with variations in the helix structure and has been concluded to depend on helix length. Thus the estimates for peptides are to be viewed with caution. The length dependence of the α-helix *CD* is unsurprising: the signal behaves as if it were for a helix approximately four residues shorter than it in fact is; this corresponds to the number of unanchored hydrogen-bonding groups.

β-sheet CD

The spectroscopic characterization of β-sheets in proteins has proved more difficult than that of α-helices as in general they are less soluble in solvents with a good UV transmission as well as the intrinsic reason that they are generally structurally (and hence spectroscopically) less well-defined: they may be parallel or antiparallel and of varying lengths and widths. Furthermore, an extended β-sheet is usually found to show a marked twist, rather than to be planar. Such tertiary structures also influence the overall *CD* spectrum.

The general characteristics of β-sheet *CD* may be taken to be a negative band between 215 nm and 219 nm and a larger positive band between 195 nm and 202 nm. The peptide and antibody spectra of Figure 4.14 are typical examples. This level of characterization of the spectrum might be deemed to be not much worse than that of α-helices were it not for the fact that the 'random coil' spectral features have their maxima at similar wavelengths and are of opposite sign from those of the β-sheet. This means that an empirical fitting program may incorrectly weight these spectra (and the β-turn components) in an attempt to better account for the wavelength and magnitude variations that occur.

A 222 nm:208 nm intensity ratio of about 0.8 has been proposed for a single-stranded α-helix and a ratio of about 1.0 for a two-stranded coiled-coil of α-helices.[130] The actual ratio depends on H-bonding and the environment of the helix. Larger 222 nm:208 nm ratios widely believed to correspond to helices in more hydrophobic environments.

Describing a structural motif as α-helical when it is only about four amino acids in length is unhelpful as the spectrum will be dominated by 'end-effects'.

As protein backbone CD signals arise from transitions of the amide bonds, it is usually more useful to express the CD in terms of concentrations of amide bonds (or amino acids) rather than protein molecules. The spectra of proteins of different sizes may then directly be compared.

The 3_{10} helix has a *CD* spectrum broadly similar to that of an α-helix, but the 222 nm band is much smaller than the 208 nm band and the positive 195 nm band is also small.

β-turn CD

The label β-turn is usually used to include all possible turns that occur, not simply the ones that enable a single β-strand to become an anti-parallel β-sheet. About one quarter of the residues in globular proteins then fall into this structural group. Despite this range of structures, a 'typical' β-turn *CD* spectrum has been identified which has a weak red-shifted negative $n{\rightarrow}\pi^*$ band near 225 nm, a strong positive $\pi{\rightarrow}\pi^*$ transition between 200 nm and 205 nm, and a strong negative band between 180 nm and 190 nm. However, the spectra of turns are really not nearly as well-defined as the previous sentence might imply.

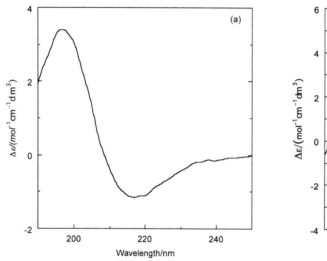 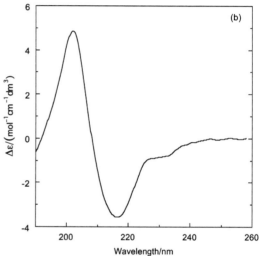

Figure 4.14 β-sheet *CD* spectra of (a) a nonapeptide GLRILLLKV in a solution containing soya bean lipid liposomes and (b) a monoclonal antibody.

Random coil CD and polyproline II helix

In an attempt to avoid the somewhat misleading implications of the label 'random coil', Woody has coined the phrase 'unordered conformation'.[133] Whatever label one uses, it is important to realise that we are generally grouping the parts of the folded protein that do not fit into one of the previously discussed categories which means that there may be ordered structures included in this category. Furthermore, even truly disordered or unfolded proteins[134] have a strong negative *CD* signal at ~200 nm, a small positive band at about 218 nm in many systems, and perhaps a very weak negative band at 235 nm.

Unfortunately for the use of *CD* as a means of identifying unfolded proteins, the very well-defined left-handed polyproline II helix[133] of collagen is indistinguishable from the general description of a random coil spectrum. This similarity of random coil and polyproline type II helix *CD* signatures suggests that the local arrangement of transition moments is similar in the two geometries.[136]

As shown by Woody and Lakey,[137] the underlying reason for the similarity of the spectrum is that the characteristic negative maximum at ~200 nm (*cf.*

Raman optical activity experiments and calculations support the occurrence of polyproline II conformations in unfolded proteins.[135]

Figure 4.15) results from the coupling of a peptide bond N-terminal to the chiral α-carbon in a polypeptide chain. Therefore the simplest peptide bonds have a preferred conformation that is essentially that of the polyproline II helix. So in the absence of other well-defined secondary structures such as the α-helix, polyproline II conformations dominate the local structure and hence the *CD* spectra of peptide chains.

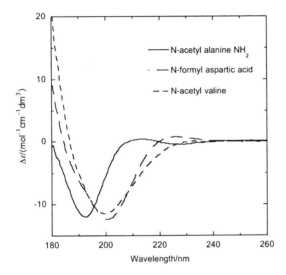

Figure 4.15 *CD* spectra of some peptide bonds illustrating the variation in positions of the ~200 nm negative band of the polyproline II conformation.[137]

A variant on the β-strand structure, referred to as β_{II} also has the same spectral signature. Frequently other information about the protein enables these three options to be distinguished. Sreerama and Woody concluded that the *CD* spectrum of the β_{II} structure *CD* spectrum owed its form to a high percentage of polyproline in the proteins of this class.[138]

Determining the percentage of different structural units in a protein

There are a range of different computer programs available for determining the percentage of different structural motifs just from the *CD* spectrum. The cost of these programs does not necessarily correlate with their reliability. Many of them are freely available on different web sites *e.g.* reference [139]. All use 'standard' reference spectra, usually of proteins where the solution phase and crystal structure are known to be the same. A measured spectrum is then decomposed so that an appropriately weighted sum of the reference spectra equals the measured spectrum. This is then converted into percentages of different secondary structure motifs.

The preceeding description of the spectral features of different motifs suggests that spectral data extending down to at least 190 nm should be used for any structure fitting analysis. A wider wavelength range provides more transitions and (usually) a more reliable answer. However, there are the inevitable qualifications to this statement. Any error in the concentration or path length translates into an error in the proportions of secondary structure that result from the fit. The contributions to the observed *CD* from side chains can also become significant, depending on how many there are. It is currently accepted that α-helix estimates from most fitting programs will be reliable. Other estimates may need to be confirmed from independent sources.

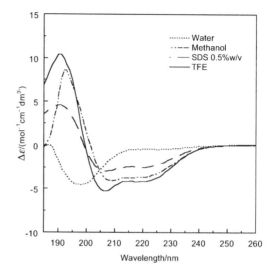

Figure 4.16 *CD* of a peptide (MSLSRRQFIQASGIALCAGAVPLKASA) in different solvents.[140]

Protein CD as a function of structure and environment

Many proteins change their structure as a function of environment and this is reflected in their *CD* spectrum. Solvent effects on short peptides can be particularly dramatic as shown in Figure 4.16. In this example, a change from approximately 50% α-helix (in TFE) to random coil (in water) is observed.

Protein aromatic region *CD*

The aromatic chromophores of protein side chains are planar and so have no intrinsic *CD*. However, their location as part of an amino acid results in an induced *CD* signal in their transitions. This is further enhanced when they are located in the chiral environment of a peptide or protein. The sign and intensity of the *CD* induced into the achiral aromatic side chains is dependent on their environment. It is a useful fingerprint of protein identity and conformation, but seldom leads to direct structural analysis. A selection of aromatic protein spectra for a range of biopharmaceutical (*i.e.* therapeutic proteins in a formulation buffer) products are illustrated in Figure 4.17. In this context it is worth noting that *CD* is one of a limited number of techniques required by regulators to show that the structure of the active protein has not changed from batch to batch.

As side chains are usually isolated from one another, their *CD* can be treated analagously to the situation when we look at the *CD* induced into a

In contrast to fluorescence spectroscopy, it is not possible to limit consideration of *CD* data specifically to the tryptophans.

ligand bound to the macromolecule. Side chain *CD* has proved to be particularly useful for probing conformational changes in a part of the protein containing one of these residues. The development of site-directed mutagenesis (whereby selected chromophores can be removed from the protein being studied) has greatly facilitated this approach. Quantitative analysis of the kind outlined in Chapter 8 has seldom proved possible due to the number of transitions from amide groups around any given residue. Progress is, however, being made by Hirst and co-workers for both side chains and ligands using a version of the matrix method approach mentioned in §4.3.[112] The challenges remain to make accurate assignments of magnitude and polarization of chromophore transition moments and accurate determination of couplings between transitions.

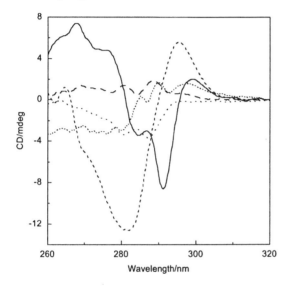

Figure 4.17 A selection of aromatic protein spectra for a range of biopharmaceutical products.[141] Concentrations of all samples: 1 mg/mL in a 1 cm path length.

Protein *CD* due to interactions between side chains

Groups that can essentially be ignored in the absorption spectrum may become significant in the *CD* if there is a magnetic dipole allowed (*cf.* Chapter 9) component to their transitions. For example, disulfide or cystine groups (covalent bonds formed between cysteine residues in different parts of a protein) have weak transitions in the aromatic region of the spectrum ($\varepsilon_{max} \sim 300$ mol^{-1} cm^{-1} dm^3, for broad bands with maxima between 250 and 270 nm). However, they are magnetic dipole allowed and may contribute significantly to the protein *CD* in this region.

When peptides with aromatic residues assemble (such as in a membrane), bands in the 230–240 nm region of the *CD* spectrum appear. Such effects are often even more apparent in *LD* spectra as shown in Figure 5.6. They arise as a result of π-π^* exciton coupling (*cf.* Chapter 8) of aromatic transitions, where one of the components occur at about 235 nm and the shorter wavelength component is masked by the backbone transitions. The unusual

CD spectra of gramicidin (*cf.* §5.4 for information on gramicidin) may in large part be due to the many (25%) aromatic residues it contains that could couple together.

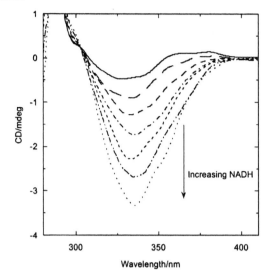

Figure 4.18 *CD* of lactate dehydrogenase (LDH, 22.5 μM) with increasing concentrations of NADH (0.1 mM, 0.15 mM, 0.2 mM, 0.25 mM, 0.3 mM, 0.35 mM, 0.4 mM).

Protein-ligand interactions and induced *CD* signals

Lactate dehydrogenase and NADH

As with DNA ligand systems, one expects a change in both the ligand and protein *CD* upon interaction. The *CD* spectrum of lactate dehydrogenase (LDH) with nicotinamide adenine dinucleotide (NADH) is given in Figure 4.18. At first sight there appears to be an induced *CD* signal, however, the ligand is chiral and the observed spectrum is simply the sum of the component parts. So care must be taken in using *ICD*.

Actin, ATP and myosin

Actin is a protein that binds to myosin during muscle contraction. Some indication of the lack of protein secondary structure conformational change resulting from of complexation is given by the comparison between the theoretical sum of the actin and myosin S1 spectra in Figure 4.19 compared with the experimental spectrum of the mixture. The experimental *CD* spectrum of actin-S1-ATP is similar to the theoretical *CD* spectrum from the sum of the components over the full wavelength range indicating there is no significant change in the component secondary structure. The true and theoretical sum spectra, however, differ enough in the 208 nm region to confirm the binding.

4.6   Oriented *CD*

Oriented *CD* (*OCD*) spectroscopy can be a very useful technique. However, it is prone to artifacts resulting from an *LD* contribution to the observed spectrum (as discussed below). A sample for *OCD* should be symmetric about a unique orientation axis; data must then be collected with the light propagating along that unique axis and data compared for rotations of the sample about that axis using a cell such as that illustrated in Figure 4.20. If the rotated spectra are the same, then these data can be believed.

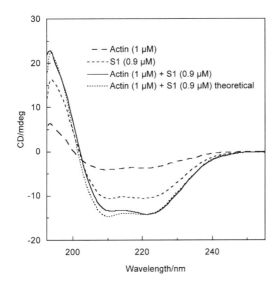

Figure 4.19 *CD* spectra of actin (1 μM); myosin S1 (0.9 μM); the mixture of actin and S1 (1 μM and 0.9 μM respectively); and the theoretical spectrum of actin (1 μM) + S1 (0.9 μM).[66] All samples in KCl (15 mM); MgCl$_2$ (0.75 mM) and MOPS (3 mM) pH 7. Spectra were collected using a 1.1 mm diameter capillary and 5 μL sample volume.[73]

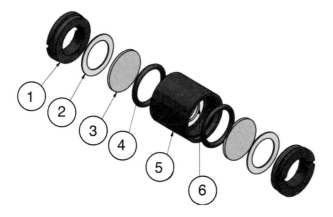

OCD samples should ideally be prepared in a cell where the humidity can be controlled.[58]

Figure 4.20 Schematic diagram of a unit to hold an *OCD* cell. Position 1: a threaded end cap; position 2: a teflon seal; position 3: a quartz window on which sample is dried; position 4: an O-ring; position 5: small hole to insert salt solution[58] to achieve required humidity level; and position 6: teflon seal.[142]

The Latin word *mellitus* means 'honeyed, sweet as honey'. Anomalously the bee venom peptide has come to be spelled melittin in most references.

The ~195 nm *OCD* band is red-shifted compared with the isotropic *CD* spectrum where the high energy π-π* band is closer to 190 nm due to more cancellation of intensity from the larger isotropic 208 nm band.

Quadrupolar terms may also contribute to the CD of oriented samples as discussed in §10.10.

To understand *OCD*, let us consider the *CD* spectrum of an α-helical peptide inserted into and perpendicular to a lipid bilayer that has been dried onto a quartz plate. If the orientation is perfect, from Figure 3.19 one can deduce that the 210 nm region of the spectrum should have no intensity as the light is propagating along the direction of its polarization. So an α-helix inserted into a membrane will have a negative 222 nm *CD* band and a positive ~195 nm band.

The *OCD* spectrum of the bee venom peptide melittin shown in Figure 4.21 indicates that the melittin has indeed inserted approximately parallel to the lipids of the bilayer. The fact that the 208 nm band is still apparent in this case tells us that the melittin is not perfectly perpendicular to the membrane normal. As with *LD*, we cannot tell whether this is because all the peptides are tilted or there is disorder in the system.

If instead of being inserted into the membrane, a helix lies parallel to the surface of the membrane, then the transitions polarized along the helix axis (208 nm in this case) will have full *OCD* intensity and those perpendicular to the helix axis will have half the maximum possible *OCD* intensity since they are oriented partly along the propagation direction and partly not (assuming perfect orientation). Thus the 208 nm band will be comparatively enhanced for surface helices rather than diminished as is the case for inserted peptides. In using *OCD* it is important, as with *LD*, to recall that orientation is unlikely to be perfect.

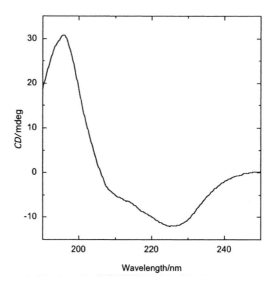

Figure 4.21 Oriented circular dichroism spectra of melittin in DMPC (dimyristoylphosphatidylcholine) at 1 to 100 peptide to lipid molar ratio. The sample was added to a quartz window in a dropwise fashion, dried in a flow of nitrogen, and left overnight at room temperature to slightly hydrate the sample. The final spectrum is the average of the four similar spectra obtained at 0°, 90°, 180° and 270°.[143]

Contributions to *CD* signal recorded on oriented samples

The above discussion is based on the assumption that what is being measured is only *OCD*. However, the apparent *CD* signal recorded using a *CD* instrument and an oriented sample, CD_{obs}, consists of the true *CD* plus several *LD*-based contributions. For a uniaxial sample with the optic axis perpendicular to the propagation direction of the light and 45° from the axis of induced birefringence of the PEM[76]

$$CD_{obs} = CD + LD\cos(2\beta_o)\sin\kappa$$
$$+ \frac{1}{6}\left[CB \cdot LD \cdot LB - CD \cdot LB^2 + \left(\frac{\ln 10}{2}\right)^2 \left(CD^3 + CD \cdot LD^2\right)\right] \quad (4.2)$$

Here κ denotes a static, spurious birefringence (expressed as retardation in radians) in the modulator or sample optics, oriented at angle β_0 to the photo elastic modulator (Figure 10.3) axis. *LD* is a linear dichroism of the sample oriented at 45° to the same reference axis. Linear birefringence (*LB*) and optical activity (circular birefringence, *CB*) in the sample contribute as well, but in most experiments the dominant term, which may exceed the intrinsic *CD* signal itself by orders of magnitude, is the second term in equation (4.2). In this term the *LD* of the sample is mixed into the recorded instrument *CD* by any non-parallel birefringence. Thus, while the higher order terms in the *LD* and *CD* may be ignored, the second term in equation (4.2) can give rise to serious artifacts should the sample exhibit a significant *LD* in some direction perpendicular to the propagation direction of the light beam at the same time as the instrument, the sample cell or modulator have a static birefringence.

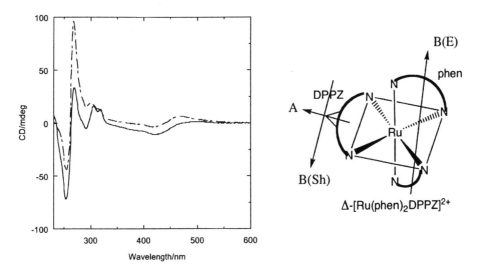

Figure 4.22 *CD* spectrum of Λ-[Ru(phen)₂DPPZ]²⁺ in water (dashed line) and in a lamellar liquid crystal (octanoate-decanol-water) at normal incidence (solid line). The difference between the isotropic and oriented *CD* is the signal that is not contributed to the oriented spectrum by the transition along the long axis (A) of the molecule, which is aligned with the direction of the propagation of light.

Thus, the second correction term of equation (4.2), and hence the recorded *OCD*, will depend on the *LD* in the sample plane and thus on how the sample is oriented. Since

$$LD(\eta) = LD_{max}\left(\cos^2 \eta - \sin^2 \eta\right) \qquad (4.3)$$

where η is the angle of rotation of the sample around the normal to the orientation plane and LD_{max} is the maximum *LD* signal of the sample occurring at $\eta = 0$, we may write the variation in observed *OCD* when rotating the sample around the normal to the orientation plane (neglecting the higher order terms of equation (4.2)) as

$$CD_{obs}(\eta) = CD + LD_{max} \cos(2\eta) \cos(2\beta_o) \sin\kappa \qquad (4.4)$$

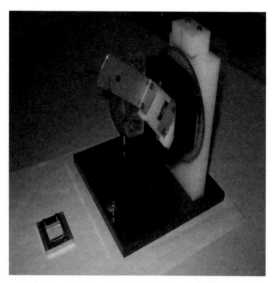

Figure 4.23 Goniometer cell holder for lamellar membrane measurements (for oriented *CD* at normal incidence, as well as, for *LD* at tilted geometry).

As noted above, in practice, the problem may be solved by measuring the *OCD* at many angles uniformly distributed in the plane of the sample because the average

$$\langle\cos(2\eta)\rangle = 0 \qquad (4.5)$$

These issues are illustrated for the light-switch complex [Ru(1,10-phenanthroline)$_2$dipyrido[3,2-a:2',3'-c]phenazine]$^{2+}$ ([Ru(phen)$_2$DPPZ]$^{2+}$ in a lamellar liquid crystal host (Figure 4.22).[144] Its *LD* spectrum is shown in Figure 6.12. A cell holder for such measurements is illustrated in Figure 6.12. The goniometer of this device allows both measurement at normal incidence (angle ω between sample plane and propagation direction of light is then 90°) and also at different angles in sample plane (angle η in equation (4.4)). The same device can also be used for measuring membrane *LD* at inclined incidence ($\omega < 90°$) as illustrated in Figure 6.12 (*cf.* equation (6.42)).

4.7 *CD* of chiral dimers

At the beginning of this chapter we illustrated in qualitative terms how a combination of a linear displacement of charge (an electric dipole moment) and a circulation of charge (a magnetic dipole moment), during the excitation process of a chiral molecule, gives rise to a circular dichroism signal. In Chapter 8 this approach is given a firm theoretical basis to show how the *CD* is proportional to the scalar product between the electric dipole transition moment (μ) and the magnetic dipole transition moment (m)

$$CD \propto \mu \cdot m \qquad (4.6)$$

This is the Rosenfeld equation which tells us that the *CD* is positive if μ and m are parallel and negative if they are antiparallel. Equation (4.6) thus predicts that for the inherently chiral, right-handed spiral-shaped hexahelicene isomer depicted in Figure 4.1, the *CD* is positive, since μ and m are parallel (irrespective of in which direction we let charge be displaced around the screw). If we observe the same molecule and charge displacements in a vertical mirror it is easy to see that m but not μ will change into the opposite direction, so that they now become antiparallel instead of parallel. According to equation (4.6) the *CD* for the mirror image enantiomer should thus be negative. For such inherently chiral chromophores for which the relative orientations of electric and magnetic transition moments may be deduced simply by visual inspection, we have a way of relating the sign of the *CD* at the wavelength of transition to the molecular absolute configuration, or in other words to tell, in the case of hexahelicene, whether we are dealing with the right or left-handed configuration.

Another similarly simple way of non-empirical assignment of absolute configuration can under favorable conditions be applied to chiral dimers as will be briefly illustrated here. As discussed in Chapter 8, if achiral chromophores are arranged in some chiral fashion relative to one another, they can exhibit circular dichroism. We conclude this chapter by a qualitative description of the, so-called, exciton *CD* of a dimer consisting of two electronically separate chromophores, each carrying an electric-dipole allowed transition. This situation is treated in detail in §8.5 and §8.6. Here the example will be given in qualitative terms and a practical scheme advised for assignment of μ and m which is useful for interpretation of *CD* spectral patterns of dimers. The same scheme may also be applied for the analysis of *LD* for dimers in oriented systems.

It is often possible to determine if an orientation of dipole moments is attractive or repulsive just by visual inspection: put a negative charge at the tail of each dipole arrow and a positive charge at the head of each arrow and see where opposite charges attract and same charges repel each other.

Exciton coupling between two dimers

Exciton coupling between two identical electric-dipole allowed transitions sitting on two chromophores that are chirally arranged relative to each other (Figure 4.24a), changes the absorbance spectrum as well as the *CD* and may be described as follows in terms of the effective electric and magnetic transition moments.

- *New transitions*: Two new 'exciton' transitions (Figure 4.24b(ii)) will appear in the spectrum: one at higher energy (shorter wavelength) and one at lower energy (longer wavelength) than the transition in the isolated (monomeric) chromophore (Figure 4.24b(i)). The one at higher energy corresponds to a repulsive (destabilizing) arrangement of the electric dipole transition moments relative to one another, the one at lower energy corresponds to an attractive (stabilizing) arrangement.

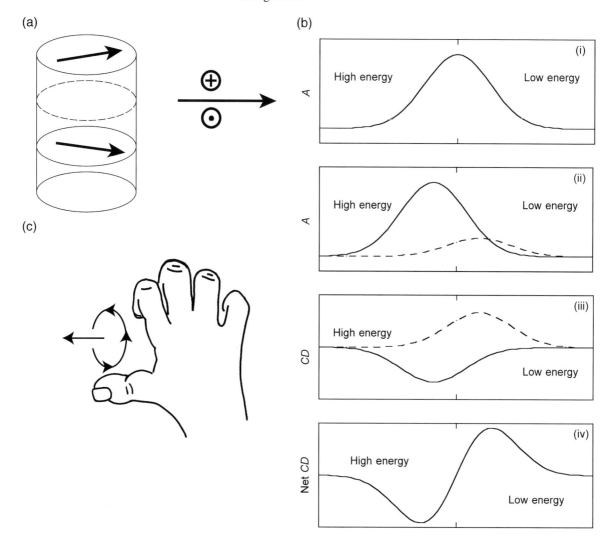

Figure 4.24 (a) Diagram illustrating exciton coupling between two chirally arranged monomeric chromophores with angle between the two transition moments of ~36°. The net dipole is shown to the right. (b) The associated absorption and *CD* spectra: (i) Monomer absorbance; (ii) dimer absorbance; (iii) dimer *CD* exciton bands, (iv) net *CD* spectrum. For fuller explanation see text. (c) Right-hand-thumb rule: if fingers are aligned with the electron flow then the thumb shows the direction of the axial magnetic dipole moment vector.

- *Total electric dipole transition moment of dimer:* The two new transitions correspond to collective charge diplacements described as the sum or difference of the two monomeric electric dipole transition moments.

$$\mu(\text{dimer}) = \frac{1}{\sqrt{2}}\left[\mu(\text{monomer 1}) \pm \mu(\text{monomer 2})\right] \qquad (4.7)$$

Here the factor of $1/\sqrt{2}$ arises to ensure the total absorption intensity is conserved. However, as a result of the + and – signs in equation (4.7), the absorption intensities of the two transitions may differ from each other.

The square of a transition moment is proportional to the absorption intensity, so the factor of $1/\sqrt{2}$ ensures the squares of the two new transition moments sum up to twice the square of the transition moment of each monomer.

For example, if the monomeric transition moments are parallel, the plus-combination will be a transition that contains all the absorption intensity of the two monomers while the minus-combination will have zero absorption (the two antiparallel monomeric dipoles exactly cancelling each other). In the spectrum, this extreme case will look like a shift of the monomer absorption band to either shorter or longer wavelength depending on whether the dipole combination with absorption intensity is, respectively, repulsive or attractive. However, if the monomeric transitions are oriented at any other angles than parallel, antiparallel or orthogonal to each other, the new transitions of the dimer will both have non-vanishing (albeit generally different) absorption intensities and be oriented in different directions.

These two transition moments, also represent what may be probed by linear dichroism spectroscopy for the coupled system. Since the two new transitions can be shown to be always polarized perpendicular to each other, their directions in an oriented sample may be an interesting source of information.

- *Total magnetic dipole transition moment of dimer:* The direction of any component of the total magnetic dipole transition moment that is collinear with the total electric dipole transition moment μ(dimer), is determined by the right-hand-thumb rule: is there any circulation of charge in a plane perpendicular to μ(dimer).

The right-hand-thumb rule involves aligning one's right hand thumb along the electric dipole transition moment direction and curling ones fingers around that direction. The direction of curl is the magnetic dipole transition moment.

- *The CD of the dimer* is now given by equation (4.6) with μ(dimer) being the total electric dipole transition moment of the dimer and m being the magnetic moment component as determined above.

CD resulting from exciton coupling between two dimers

Let us apply this set of rules to a chiral dimer as shown in Figure 4.24a. This dimer might, for example, be two identical DNA bases (*cf.* Figure 8.10) stacked on top of each other in a right-handed helix (with 36° dihedral angle between), but it could also represent general cases of chirally twisted bi-aryls *etc*. Figure 4.24b is a book-keeping diagram showing (from the top) (i) the absorption of the isolated monomeric transitions (corresponding to the arrows drawn for each of the monomers in Figure 4.24a), (ii) the absorption bands of the two new (exciton) transitions, (iii) the deduced *CD* components at higher and lower energy, and (iv) the expected net *CD* spectrum. The derivation of Figure 4.24b follows from the step-by-step process given below.

(1) Draw the directions of the monomeric transition moments and put heads on the arrows (the choice of direction at this stage does not matter) as in Figure 4.24a.

(2) Put a minus charge at the tail and plus charge at the head of each arrow. Determine whether the arrangement is repulsive or attractive. In Figure 4.24a, with a small angle (36°) between the dipoles, we conclude a repulsive arrangement is illustrated; we have thus illustrated the case of the higher energy (shorter wavelength, solid curve) relative to the monomer transition in Figure 4.24b(ii).

(3) Form the sum of the monomeric transition dipole vectors, μ(dimer), as illustrated by the longer arrow to the right of Figure 4.24a. (This is thus the polarization of the short-wavelength exciton transition.)

(4) Look for circulation of charge in plane perpendicular to μ(dimer) as shown in Figure 4.24 using the right-hand-thumb rule. Your thumb points in the opposite direction from the μ(dimer).

(5) μ and m are antiparallel: thus, the *CD* is negative.

(6) Now change the direction of one of the monomer dipole arrows in Figure 4.24a. You will end up with the other dimer transition—corresponding to the minus sign in equation (4.7). You will find that this combination is an attractive one (longer wavelength), with shorter μ(dimer), as shown by the smaller (dashed) absorption band to the right of the monomer transition on the second line of Figure 4.23b (ii). In this case m is parallel with μ(dimer), so the *CD* is positive. While μ is shorter than in the case of the plus combination, m is bigger. It can be shown theoretically (*cf.* §8) that $\mu \cdot m$ for the two exciton transitions should always have opposite signs but the same magnitude, and should give rise to a symmetric bisignate *CD* pattern shown in Figure 4.24b(iv).

5 Advanced *LD* methods for biological macromolecules

LD can be applied to a wide range of systems. This chapter contains illustrations of how LD can be used for structural, kinetic and mechanistic studies of various biomacromolecular systems that can be macroscopically oriented (mainly in flow).

Linear Dichroism and Circular Dichroism: A Textbook on Polarized-Light Spectroscopy
By Bengt Nordén, Alison Rodger and Timothy Dafforn
© B. Nordén, A. Rodger and T. Dafforn, 2010
Published by the Royal Society of Chemistry, www.rsc.org

5.1   Introduction

The applications outlined in Chapter 3 are fairly direct applications of *LD* to give geometric information from spectra, either qualitatively or using equations (3.3) or (3.6). This chapter contains a collection of examples which are slightly more complicated for a variety of reasons including: the need for data from other techniques to interpret the *LD* data (*e.g.* requiring *CD* data to determine protein secondary structure); kinetics processes are operative; the system is a complex of molecules (*e.g.* DNA/protein complexes; carbon nanotube/DNA complexes; bacteriophage); or the experiment requires unusual equipment (synchrotron radiation *LD*).

5.2   DNA-ligand absorbance, *LD* and *CD*

In Chapters 3 and 4 we considered what *LD* and *CD* spectra might be expected for DNA-ligand systems. It is frequently convenient to complement one set of data with the other. For a ligand whose transition polarizations have been assigned, the *LD* gives the orientation of the ligand with respect to the helix axis. The *CD*, especially for an intercalated ligand, enables one to determine the orientation of a ligand in the base pair pocket. This is illustrated in Figure 5.1.

> The discussion of intercalator *CD* is all in terms of transition moment directions. One needs to know transition moment polarizations to assign intercalation orientations.

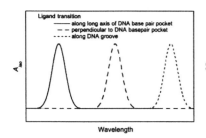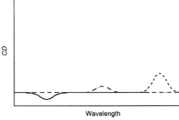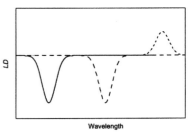

Figure 5.1 Schematics of the spectra expected for the absorbance, *CD* and *LD* spectra for ligands bound to DNA. The left hand spectrum in each box is for an intercalator with a transition polarized along the long axis of the base pair pocket. The middle spectrum in each box is for an intercalator with a transition polarized perpendicular to the long axis of the base pair pocket. The right hand spectrum in each box is for a groove binder with a transition moment polarized along the groove.

5.3   Liposome orientation

> Cylindical micelles made of amphiphilic molecules such as cetyltrimethylammoniumbromide (CETAB) can also be aligned by flow. The orientation of chromophoric molecules, bound in their membrane-like interior or surface, may be studied by *LD* with respect to orientation and interactions.[145]

The physical basis for the alignment that underlies the use of *LD* for membrane proteins in solution is the observation that spherical liposomes become ellipsoidal in shear flow thus orienting with their long axes arranged circumferentially around the *LD* cell as discussed in §2.3, §3.7 and §5.4. If we know the degree of alignment of the membranes then more information may be gleaned from *LD* measurements. It is for this reason that all the examples of §5.4 include a probe molecule. With bacteriorhodopsin, the probe is an intrinsic part of the protein, but more generally, an external probe has to be added for determining the degree of membrane orientation (orientation factor *S* of equation (3.6)).

Ideally, a probe molecule for this type of application should have a high extinction coefficient at wavelengths away from the peptide/protein signals, have a low extinction coefficient in the peptide/side chain region, and not affect the interaction of the peptide/protein with the membrane. No universal probe has yet become apparent.

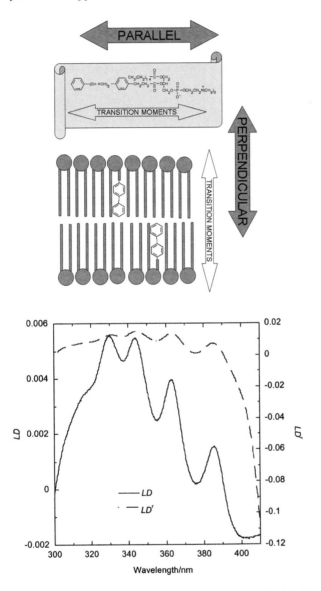

One way to make liposomes is first to dissolve the lipids in chloroform (spectrophotometric grade) at ~10 mg/mL. Probes or analytes can be added from solvents miscible with chloroform such as ethanol. The solvent is then removed, *e.g.* by a stream of nitrogen, leaving the lipid as a thin film around the sample vial. The remaining solvent is removed by leaving the lipid film under vacuum. The required aqueous buffer is then added to the vial containing the film to give a total lipid concentration of ~5 mg/mL. The aqueous lipid suspensions may then be subjected to 5 freeze-thaw cycles (between a dry ice-ethanol bath and a 50 °C water bath) with manual agitation during thawing to remove the very small particles. If the lipid samples are then extruded ~13 times (usually while being held above the gel to liquid phase transition temperature) through a 100 nm polycarbonate membrane filter, liposomes of average size about 120 nm are produced.

Figure 5.2 (a) Schematic of β-DPH-HPC orientation in stretched film (positive *LD*) and in lipid membranes (negative *LD* signal). (b) *LD* and *LD*^{r} spectrum of β-DPH-HPC in a stretched PE film. Data were collected as described in §2.3.

After testing a variety of candidates[146] β-DPH-HPC (2-(3-(diphenyl-hexatrienyl)propanoyl)-1-hexadecanoyl-*sn*-glycero-3-phosphocholine) and retinoids, particularly retinoic acid, seem at present to be the best candidates.[54, 147, 148] Figure 5.2 shows the PE film *LD* (§2.3) spectrum of

DPH-HPC. Its positive *LD* signals mean that the 340 nm region transition moments are along the long axis of the probe molecule. Thus, equation (3.7) leads one to expect that, when β-DPH-HPC is inserted into a membrane parallel to the lipids (and to the membrane normal), it will have a negative signal. An advantage with this probe is that it is covalently attached to a phospholipid.

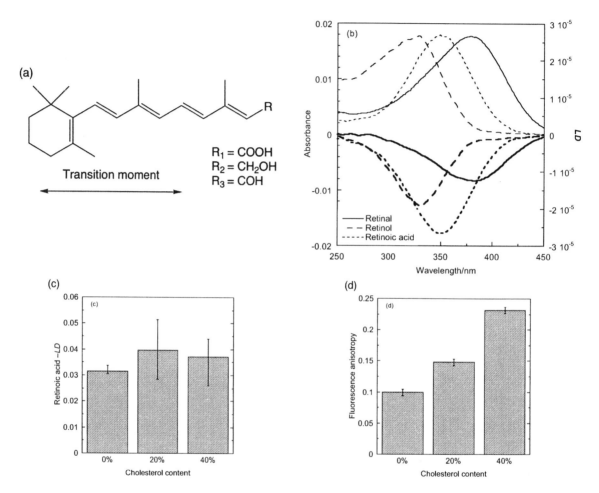

Figure 5.3 (a) Structure of retinoic membrane probes retinoic acid (R_1), retinol (R_2), and retinal (R_3). The absorbing transition moment is oriented along the long axis of the molecule.[148] (b) Absorbance (thin lines) and *LD* (thick lines) spectra of retinoids in POPC (palmitoyloleoyl phosphatidylcholine) liposomes showing positions of their absorption bands and their relative degrees of alignment. The negative *LD* indicates that retinoids insert parallel to the membrane normal. (c) *LD'*, of retinoic acid in POPC liposomes with increasing molar concentrations of cholesterol. (d) Fluorescence anisotropy of the membrane fluidity probe DPH in POPC liposomes with increasing concentration of cholesterol.

Retinoic acid, retinal and retinol (*cf.* Figure 5.3a) have been used to study systematically membrane orientation as function of phospholipid composition and cholesterol concentration.[148] The retinoids have confined absorption bands in the 300–400 nm region (Figure 5.3a), often well separated from protein absorption, and can be added separately to a sample with lipid vesicles and peptide or protein to determine the orientation parameter S in an individual experiment. This then allows for accurate calculation of peptide chromophore insertion angles.[149]

It is important to realize that the orientation parameter, S, reports on both the degree of vesicle alignment (deformation and orientation in the flow) and on the local lipid order within the membrane. This is well-illustrated in Figure 5.3c and Figure 5.3d which show that the magnitude of the retinoic acid LD^r in cholesterol-containing liposomes is only modestly affected by increasing cholesterol concentrations despite the fact that the lipid order (or membrane fluidity) monitored from the fluorescence anisotropy of 1,6-diphenyl-1,3,5-hexatriene (DPH)—a common probe of membrane fluidity—increases monotonically as expected. This is due to an increased bending rigidity of the liposomes resulting in less deformation and less overall alignment, reflected in the LD.

5.4   Membrane peptides and proteins in liposomes

As discussed in §3.7 membrane proteins comprise ~30% of the proteins expressed in a cell, yet the structural data for them is only a small fraction of that for soluble proteins because they generally require lipid environments in order to maintain both their structure and function. LD is proving an ideal technique to follow the insertion, orientation and kinetics of structural changes of membrane peptides and membrane proteins when they are mixed in aqueous solution with liposomes, a model membrane bilayer system (*cf.* §2.3).

The LD signals expected for an α-helical and a β-sheet peptide in liposomes are described in §3.7. In this chapter, two more complicated examples are presented. The bacteriorhodopsin case discussed below shows how LD can be used to compare crystal and solution data; the gramicidin case illustrates the use of a probe to monitor liposome integrity (if the probe signal disappears then one may conclude that the structural integrity of the liposomes has been compromised) and also peptide insertion kinetics.

LD is ideally suited to indicate whether a peptide bound on the surface of a membrane or inserted into it, as long as we know the peptide secondary structure in the same environment from CD (*cf.* Chapter 4). For example, as discussed in §3.7 for aurein 2.5, if a peptide folds into an α-helix on the surface of the membrane, its 222 nm n-π^* transition which is perpendicular to the helix axis (Figure 3.19) will give rise to a negative LD. If the helix then inserts, the LD at 222 nm will change to positive because it then points in the flow direction half of the time. If the peptide folds and inserts in one simultaneous step, one would see a random coil circular dichroism spectrum converting to a helical one on exactly the same time scale as the appearance of the positive LD. An ideal antibiotic peptide would insert into and disrupt bacterial membranes but not have that effect on host cell membranes; so being able to study different lipids is an additional attraction of LD.

Bacteriorhodopsin

Bacteriorhodopsin (BR) is a membrane protein found in the purple membrane of halobacteria;[150] it is a 248 residue protein and includes a covalently bound retinal (a vitamin A derivative) chromophore (Figure 5.4). It is undoubtedly one of the most studied membrane protein. BR captures light energy (using the retinal chromophore) and uses it to move protons across the membrane out of the cell. The resulting proton gradient is subsequently converted to chemical energy by the cell. Each BR has seven trans-membrane helices, three of which in the X-ray crystal structure have their axis at ~70° to the lipid long axis and the remaining four are parallel to the lipid long axis.[151] The long axis of the retinal lies at ~69° to the lipid long axis in the structure. A flow *LD* spectrum of BR inserted into liposomes is shown in Figure 5.4. The 570 nm peak is due to a long axis polarized transition of the retinal chromophore; the broad peak in the near UV region (260–290 nm) is due to the transitions of the protein aromatic side chains (dominated by the indole chromophore of the tryptophan residues); the peak observed in the far UV region (220–230 nm) is largely due to peptide $n \rightarrow \pi^*$ transitions of the amide groups although tryptophans and tyrosines also exhibit distinct $\pi \rightarrow \pi^*$ transitions in this region of the spectrum.

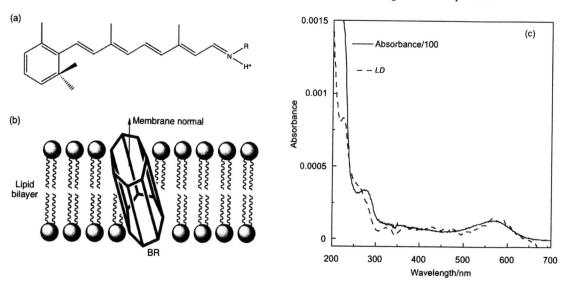

Figure 5.4 (a) All-trans retinal converted to the Shiff base. (b) A schematic illustration of bacteriorhodopsin (BR) inserted into the liposome bilayer. (c) Absorbance and flow *LD* spectra of BR (0.2 mg ml^{-1}) inserted into a lipid bilayer (0.5 mg ml^{-1} soya bean phosphatidyl choline liposomes).[147]

The key question is whether the secondary structure orientations derived from the X-ray crystal structure are appropriate for BR in a solvated lipid environment. If we assume that the retinal is at an angle of 69° to the lipids as in the crystal, equation (3.6) then gives $S \sim 0.05$. It follows from the *LD* magnitudes that the tryptophan transitions are oriented so that the angles they make to the membrane normal (Figure 2.7) are different for different transitions: $\beta(\text{L}_\text{a}, 270 \text{ nm}) \sim 60°$ and $\beta(\text{L}_\text{b}, 287 \text{ nm}) \sim 65°$. This is consistent with the X-ray structure which shows that the retinal is sandwiched by tryptophan residues.[150] The protein backbone *LD* spectrum, however, shows a positive maximum at 220 nm and a negative contribution (which looks like a positive minimum) at ~213 nm, from which it follows (equation (3.7)) that the $n{\rightarrow}\pi^*$ transition (which is polarized perpendicular to the α–helix long axis, Figure 3.19) is ~58° from the average lipid direction. Thus the average orientation of the transmembrane helices is ~30° from the membrane normal (or lipid long axis). This value suggests that the protein is less rigidly held in a liposome than when dried or crystallized — which is not entirely surprising.

Gramicidin insertion

Many naturally occurring peptides show antibiotic properties with mechanisms of action that avoid the current drug resistance mechanisms of bacteria. Some of them act by inserting into the bacterial cell membrane and disrupting its function in some way, the most dramatic effect causing the cell membrane to lose its structural integrity so it no longer encapsulates the cytoplasm. Gramicidin D is a naturally-occurring mixture of gramicidins A (~85%), B (~10%) and C (~5%), each discriminated by changes in amino acid sequence at position 11 (denoted X), being tryptophan, tyrosine and phenylalanine respectively:[152] HCO-L-Val-Gly-L-Ala-D-Leu-L-Ala-D-Val-L-Val-D-Val-L-Trp-D-Leu-L-X-D-Leu-L-Trp-D-Leu-L-Trp-NH(CH$_2$)$_2$OH

Gramicidin inserts into bacterial membranes and kills bacteria by transporting ions and hence dissipating ion gradients which the bacterium generates across the membrane. Although unusual in being a mix of alternating D- and L- isomers, gramicidin is widely used as a model ion channel peptide.

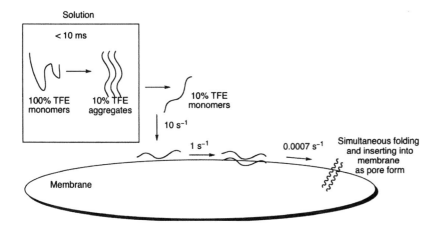

Figure 5.5 Schematic illustration of the mechanism of insertion of gramicidin into soya bean liposomes from 100%TFE. The processes in the box occur in the dead-time of a stopped flow instrument.[153] The final folded form in this case is the non-conducting channel.

The kinetics of gramicidin's insertion has mainly been studied by fluorescence and *CD*. Fluorescence gives an indication of the hydrophobicity of the environment of the tryptophans in the peptide and also shows when the scattering of the sample changes (since scattered photons register as an unstructured fluorescence spectrum), which in turn indicates particle size changes. *CD* gives a direct read-out of any changes in secondary structure of the peptide such as its folding into a dimer of left-handed β-helices (ion-channel form) or into an intertwined dimer (non-conducting form). However, neither technique indicates whether the peptide has actually inserted into the membrane. Changes in the gramicidin fluorescence signal had in fact previously erroneously been assumed to indicate insertion until a combined *CD* and *LD* study showed that gramicidin first inserts its tryptophan residues into the membrane; this is later followed by the simultaneous folding and insertion of the backbone as illustrated in Figure 5.5.[153]

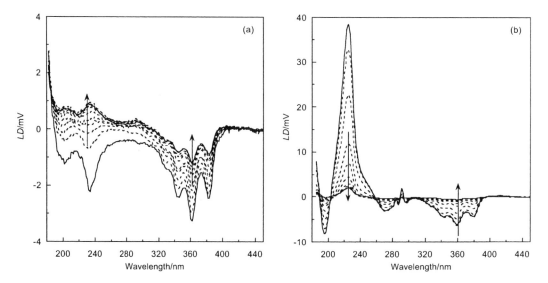

Figure 5.6 Kinetics of insertion of gramicidin D into DPPC/β-DPH-HPC (100:1 w/w) liposomes at (a) 30 °C and (b) 50 °C. The signal from 300–400 nm is from the β-DPH-HPC lipid which acts as a probe of liposome orientation. Data were collected at ASTRID[154] synchrotron radiation source on samples with final concentrations of 1 mg/mL lipid, 0.01 mg/mL β-DPH-HPC and 0.1 mg/mL gramicidin in 10% (v/v) TFE using a micro-volume *LD* cell (Figure 2.6). Spectra were collected at 12 minute intervals. *LD* measured in mV; $LD = 3.3 \times 10^{-4}$ per mV.

Figure 5.6 provides a clear illustration of the effect of the membrane structure on a peptide. When gramicidin was added to 1,2-dipalmitoyl-sn-glycero-3-phosphocholine (DPPC) liposomes at 30 °C (when the membrane is in its more rigid gel phase below the phase transition temperature of 41 °C[155]) a negative *LD* signal around 234 nm is observed at time $t = 0$ (Figure 5.6a). This shows that the polarization of the transition(s) at this energy is pointing more parallel than perpendicular to the membrane normal (equation (3.7)). The 234 nm negative signal is almost certainly dominated by the tryptophan B_b band indicating that the tryptophan average long axis is initially preferentially oriented parallel to the membrane normal. After two hours the tryptophans have reoriented so that their long axes are perpendicular to the membrane normal. The small signals at low wavelength are consistent with a lack of structure in the peptide backbone and/or a lack of orientation of the peptide backbone relative to the lipid bilayer. This may

Note the absence of aromatic signals around 270–290 nm in Figure 5.6a. This region of the spectrum has *Aiso* that is ~10% of the intensity of the B_b band. Its zero *LD* is consistent with the tryptophans being tilted so their short axis is perpendicular to the membrane surface with the transition moments lying close to the magic angle of 54.7°, where $3\cos^2\beta = 1$ (equation (3.6)).

be a result of the closely packed palmitoyl chains in the lipid excluding the peptide from insertion into the bilayer.

When the experiment with DPPC/β-DPH-HPC was repeated at 50 °C (*i.e.* above the phase transition temperature of the membrane), the gramicidin *LD* signal is much larger than that observed at 30 °C (Figure 5.6b). The order of magnitude of the probe signal, however, is similar at 30 °C and 50 °C, showing that the change is due to the peptide-lipid interaction not to a changed orientation of the liposomes. The magnitudes of the *LD* signals from both the peptide and the lipid probe decrease with time in a sigmoidal profile with a $t_{1/2}$ of around 30 minutes. In this case there is no evidence of change in peptide orientation within the membrane during the experiment so we conclude that the insertion is fast and the peptide then disrupts the liposomes (causing the vanishing *LD*).

The tryptophan transition polarizations are given in Figure 3.19. The aromatic L_b band with vibronic contributions around 280–290 nm[96] is large in Figure 5.6b indicating that there is significant orientation of the tryptophans (in contrast to the 30 °C experiment).

Structure of membrane-inserted gramicidin

Figure 5.7 shows *LD* and *CD* spectra of gramicidin inserted into a DOPC (dioleoyl phosphatidylcholine) liposome, before and after equilibration at 50 °C.[148] The *LD* spectra are dominated by the four tryptophans in gramicidin: the two peaks with positive signature close to 290 nm are due to the L_b transition, the negative peak in the 260–300 nm region is due to the L_a transition and the strongly positive peak at 225 nm is dominated by the B_b transition (Figure 1.6 and Figure 3.19). The average orientations of the tryptophans in gramicidin have been determined by resolving the contribution of the overlapping L_a and L_b transitions to the LD^r in the 260–300 nm region of the spectrum (*cf.* Chapter 6) after S had been determined using retinoic acid as membrane probe. The resulting tryptophan orientations are given in Table 5.1.

Table 5.1 Average orientation of tryptophans in the non-channel and ion-channel form of membrane-inserted gramicidin. Angles β are with respect to the membrane normal (Figure 2.7).

	L_a		L_b		B_b
	LD^r	β	LD^r	β	β
Non-ion channel	−0.047	41°	0.059	78°	~75°
Ion channel	−0.041	44°	0.068	80°	~75°

The quite different *CD* spectra in Figure 5.7 indicate substantial conformational differences between the two types of gramicidin dimers, in the initial and final states. This is in agreement with the initial formation of a postulated, kinetically trapped intertwined dimer structure, followed by a slow conversion to a thermodynamically more stable head-to-head stacked β-helical dimeric ion-channel.[156] By contrast, the corresponding *LD* spectra do not evidence any major structural differences between these two forms of gramicidin. The analysis of *LD* spectra described above showed that the variations observed in the *CD* spectra must be due to subtle re-orientations of tryptophan residues. Since the gramicidin *CD* spectrum is mainly a result of excitonic interactions between the tryptophan chromophores, it is very sensitive to their mutual orientations and spatial separation and the change in *CD* becomes extreme in comparison to the structural rearrangement that actually occurs.

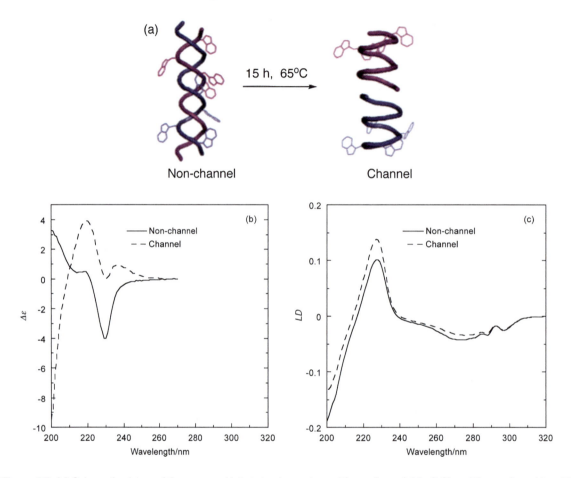

Figure 5.7 (a) Schematic picture of the proposed intertwined non-channel form of gramicidin (left) and the confirmed head-to-head left-handed single-stranded helical dimer (right) ion-channel form of gramicidin. (b) Circular dichroism spectra and (c) linear dichroism spectra of gramicidin in DOPC liposomes incorporated at a peptide-to-lipid molar ratio of 1:50 recorded before and after heating at 65°C for 15 hour to convert the non-channel form to the ion-channel form.[158]

The four tryptophans, adopt, for both types of gramicidin dimer, a preferred membrane-interface alignment, similar to the unconstrained orientation distribution of unsubstituted indole (Figure 5.7).[157] Based on this structural relationship it has been proposed that the 'intertwined' gramicidin model does not apply to true lipid bilayer environments where the tryptophan-interface interactions will energetically stabilise conformations having the tryptophans positioned close to the polar head groups. Intertwined conformations have been observed in solvents with lower dielectric constant than the membrane interface has.[158]

Tryptophan orientation in trans-membrane and surface-oriented peptides

Aromatic residues are present at higher levels in membrane proteins than with aqueous soluble proteins. They seem to play an important role as membrane anchors. Recent studies of tryptophan orientations in membrane-bound peptides indicate a common orientation mode for tryptophans attached to the ends of trans-membrane segments, whereas the tryptophans of surface-oriented α-helical peptides adopt distinctively different orientations.[157,158]

5.5   DNA superstructures

Since *LD* is a quadratic function of the cosine of angle α (equation (3.3)) there is a simple way to calculate the effect of averaging the orientation of transition moments in supramolecular structures using a notation known as the 'ensemble formula'.[159] The *LD* of any supramolecular aggregate with local uniaxial symmetry (*e.g.* a helix) is obtained as the sum of the squares of the absorption vectors projected on to the orientation axis of the aggregate minus the sum of the squares of the projections perpendicular to this axis. The ensemble formula can be applied to gel networks, liquid crystals or membranes built up of (locally uniaxial) subsystems, or to helices that are wound into superhelices, including supercoiled DNA in plasmid supercoils or in chromatin. For such supercoiled DNA structures, the *LD* may be described as a function of the pitch angle, Ψ_i, of each superhelix:[159]

$$LD^r = \frac{3}{2} S\left(3\cos^2 \alpha - 1\right)\left(\frac{1}{2}\right)^N \prod_{i=0}^{N}\left(3\cos^2 \Psi_i - 1\right) \tag{5.1}$$

where S is the usual orientation parameter (in this case of the superstructure) and α is the angle the transition moment of interest makes with the local orientation axis, N is the order of supercoiling: $N = 0$, no supercoil ($\Psi_0 = 0$), $N = 1$ a first supercoil, with pitch angle Ψ_1; $N = 2$ a supercoiled supercoil with pitch angle Ψ_2; and so on. The effect of increasing the order of supercoiling is to decrease the total *LD* amplitude of the structure (since supercoiling requires $\Psi_i > 0$). If $\Psi_i > 54.7°$ (very tight supercoiling), then $(3\cos^2\Psi_i - 1)$ is negative and the sign of the net *LD* can oscillate from positive to negative with every additional supercoiling. In addition, since supercoiling generally makes a particle shorter, the effective orientation will probably be worse and the *LD^r* further decrease by a decreased S (though the DNA does also get stiffer which increases S).

Consider a stretched-out DNA double helix. If a single supercoil is introduced into it so that the local double helix tilts at an angle Ψ_i, then the transition moments are tilted from the main supercoil axis. The resulting averaging introduces a factor of

$$\left(\frac{3\cos^2 \Psi_i - 1}{2}\right)$$

Each further supercoil has the same effect.

5.6   Protein-DNA interactions

During the life of a cell, a number of processes occur that require interactions between proteins and nucleic acids. Some of these affect DNA structure and thus *LD* is a useful tool for their study. *LD* has proved to be particularly well-suited to the study of non-specific DNA-protein interactions as the orientation properties of the complexes formed are generally different from those of the separate constituents. Furthermore, as with DNA-drug systems, one key advantage of *LD* is that if a globular protein binds to a polymeric DNA molecule (or conversely) the protein is invisible until it is bound. Thus *LD* titrations are useful for following formation and stoichiometry of such complexes.

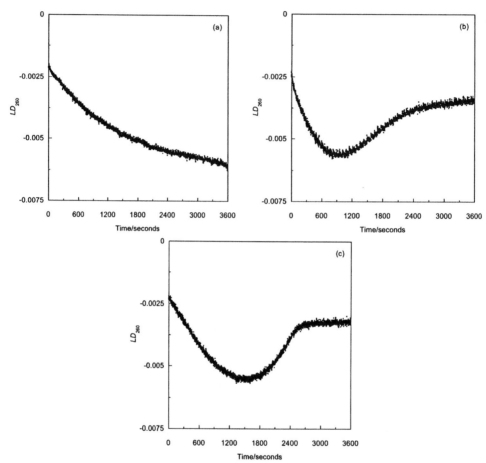

Figure 5.8 *LD* of restriction digests of a circular, super-coiled plasmid (pCDNA3.1 derivative (Invitrogen, Paisley, U.K.), 6882 base pairs in length), showing changes in DNA length (*LD* magnitude) as a function of time. (a) Plasmid digested with *Eco*RI, which has a single cut site in the DNA sequence. (b) Plasmid digested with *Eag*I, which has two cut sites in the DNA sequence. (c) Plasmid digested with *Bst*Z17I, which also has two cut sites in the DNA sequence.[160]

Despite the presentation in most biochemistry text books of nucleic acid systems as necessary but mechanistically rather uninteresting introductions to the more central parts of cell function, recent discoveries have revealed that nucleic acid structures display a variety of previously unsuspected interactions and may indeed play a centre-stage role in regulating many biomolecular interactions including the main transcription/translation steps leading to the formation of proteins. This is, for example, the case for the postgenetic roles of small RNAs (previously treated as debris) as well as the regulatory expression machinery involving chromatin. The effect of restriction enzymes is discussed below as are chromatin and two recombination proteins (bacterial RecA and human Rad51).

Restriction enzymes: unwinding of DNA supercoiling

Most plasmid DNAs in bacteria exist as supercoiled circular structures. Cells have a range of enzymes designed to control DNA structure. One such example is the type II restriction endonucleases which are enzymes that recognize and cleave specific DNA sequences. They have evolved to protect prokaryotic organisms from invasion by foreign DNA: the host's DNA is protected by a specific DNA methylation pattern (N4 or C5 methylation of cytosine or N6 methylation of adenine) whereas the foreign DNA is not so protected and is therefore susceptible to cleavage (restriction) by the host enzymes.[161] There are thousands of restriction enzymes that have been identified to date. They are also extremely important tools for the molecular biologist as they are used in many techniques that require DNA manipulation and have applications in, for example, diagnostics and recombinant protein production.

The normal methodology in molecular biology experiments is to leave the DNA and the restriction enzyme incubating at a chosen temperature for long enough to hope it has all reacted. This is sometimes checked by gel electrophoresis and sometimes not. There are other methods of monitoring the reaction but they involve either extra chemical labelling of the DNA with a fluorophore[162] or require the DNA to become single stranded after cleavage.[163] LD has proved to be the ideal technique to probe the effects of a restriction enzyme on a circular supercoiled plasmid DNA taking it from supercoiled to linear DNA and then to shorter linear pieces of DNA (if the DNA has multiple restriction sites). As the above discussion leads us to expect, we see first an increase in LD, as the DNA supercoiling decreases and the DNA lengthens, then a decrease if it is chopped into shorter pieces.

Three examples of the effect of restriction enzymes on DNA are shown in Figure 5.8.[160] The enzyme *EcoRI* has a single recognition site in the plasmid and as such is expected produce a linear DNA molecule. Figure 5.8 (a) shows an increase in the magnitude of the negative $LD_{260 \text{ nm}}$ which is consistent with an increase in orientation after linearization of the plasmid, as predicted. The effect of an enzyme with two cleavage sites in the plasmid DNA is shown in the LD traces of Figure 5.8b and c. For both enzymes *Eag*I and *Bst*Z17I, there is an initial increase in the magnitude of the negative LD signal at earlier time showing the linearization of the plasmid followed by a decrease when the second cut takes place. Intriguingly the rate of the second cut with *Bst*Z17I is faster than the first cut.

Chromatin structure

Chromatin is the DNA-histone protein complex that accommodates the DNA in the eukaryotic cell nucleus. The human haploid genome contains some 3×10^9 DNA base pairs corresponding to a linear length of DNA of about 1 m divided in 23 chromosomes. To accommodate these giant pieces of DNA in an eukaryotic cell nucleus of about 5 μm in diameter, efficient packing is required: this packing takes place at a number of levels—each could be represented by one factor in the superhelix product describing *LD* in equation (5.1).

(a)   (b)

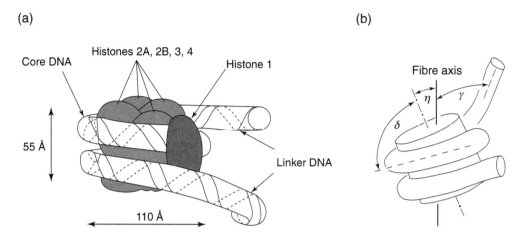

Figure 5.9 (a) The nucleosome: DNA wrapped around a core of histone proteins. (b) Definition of orientation angles in equations (5.3) and (5.4) for DNA pitch, nucleosome tilt and linker DNA angle.

The first level of packing consists of 145 base pairs of DNA wound 1.8 turns around a complex composed of eight basic histone proteins (two each of histones H2A, H2B, H3 and H4). At the point where DNA enters and exits this protein core some additional 20 base pairs are associated with a fifth histone, H1. This whole unit is termed a nucleosome (Figure 5.9). The nucleosomes are the repeating units of the nucleic acid-protein complex known as chromatin. They are linked by a length of DNA whose length varies between chromatins of different origins, cell tissues and possibly also between chromatins having varying transcriptional activity.

In vitro at low ionic strengths (<10 mM NaCl) chromatin is found in an extended 'beads-on-a-string' (the beads are called 'chromatosomes') configuration with the beads having diameters of typically 10 nm (as judged from electron microscopy). At higher ionic strengths these structures are condensed into the next level of folding, a compact fibre of about 30 nm in diameter. Despite almost 30 years of research the structure of chromatin and its molecular determinants have remained enigmatic, particularly in the condensed fibrous state. A recent low-resolution X-ray structure of a tetranucleosome reveals a concrete well-defined structure, although it is unclear to what degree it is representative of the 30 nm fibre structure.[164]

The *LD* of a chromatin sample will consist of a weighted average of contributions from the linker DNA and from the beads:

$$LD^r (\text{chromatin}) = fLD^r (\text{linker}) + (1 - f)LD^r (\text{bead}) \qquad (5.2)$$

where f is the fraction of DNA bases in the linker segments ($f \sim 0.1$). To a first approximation the linker region has approximately 1 supercoil (so $N = 1$ in equation (5.1)) and the bead has approximately 1.8 supercoils (so $N \sim 2$). Thus from equation (5.1) it follows that

$$LD^r \text{(linker)} = \frac{3}{4} S \left(3\cos^2 \alpha - 1 \right) \left(3\cos^2 \gamma - 1 \right) \tag{5.3}$$

$$LD^r \text{(bead)} = \frac{3}{8} S \left(3\cos^2 \alpha - 1 \right) \left(3\cos^2 \delta - 1 \right) \left(3\cos^2 \eta - 1 \right) \tag{5.4}$$

where α is the angle of base orientation relative to local DNA helix axis ($\alpha \sim 86°$), γ is the tilt angle representing the static or dynamic bending of the linker double helix, δ is the pitch angle for the chromatosomal DNA ('bead') and η is the angle between the superhelix axis of the chromatosome and the chromatin fibre axis as illustrated in Figure 5.9b. ($\pi/2-\eta$) is the chromatosome tilt angle.

> The DNA transitions of interest are all π-π^* transitions of the bases.

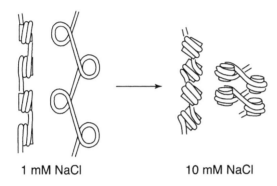

1 mM NaCl   10 mM NaCl

Figure 5.10 Cross-wise linker model of the 30 nm chromatin fibre consistent with *LD* and X-ray data for the orientation of linkers and chromatosomes.[10]

The first flow linear dichroism spectrum of condensed chromatin showed a positive *LD* peak at 260 nm which changed to negative with decreasing ionic strength as the complex unfolded and the strongly negative DNA contribution from linker dominated.[165] An interesting warning arises from later work on chromatin: apparently inconsistent data from different orientation techniques simply reflected different orientations induced into the DNA by flow and electric field, rather than any perturbation of the molecular structure by the applied fields. For example, chromatin partly condensed by Mg^{2+} ions when oriented by flow has its orientation determined by the hydrodynamic properties of the entire fibre, whereas the electric field orientation is dominated by counterion polarization along the linker parts of DNA.[166] Based on *LD* data and topological and energetic arguments a 'cross-wise-linker' arrangement has been suggested for the chromatin fibre (Figure 5.10).

5.7   Site-specific *LD* spectroscopy: DNA binding of recombination enzymes

A class of proteins that has been extensively studied due to their important biological roles are recombination enzymes, such as RecA (in *E. coli*) and Rad51 (in eukaryotes including human). While crystals of the pure proteins have been obtained and studied using X-ray diffraction, their mechanistically more interesting DNA complexes have largely escaped structural examination as their filamentous characters make them in practice impossible to crystallize. Such filamentous structures, however, may be easily aligned in shear flow providing excellent targets for *LD* spectroscopy.

Flow *LD* is thus probably the most sensitive tool for monitoring the formation and determining the stoichiometry of DNA-RecA complexes.[167] *LD* also provided early evidence of a preferred perpendicular DNA base orientation in the fibres using 'etheno-modified' DNA bases which are chromophores with absorption bands outside the protein absorption region. Reference[168] provides many good examples of how flow *LD* spectra can provide information about composition of RecA-DNA complexes, although normally not any detailed conclusions regarding their structure.

Limitations of *LD* spectroscopy

It is sometimes possible to analyze DNA and nucleotides in the presence of proteins, as illustrated for FtsZ in Figure 3.23, however, often the protein shields the nucleobase absorbance around 260 nm for both DNA and any bound cofactors, *e.g.* ATP for RecA.

Some general limitations of *LD* spectroscopy may be summarized by saying that orientation information is limited due to one or several of the following causes:

(i)   *Angular distribution.* Spectroscopically identical chromophores may have different orientations due to dynamics (lack of strong orientation constraints) or to structural variations—in both cases leading to widened angular distributions about which $< \cos^2 \alpha >$ may provide limited information.

(ii)   *The angular sign ambiguity*, $\pm \alpha$, *when solving* $< \cos^2 \alpha >$ *for* α. In certain cases of low symmetry and with additional information about directionality the sign may be concluded. An example is the ruthenium complex of Figure 5.14 where knowledge that the DPPZ moiety is intercalated provides the information that an internal coordinate axis of the complex is pointing inwards on the DNA, in turn allowing the roll sign to be determined from an off-diagonal element of the orientation tensor (a transition moment directed along the magic angle of the complex).

(iii)   *Azimuthal ambiguity.* For an *LD* experiment, a linear chromophore may be characterized with respect to a unique orientation axis (*e.g.* a fibre axis) by an angle; a planar but otherwise asymmetric chromophore similarly needs two angles (roll angle around a certain rotation axis and tilt with respect to the same rotation axis) to define its orientation with respect to the axis. In most *LD* experiments, due to the uniaxial symmetry of orientation, all orientations around the compass will have the same *LD* and, thus, *LD* contains no information about the azimuthal angle.

Ambiguities of this kind are not unique to *LD*. Other spectroscopic techniques often need extra information to make interpretation feasible. Even the most exact of all structural tools, X-ray diffraction, needs assumptions or auxiliary information (about phase angle) to translate a diffraction intensity pattern into an electron-density map.

In the particular case of proteins, an added complication is that the *LD* spectrum of the protein fibre or a protein-DNA complex in the near-UV protein absorption region derives from an average over all aromatic residues.

It is generally rather non-specific regarding secondary/tertiary structures as well as the overall orientation of the protein. A methodology called 'Site-Specific Linear Dichroism' (*SSLD*) has opened up novel applications of *LD* to extract more specific orientation information about protein structure.

Principles of site-specific *LD* spectroscopy

The principle of the *SSLD* methodology is to highlight a selected single chromophore by a molecular replacement technique so that the altered optical properties of the whole oriented system may be directly related to the specific change. *SSLD* can be implemented by replacing one of the aromatic residues (say a tryptophan) with an optically 'invisible' residue (say threonine) and thereafter comparing the *LD* of this mutant with that of the wild type protein. Provided the orientations and structures of the DNA complexes are not significantly changed by the residue replacement, the difference *LD* spectrum will be that of the replaced residue and may thus provide specific information about its orientation in the protein. In this way *SSLD* can give orientation data for a large number of chromophores. It could in principle be used to solve three-dimensional structures in systems that for some reason are not amenable to X-ray crystallography or NMR study. Such situations are the rule rather than the exception with fibrous nucleo-protein complexes or other large structures containing flexible fragments or else non-repetitive arrangements that may prevent crystallization, including most membrane proteins. With this methodology membrane proteins may, *e.g.*, be studied in their native environment with the additional potential of following structural changes in response to various chemical, electric or photo-physical perturbations. This approach was adopted in the RecA study discussed below.

RecA

The principle of *SSLD* was introduced in a study of RecA for which the *LD* spectrum of the wild-type protein complexed with DNA was compared with that of a mutant protein which had been engineered to have a threonine residue replace one of the two tryptophans that are natural in the RecA protein. Since threonine is optically transparent in the UV region where tryptophan absorbs light, the difference *LD* spectrum between the two complexes represents the *LD* spectrum of the mutated tryptophan. From knowledge of the transition moment directions of indole, the chromophore responsible for tryptophan's near-UV absorption, one may use the *LD* data to determine the orientation of the indole (in terms of two angles) relative to the fibre axis of the RecA-DNA complex.[169] In practice, one may need to account for different orientation parameters for the two fibres so we work with

$$SSLD(S = 1) = \frac{LD(\text{RecA wild type})}{S(\text{RecA wild type})} - \frac{LD(\text{RecA mutant})}{S(\text{RecA mutant})} \quad (5.5)$$

From the detailed shape of the reduced *SSLD* spectrum, *i.e. LD* divided by the corresponding *A*, and considering the mutual overlap of the three tryptophan transitions (L_b, L_a and B transition moments absorbing at 285, 265 and 220 nm), the orientation of the substituted tryptophan can be calculated in terms of the angles of the long-axis and the plane relative to the the fibre axis of the RecA-DNA complex.[169]

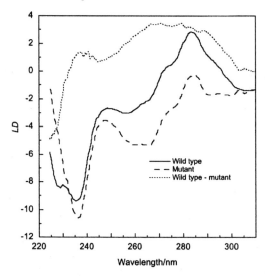

Figure 5.11 The *SSLD* spectrum of a selected tryptophan (trp291) in a flow-oriented RecA-dsDNA fibre is obtained by subtracting the *LD* of the corresponding threonine mutant complex from that of the wild-type complex. The positive and negative *LD* signals seen in the differential spectrum represent, respectively, the L_b, L_a and B transition moments absorbing at 285, 265 and 220 nm in the tryptophan chromophore (Figure 3.19).[169]

For the wild type RecA fibre where A_i denotes the absorbance of the *i*th transition of the fibre and α_i is the angle the transition moment makes with the fibre axis:

If more than one transition of a replaced chromophore overlap then the spectra need to be decomposed as discussed in §6.7.

$$LD(\text{RecA wild type}) = \frac{3}{2} S(\text{RecA wild type})$$

$$\times \left\{ A_1(\lambda)\left(3\cos^2\alpha_1 - 1\right) + \sum_{i=2}^{N} A_i(\lambda)\left(3\cos^2\alpha_i - 1\right) \right\} \quad (5.6)$$

For the mutant RecA fibre where transition one has been replaced by something that has no spectroscopy in the region of interest

$$LD(\text{RecA mutant}) = \frac{3}{2} S(\text{RecA mutant})\left\{ 0 + \sum_{i=2}^{N} A_i(\lambda)\left(3\cos^2\alpha_i - 1\right) \right\} \quad (5.7)$$

If more than one transition is lost then more zeroes appear in equation (5.7). Combining these equations and considering LD^r at the appropriate wavelength allows direct determination of the angle α_1 characterizing the orientation of the substituted residue in the structure.

A low level structure of the RecA-DNA fibre was determined from *SSLD* spectra of eight substituted aromatic amino acids (Figure 5.12). The structure has the DNA molecule in the centre, with the RecA outside.[170] The orientation of some of the amino acids was altered upon binding to the DNA.

The recent crystal structure of a RecA-DNA complex, with protein units artificially tethered together by covalent linkers,[171] together with the available *SSLD* data and molecular modelling provides a basis for answering questions regarding the function of this recombinase.

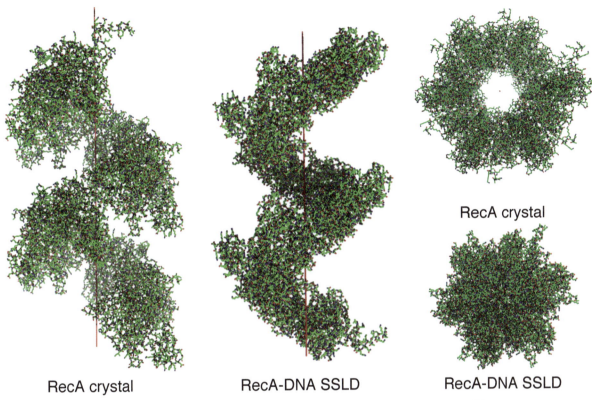

RecA crystal

RecA crystal   RecA-DNA SSLD   RecA-DNA SSLD

Figure 5.12 (a) Structure of RecA[172] and a RecA-DNA complex modelled using *SSLD* as in reference [170].

There are several conditions that have to be fulfilled in order to justify application of *SSLD*. First of all, the internal structure of the mutated protein as well as its arrangement in the DNA complex must not significantly deviate from those of the wild-type complex, the structure which we want to determine. Secondly, in order to solve equation (3.3) for angles α, we need first to assess a value of the orientation parameter S. In this instance, this was done using small-angle neutron scattering on the same sample in a niobium Couette cell (transparent to the neutron flux).[173]

Human Rad51-dsDNA complex

The human recombination protein Rad51 provides an example of a fruitful combination of *SSLD* with molecular modelling to determine the three dimensional structure of a fibrous protein complex. In eukaryotes, from yeast to human, Rad51 is involved in homologous recombination. Rad51 is functional when it assembles into long fibrous structures around DNA. Thus the 'active' form of Rad51 is too large for NMR and too flexible to form

Homologous recombination is the process of exchanging DNA strands of homologous sequences. It is important for both the repair of damaged DNA and the maintenance of genomic diversity.

crystals for X-ray diffraction.

For human Rad51, only the structures of two disconnected pieces of a monomer are available: a crystal structure of the Rad51 ATP-binding domain and an NMR structure of Rad51 N-terminal domain. By combining these structures, homology modelling of a crystal structure of a yeast Rad51 fragment, and data from *SSLD* studies where some eight tyrosine residues were replaced, it has been possible to put together a 'Frankenstein model' of the Rad51 protein assembly as it occurs in the fibrous complex with DNA (Figure 5.13).[174]

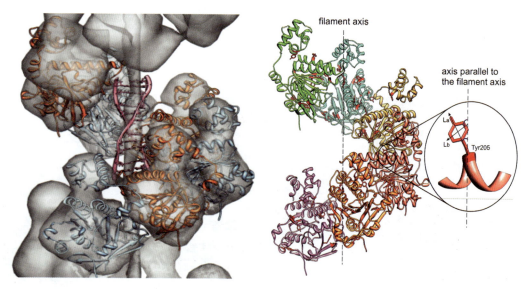

Figure 5.13 Structure of Rad51 in its active filamentous form derived from *SSLD* data collected for double stranded DNA complexes with eight Rad51 mutants in which one tyrosine at a time has been replaced by a phenylalanine.[174]

To assemble the whole structure of the Rad51 filament one must know how the monomers are oriented in space in the protein fibril—this is the key role of the *SSLD* data. In order to define precisely an object's orientation in space, three angles are needed. Two of three angles were determined from minimisation of the misfit functional between experimentally measured angles (*SSLD* data) and angles of trial filament structures. The resulting model was then refined to determine the third angle and the radius of a filament (based on the fact that two monomers have to be connected to each other). The net result is the three dimensional structure of human Rad51 in its active filamentous form (Figure 5.13).

An indirect *SSLD* method was used in the case of the human Rad51-double stranded DNA complex to avoid an exact determination of the orientation parameter, *S*, by a normalization procedure in which the *LD* values at one or several so-called 'magic' wavelengths are compared. At certain wavelengths (around 200, 215 and 250 nm for Rad51) in the far UV region a tyrosine and (parallel) phenylalanine transition have coinciding absorption coefficients. Thus if a tyrosine is replaced by a phenylalanine, we expect the *LD* intensity (after normalization with respect to the orientation factors *S*) of wild-type and modified nucleoprotein filaments to be the same at the 'magic wavelengths'. So we proceed by normalizing the wild-type and modified *LD* spectra to have the same *LD* intensity at a 'magic wavelength'.

5.8   Orientation of metal complexes bound to DNA

The DNA binding of chelate compounds of polycyclic heteroaromatics with transition metals has received great attention over the years not least because of the often remarkable photophysical properties of such systems. The chiral three-bladed propeller complex [Ru(1,10-phenanthroline)$_3$]$^{2+}$ (Figure 1.11, Figure 8.6) has two enantiomers, Δ and Λ, with respectively right-handed and left-handed propeller shapes, which make diastereomeric combinations with the chiral DNA. If one of the three phenanthrolines in the [Ru(phen)$_3$]$^{2+}$ complex is replaced by the dipyridophenazine ligand DPPZ (Figure 4.2, Figure 5.14, Figure 6.8) the complex is found to bind to DNA by intercalating this large aromatic moiety. As a result of protection of the DPPZ aza lone-pairs by the surrounding nucleobases, [Ru(1,10-phenanthroline)$_2$DPPZ]$^{2+}$ shows a brilliant luminescence which can thus be used to follow intercalation and dissociation kinetics of this kind of complex.

The first observation of DNA binding using *LD*, supported by *CD* showing enantiopreferential binding, of an analogous *tris*-chelate iron complex was made more than 30 years ago.[175]

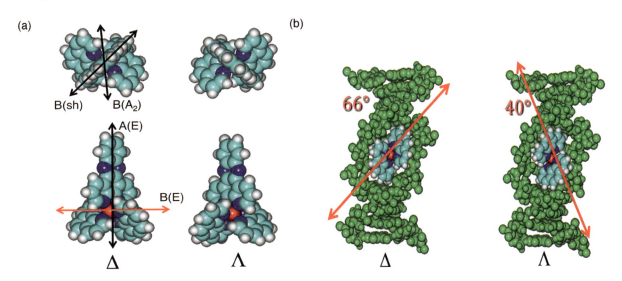

Figure 5.14 (a) [Ru(1,10-phenanthroline)$_2$DPPZ]$^{2+}$, including its transition moment directions. (b) Models of Δ- and Λ-[Ru(DPPZ)(1,10-phenanthroline)$_2$]$^{2+}$ complex fitted into the minor groove of DNA and with the DPPZ moiety intercalated.[176, 177]

The [Ru(1,10-phenanthroline)$_2$DPPZ]$^{2+}$ complex and numerous derivatives, including many binuclear compounds, have been extensively studied by photophysical methods with respect to their interactions with DNA. Flow *LD* clearly showed some interesting properties of the complexes with DNA and for a number of cases allowed the orientation angle α be determined together with its sign.[176, 177] We shall here briefly indicate how this is done.

The absorption, *CD* (Figure 4.22) and *LD* (Figure 6.12) spectra of [Ru(1,10-phenanthroline)$_2$DPPZ]$^{2+}$ in the 300–400 nm region are dominated by several intense charge-transfer transitions which overlap to different extents. The four main components are shown in Figure 5.14. The film *LD* spectrum of Figure 5.15 indicates the phenanthroline in-ligand transitions are polarized along the phenanthroline long axis. DPPZ has a long axis transition at 350 nm (polarization A(E) in Figure 5.14a and a *z*-polarized peak in Figure 6.12).

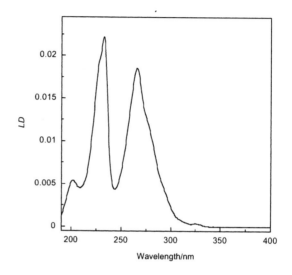

Figure 5.15 PE film *LD* spectrum of 1,10-phenanthroline prepared as described in §2.3.

From the flow *LD* spectra of [Ru(phen)$_2$DPPZ]$^{2+}$ enantiomers with DNA shown in Figure 5.16[176] it can be seen that the two enantiomeric forms of the complex have a similar sign pattern, but they do behave quite differently with Δ having only a very small positive signal near 400 nm whereas Λ has a large positive *LD* there. A first intuitive conclusion might be that the two enantiomers adopt quite different binding geometries. However, it turns out that the variations are due to subtle changes in the rotation of the complex around an axis that we may call the intercalation axis or the DNA-base tilt axis of Figure 3.5—the long-axis of the DPPZ ligand denoted A(E) in Figure 5.14a. The very similar negative 360 nm *LD* for both complexes is due to an A polarized transition perpendicular to the DNA helix axis.

Figure 5.14b shows models of the [Ru(phen)$_2$DPPZ]$^{2+}$ complex fitted into the minor groove of DNA and having the DPPZ moiety intercalated. The two enantiomers, Δ and Λ, must place their non-intercalated phenanthroline wings in different positions in the groove. As indicated in the model, the plane of the DPPZ ligand is not perfectly perpendicular to the DNA helix axis but turned slightly clockwise (when viewed from the minor groove) around the DPPZ long-axis when looking into the groove. This turn, or 'twist', is characteristic of all [Ru(phen)$_2$DPPZ]$^{2+}$ derivatives. It varies in magnitude from 5° to 15° and can be determined to within 1° from the B(E) transitions which lie close to the magic angle, so small changes in the orientation of this axis significantly affect the *LD*. Intriguingly both enantiomers tilt in the same direction, indicating it is due to intrinsic characteristics of the DNA structure rather than of the metal complexes.

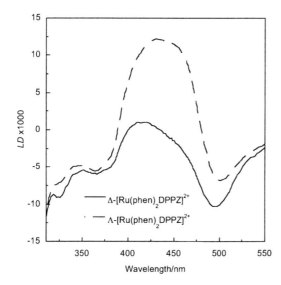

Figure 5.16 Flow *LD* spectrum of Δ- and Λ-[Ru(phen)$_2$DPPZ]$^{2+}$ bound to DNA.

A possible explanation of the twist may be that the metal complexes induce a local conversion of B-DNA into A-DNA. As may be seen from Table 3.1, the A form structure is characterized by base pairs tilted about 20° along the long-axis of the base pairs. This base tilt corresponds to the same direction of tilt as induced by [Ru(phen)$_2$DPPZ]$^{2+}$ when it intercalates from the minor groove. If instead, it had been inserted from the major groove of A-form DNA, a negative tilt of the complex would have been expected.

5.9   Bacteriophage, a viral pathogen of bacteria

LD methods can be applied to any biomolecule that assembles as a fibre or other asymmetric moiety, perhaps the extreme being that of bacteriophages. Bacteriophages are viruses that infect bacteria. The structure of these viruses is well understood and consists of an external protein coat enclosing a circular piece of single stranded DNA that encodes the bacteriophage genome. Filamentous bacteriophage are a subset of these viruses that have a high aspect ratio, *i.e.* they are much longer than they are wide. Perhaps the best studied and understood filamentous bacteriophage is M13 which is approximately 800 nm in length but only 8 nm in diameter. Filamentous bacteriophage are very rigid having a persistence length in excess of 1000 nm.[178] This makes bacteriophage an ideal object for study by *LD*.

LD spectra of bacteriophage have very high amplitudes by virtue of the high alignment potential of the particles.[179] The *LD* signals are further enhanced at higher concentrations because the bacteriophage themselves are able to auto-assemble into aligned structures: at high concentrations bacteriophage enter a liquid crystalline phase in which they align with respect to one another to form either smectic or nematic phases. These phases produce a very strong *LD* signal even when they are simply lain down on quartz plates.

LD studies of bacteriophage M13 at lower concentrations in solution using a Couette cell to induce flow alignment provide a complex spectrum that contains both protein and DNA signals as illustrated in Figure 5.17. At low wavelengths two peaks can be observed with a negative signal in the 190 nm region, positive maxima at 206 and 226 nm and a positive minimum, or more likely a negative maximum, between them at 217 nm. The major part of the M13 coat is made up of approximately 2600 copies of the same polypeptide which contains a single tryptophan; it also has 7200 DNA bases all contributing to the *LD* signal.

The bacteriophage geometry in a flow *LD* experiment is analogous to that of membrane proteins in liposomes, so from Figure 3.21 we deduce that the low wavelength pattern is due to the largely α-helical proteins being oriented more parallel than perpendicular to the phage long axis. The 226 nm positive *LD* is most likely due to tyrosines and tryptophans since DNA bases have minimum absorbances here (Figure 1.4). The higher wavelength data are dominated by signals from the single stranded DNA and the tryptophan. Evidence of the DNA signal can be seen as a negative maximum at 250 nm and positive signals above this.

The low wavelength component of the *π-π** transition in an α-helix is polarized perpendicular to the helix (Figure 3.19).

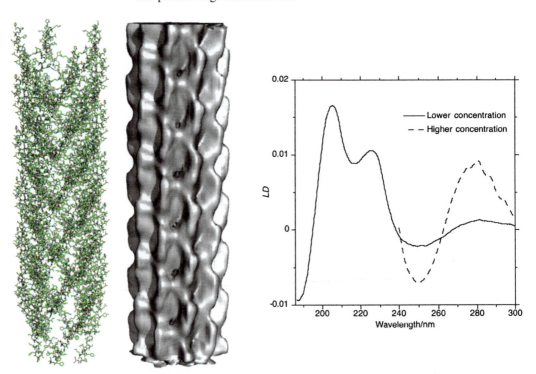

Figure 5.17 M13 bacteriophage atomic level and electron microscopy reconstruction[180] together with *LD* spectra.[181]

Clack and Gray used bacteriophage *LD* spectra to deduce that the tilt of the DNA bases within this phage is greater than 60°.[179] The long wavelength tryptophan signal is very well resolved as a result of the organisation of the tryptophans within the bacteriophage coat with four minor positive maxima at 274, 281, 286 and 293 nm. As illustrated in Figure 3.19, the long wavelength tryptophan transitions are oriented at about 45° to the long axis of the indole group. That the 285 nm region is positive tells us this direction is preferentially oriented along the long axis of the bacteriophage particle.

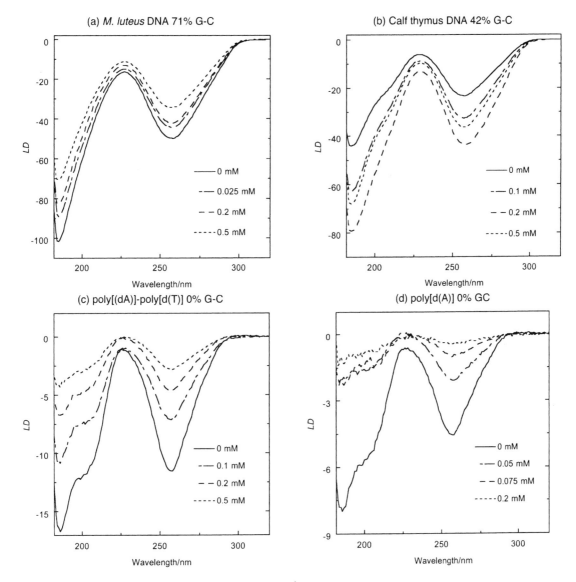

Figure 5.18 *LD* spectra of DNAs (in mV where $LD = 3.3 \times 10^{-4}$ per mV) as a function of NaF concentration (indicated on each figure) of (a) *Micrococcus luteus* (200 μM), (b) calf thymus (200 μM), (c) poly[d(A)]-polyd(T)] (200 μM), and (d) poly[d(A)] (200 μM) in a 0.5 mm path length Couette flow cell at room temperature. Data were collected on a synchrotron where the quartz cut-off for sample capillary defines the low wavelength limit to be 182 nm in these spectra.[70] *LD* measured in mV; $LD = 3.3 \times 10^{-4}$ per mV.

5.10　　Synchrotron radiation linear dichroism of DNAs

Stray or scattered light of any wavelength that is incident on the PMT is counted in the same way as photons transmitted through the sample. Thus the high tension voltage can seem 'safe' when in fact the number of photons of the correct wavelength incident on it is very low.

Circular dichroism spectroscopy is widely used to assess the secondary structure of proteins as discussed in §4.5; the value of the wide wavelength range achieved at synchrotron sources is apparent in an ever-increasing suite of applications discussed in references [182, 183]. An increase in the quality of *LD* data has recently also been achieved by the use of synchrotron radiation (SR). A synchrotron produces intense light over a wide range of wavelengths, including in the ultra-violet region. This has two distinct advantages over the sources normally used in bench-top spectrometers (usually a xenon lamp): (i) the level of noise on the data is greatly reduced, thus data can be collected much more quickly and (ii) lower wavelengths are routinely available.

Most bench-top spectrometers have a practical lower limit of between 180 nm and 190 nm, though, depending on the nature of the sample (including its absorbance and propensity to scatter light), this can be as high as 220 nm. A limitation that is due to a low number of low wavelength photons entering the detector from a xenon lamp below 200 nm.

SR *LD* as a DNA structure probe

The value of an SR source was illustrated in Figure 2.17 where the *LD* spectra of *Micrococcus luteus* DNA collected with the ASTRID synchrotron *LD* system, a Biologic MOS-450 and a Jasco J-815 are illustrated. The *SRLD* and Jasco spectrum are very similar showing that bench-top machines can give good low wavelength data with an ideal sample. As discussed in §2.6 the MOS-450 spectrum looks reasonable until about 185 nm, however, the other spectra make it clear that the signal is seriously attenuated. In this case the instrument is not optimized for low wavelength performance and there was little indication from the high tension voltage that the data were unreliable.

As the data in Figure 5.18 show, being able to collect essentially noise-free data down to 180 nm enables differences in DNA structures to be observed whereas the longer wavelength bands show little variation with conditions. Figure 5.18 shows that the overall effect of increased salt in solution with DNA is to reduce the backbone electrostatic repulsion causing the DNA to bend. However, the salt also stabilises and hence stiffens the DNA by reducing transient single stranded regions of the DNA—this increases the *LD*. In most of the DNAs of Figure 5.18, the bending effect dominates. However, for calf thymus DNA, we first see the stabilizing effect dominating and then the bending. The 200 nm shoulder in the A-T rich DNAs disappears as the salt concentration increases. Conversely the shoulder is enhanced if the temperature is increases, and hence the helix destabilized.[70]

SR *LD* to initiate photochemical reactions

YOYO's most stable binding mode is by bisintercalation with the linker in a groove of the DNA; its next most stable mode is a groove bound mode.[184]

The intensity of a synchrotron source can also be used to initiate photochemical reactions as illustrated in Figure 5.19. The long wavelength band of YO (a cyanine dye) and YOYO (a YO dimer, Figure 5.19) is due to a single long axis polarized transition.[185] The DNA-base *LD* is negative below 300 nm as is the *LD* of the bisintercalator YOYO from 430–500 nm. Upon irradiation by the intense synchrotron source, the YOYO cleaves single strands of the DNA backbone resulting in the DNA becoming less rigid (when one backbone strand is cleaved) and then shorter (once two nearby strands are cleaved). Both of these processes have the effect of reducing the magnitude of the DNA *LD* signal.

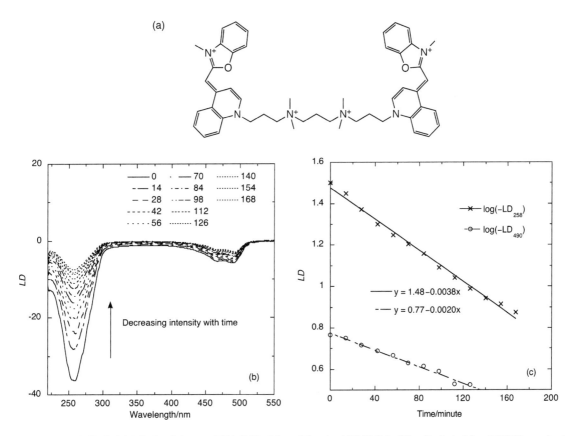

Figure 5.19 (a) YOYO. (b) *LD* of calf thymus DNA (200 μM) and the dye YOYO (10 μM) collected at the ASTRID synchrotron. The time label is that at the start of each spectrum. (c) $\log_{10}(-LD)$ *versus* time. *LD* measured in mV; $LD = 3.3 \times 10^{-4}$ per mV.

A typical data set for calf thymus DNA and YOYO at a ratio of 10:1 base:intercalator (so in the concentration regime where we can assume all added YOYO is bound intercalatively to DNA[185]) is shown in Figure 5.19b. Both the DNA and YOYO regions of the spectrum show that the reaction is first order (Figure 5.19c). However, the rate constant deduced from the DNA region of the spectrum is 0.23 min^{-1} and that from the YOYO region is 0.12 min^{-1}. Thus the DNA is losing its orientation more quickly than the YO chromophores. This is consistent with the YO chromophores stiffening the DNA locally about their binding site and/or with a small percentage of groove bound molecules actually catalysing the cleavage and no longer binding once they have cleaved the DNA—thus no longer contributing a positive *LD* signal.

5.11   DNA oriented on carbon nanotubes

Carbon nanotubes (CNTs), a comparatively recently recognized allotrope of carbon,[187] are remarkable materials which are widely studied because of their electronic properties and have a range of potential applications as semi-conductors, catalysts, optical devices *etc*. Nanotubes are well-ordered hollow graphitic nanomaterials which vary in length from several hundred nanometers to several micrometers and have diameters of 0.4 to 2 nm for single-walled carbon nanotubes (SWCNT). The literature reports a range of molecules binding to SWCNTs, but characterizing the interaction, particularly in solution phase, is challenging.

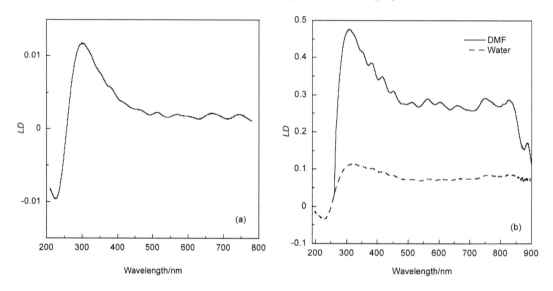

Figure 5.20 (a) *LD* spectra of SWCNTs (1 mg/mL, 500 µm path length) solubilized by sonicating in aqueous SDS (9 mM).[186] (b) *LD* spectra of ammonium functionalized SWCNTs in water and DMF aligned in a weak magnetic field (~1 T).[47]

The *LD* of CNTs (formed by arc discharge and purified by refluxing with nitric acid) gives rise to a large negative *LD* signal with maximum at 225 nm and a smaller positive *LD* at ~300 nm (Figure 5.20).[188, 189] The sign of this signal means that the dominant transition polarization of these SWCNTs lies at more than 54.7° from the nantotube long axis — so across the tube.

The *LD* spectrum of ammonia functionalized SWCNTs oriented in a magnetic field are shown in Figure 5.20b. The positive signals above 260 nm show the electronic transitions are polarized parallel to the nanotubes axis, whilst the negative signals below 260 in water show the presence of some high energy transitions transverse to that axis. Note the wiggles in the DMF spectrum indicate the presence of some 15 densely spaced transitions.[47]

Assuming that the SWCNTs have cylindrical symmetry about their long axes, then it follows that the ligand-binding geometry on a SWCNT is best defined in terms of the normal to the cylinder surface and equation (3.7) which was originally developed for shear distorted liposomes. One molecule that is widely reported to bind to CNTs and indeed to help solubilise them is DNA.[190] The *LD* signal, when that due to the CNT itself is subtracted from

The geometry of a SWCNT system is the same as that of a liposome (*cf.* §3.7 for membrane systems).

that of a solution of SWCNTs plus double-stranded calf thymus DNA that is too short to have a signal of its own is shown in Figure 5.21a. There is a single negative band at the DNA absorbance maximum; it is similar in shape to a normal DNA *LD* spectrum such as that of Figure 3.2. By way of contrast, the *LD* spectrum for single stranded DNAs (if they bind to the CNTs) is usually a couplet of bands whose position depends on the sequence of the DNA as illustrated in Figure 5.21b.

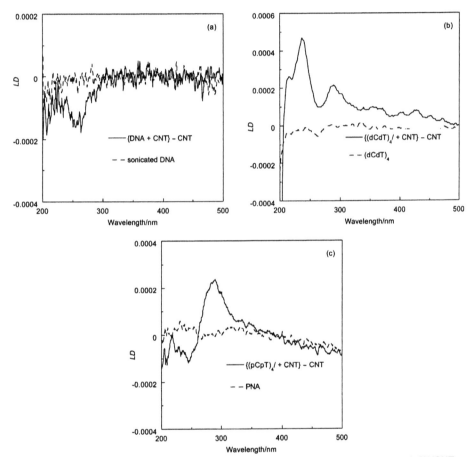

Figure 5.21 (a) Difference *LD* spectrum of SWCNT/sonicated double-stranded DNA complex and SWCNT compared with sonicated DNA (all solutions 0.033 mg/mL). (b) Difference *LD* spectrum of single-stranded DNA (dCdT)$_4$/SWCNT and SWCNT compared with (dCdT)$_4$ (all solutions 0.1 mg/mL). (c) Difference *LD* spectrum of single-stranded peptide nucleic acid (pCpT)$_4$/SWCNT and SWCNT compared with (pCpT)$_4$ (all solutions 0.1 mg/mL). All solutions were in SDS (9 mM) and NaCl (20 mM). SDS/NaCl baselines were subtracted from all spectra.[188, 189]

An intriguing twist on the DNA/CNT story is that DNA and its neutral 'analogue' PNA show opposite *LD* signals for (dTdC)$_4$ and (pTpC)$_4$ (Figure 5.21). Thus whatever the binding mode of DNA on the SWCNT, PNA is oriented quite differently. A possible orientation of DNA bases in the DNA-CNT complex that is consistent with the *LD* is illustrated in Figure 5.22. The PNA backbone has even more flexibility than the DNA backbone and its *LD* signs would suggest it wraps around the SWCNT so that the PNA bases lie

Single stranded oligo DNAs nearly all show at least small *LD* signals indicating that they can join up to form longer polymers in solution. In the presence of CNTs to which the DNA binds, the signal increases.

approximately perpendicular to those illustrated for DNA. A means of achieving this would be for it to wrap with the opposite sense which can be visualized by inserting the SWCNT with its axis horizontal rather than vertical in Figure 5.22.

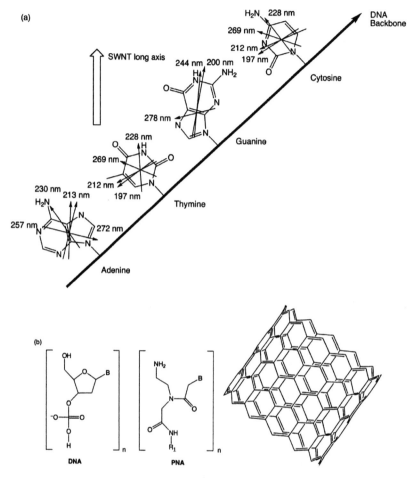

Figure 5.22 (a) Schematic of arrangement of DNA bases on a CNT surface that is consistent with available *LD* data. (b) Chemical structures of DNA (left) and PNA (right) backbones and a (13,0) zigzag CNT. B denotes base.

5.12 Peptidoglycan layer of bacterial cells

Like eukaryotic cells, bacteria have a barrier separating the interior of the cell from the outside world. This allows transport in and out of the cell to be regulated, and provides some measure of protection from outside attacks. In the case of bacteria, working from the inside to the outside, the first part of the cell barrier of a bacterium is the plasma membrane, a bilayer of phospholipids and proteins that encloses the cytoplasm of the cell. Most bacteria also have a layer of murein, or peptidoglycan, surrounding the plasma membrane. The thickness of the peptidoglycan layer varies between Gram-positive and Gram-negative bacteria with Gram-positive bacteria having a thicker layer of peptidoglycan on the outside of the cell and Gram-negative bacteria having a thinner peptidoglycan layer with an outer membrane surrounding it.

The peptidoglycan layer forms a 'net' around a bacterial cell and helps maintain the structure and shape of the cell.[191] The inhibition of synthesis of this layer is a common *modus operandi* of current antibiotics since in the absence of the peptidoglycan layer the cell autolyses and dies due to its internal osmotic pressure. As the name implies, the peptido-glycan layer is a mixed peptide/sugar complex. The peptidoglycan sacculi from the MC6R41 strain of *E. coli* (which has a defective FtsI gene that causes it to divide less often at higher temperatures) are rod shaped, making them easily oriented and so excellent subjects for linear dichroism as illustrated in Figure 5.23.

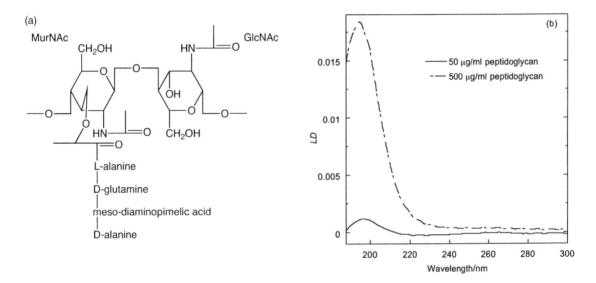

Figure 5.23 (a) A primary structure of peptidoglycan, with both types of sugar and a tetrapeptide chain extending from the MurNAc component. The ε-amino group on the meso-diaminopimelic acid is replaced when the peptide is cross-linked. GlcNAc denotes *N*-acetylglucosamine; MurNAc denotes *N*-acetylmuramicacid. (b) The *LD* spectra of two concentrations of peptidoglycan as indicated in the figure. Path length was 0.5 mm.

Lysozyme is an enzyme that disrupts the murein layer by hydrolysing the β-1–4 link between the GlcNAc and MurNAc sugars on the glycan strands.[192] The peptidoglycan net then becomes disconnected and separates. *LD* is the ideal technique to follow this process as illustrated in Figure 5.24 where the reduction of *LD* as the peptidoglycan chains gets smaller is apparent.

5.13   *LD* of flow-oriented polymers

In previous sections, DNA *LD* was qualitatively interpreted to probe the orientation of DNA bases. To conclude this chapter we briefly consider this type of experiment a little more rigorously. The simplest orientation situation for a polymer is that of uniaxial orientation. In the commonly used streaming devices of Couette flow type (Figure 2.6), however, the orientation is strictly not uniaxial, but biaxial. The LD^r can be still factorized into an optical factor, O, and an orientation factor, S, as summarized in equation (3.6). S depends on polymer stiffness, solvent viscosity, flow speed *etc*. Including the possibility of overlapping absorption bands of different transitions, which is a rule rather than an exception with most biopolymers, we have the following general formula:

$$LD^r = \frac{3}{2}S\left\{\frac{\sum_i A_i(\lambda)(3\cos^2\alpha_i - 1)}{\sum_i A_i(\lambda)}\right\}$$

(5.8)

with A_i and α_i denoting, for transition i, the relative absorption contribution and the angle of the transition moment with respect to the local polymer axis.

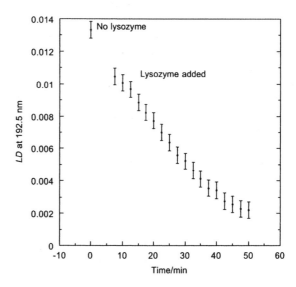

Figure 5.24 $LD_{192.5\ nm}$ of peptidoglycan (0.48 mg/mL) without lysozyme and with lysozyme (0.0097 mg/mL) as a function of time.

The most commonly used variant of equation (5.8) is equation (5.9) introduced by Matsuoka and Nordén[82] which expresses the LD^r of DNA in terms of tilt (θ_X) and roll (alternatively called 'twist') (θ_Y), (as defined in Figure 3.5, Figure 3.6, and Figure 5.25) of each base (or base pair) and an angle (δ_i) specifying the transition moment direction for the ith transition in the base plane:

$$LD^r = \frac{3}{2}S\frac{\left\{\sum_i F_i\varepsilon_i(\lambda)\left[3(-\sin\delta_i\sin\theta_X + \cos\delta_i\cos\theta_X\sin\theta_Y)^2 - 1\right]\right\}}{\sum_i F_i\varepsilon_i(\lambda)}$$

(5.9)

where F_i is the fractional content of transition i (it depends on base composition), summation is taken over all transitions, and $\varepsilon_i(\lambda)$ is the extinction coefficient of transition i at wavelength λ. Equation (5.8) and its analogues have been used to determine orientation of DNA bases within a polymer.[26, 193]

Conversely, equation (5.9) has also been applied to compute the theoretically anticipated LD^r curves for natural, mixed-sequence DNA, including the effects of band overlap and also of mutual perturbations (exciton coupling) between different transitions. Using $\theta_X = -2.1°$ and $\theta_Y = -4.0°$ (the crystal parameters of Arnott and Hukins[194] which are almost the same as those in Table 3.1 for B-form DNA).[82] As a consequence of θ_X and θ_Y both being close to zero, a nearly constant LD^r of $-3/2S$ is predicted, irrespective of transition moment directions within the bases (δ) and of exciton interactions. An 'effective' angle of 86° (which should include the effects of overlap *etc.*) was concluded to be an optimal value to be used when determining the orientation factor S from experimental LD of B-DNA.

At variance with the predicted constancy, the experimental flow LD^r of DNA (*e.g.* Figure 3.7) exhibits variations providing evidence that the planes of the nucleotide bases are not so perpendicular to the helix axis as the crystal data indicate but show some systematic inclination. Note that an isotropic dynamic wobbling giving equal inclinations in all directions is not enough to explain a wavelength dependence since no in-plane transition moment will be oriented more perpendicular than any other. Qualitatively the LD^r is consistent with a bigger base tilt (larger θ_X). Careful analysis of the LD^r variations observed in the vacuum ultraviolet region indicate that the base planes of B-DNA in solution lie at angles of about 80° or even less to the helix axis.[20]

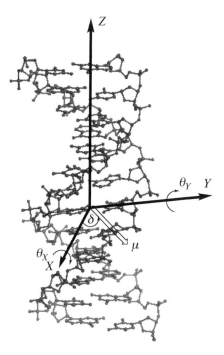

Figure 5.25 Geometry for equation (5.9).

By contrast to B-DNA, the A conformation (for $\theta_X = 19.3°$ and $\theta_Y = -3.2°$) is predicted to display a significant variation of LD^r as a function of wavelength, being some 10% more negative at long wavelength (290 nm) and less negative at shorter wavelength (240 nm). The experimental flow LD^r of DNA in 78% alcoholic solution shows variations that are in qualitative agreement with a bigger base tilt but still not fitting either of the theoretically predicted A- or B-form spectra.[35, 82]

The agreement between predicted and experimental LD^r spectra is markedly better for DNA that has been oriented in a humid matrix of PVA: at 100% humidity (B-DNA) LD^r is constant between 250–295 nm, while at 75% humidity (A-DNA) it varies by some 15% (being most negative at 295 nm) in agreement with the theoretical prediction if exciton interactions were included.[35, 82] This indicates that part of the deviation observed with flow-oriented B-form DNA is due to anisotropic bending dynamics of the biopolymer in solution. In such a case, the bending dynamics is not related to the overall stretch of DNA, since the same wavelength dependence is observed also at very high shear rates (62 000 s^{-1}).[195] Also, the homogeneous bending angle per base pair is small, of the order of 5°, for a persistence length of 60–80 nm.[196]

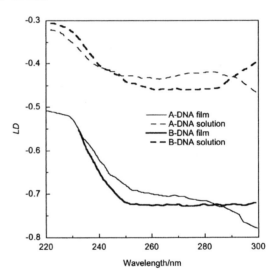

Figure 5.26 LD^r profiles of double-stranded natural A-form and B-form DNA: flow LD of DNA in solution (78% ethanol for proposed A-form, and water for B-form) and LD of DNA in stretched humid films (75% humidity for A-form and 100% humidity for B-form).[35, 82]

Both in aqueous and alcoholic solutions, as well as in the humid stretched matrices (Figure 5.26), the LD^r exhibits a distinct less negative feature at 230 nm which has been proposed to be due to n-π* transitions polarized perpendicular to the planes in the DNA bases. Such out-of-plane polarized transitions are expected in heterocyclic aromatic chromophores with aza lone-pair electrons and have been detected, using LD, in pyridine, pyrazine and s-triazine in stretched polyethylene films.[8, 197] To account for out-of-plane intensity equation (5.9) has to be modified by adding another term

$$\sum_o F_o \varepsilon_o(\lambda)\left(3\cos^2\theta_X \cos^2\theta_Y - 1\right) \tag{5.10}$$

to include the effect of any out-of-plane absorption bands $\varepsilon_o(\lambda)$. The strong influence of this term illustrates how *LD* can be extremely sensitive to contributions that deviate orthogonally from the main-stream polarizations. For both A- and B-DNA the feature can be roughly fitted by introducing an out-of-plane weak absorption band at 230 nm having $\varepsilon_o(\lambda)_{max}$ $= 250$ mol^{-1} dm^3 cm^{-1}, an intensity of the appropriate magnitude for an n-π^* transition.[35, 82]

LD of high symmetry ligands bound to a polymer

When we consider the situation of a ligand bound to DNA (or another polymer) then the optical factors of interest belong to the ligand and the orientational factor belongs to the DNA. If the ligand has a 3-fold or higher rotational symmetry (*e.g.* [Fe$_2$(LL)$_3$]$^{4+}$ Figure 3.10) then for transitions polarized along the molecular symmetry axis we may simply use equations (3.5) and (3.6) to give

$$LD^r = \left\{\frac{3}{2}\left(3\langle\cos^2\alpha\rangle - 1\right)\right\} \times \left\{\frac{1}{2}\left(3\langle\cos^2\theta\rangle - 1\right)\right\} \tag{5.11}$$

where α is the angle between the ligand's symmetry axis and the DNA axis.

For transitions polarized perpendicular to the ligand symmetry axis, the observed *LD* is the average of the *LD* intensity in the plane perpendicular to the symmetry axis, thus

$$LD^r = -\left\{\frac{3}{4}\left(3\langle\cos^2\alpha\rangle - 1\right)\right\} \times \left\{\frac{1}{2}\left(3\langle\cos^2\theta\rangle - 1\right)\right\} \tag{5.12}$$

6 Linear dichroism of small molecules

The focus of this chapter is on the LD of small molecules, though nearly all the content also applies to biomacromolecules. More detailed information can often be extracted for small molecules than from macromolecules because of their well-defined structures.

Linear Dichroism and Circular Dichroism: A Textbook on Polarized-Light Spectroscopy
By Bengt Nordén, Alison Rodger and Timothy Dafforn
© B. Nordén, A. Rodger and T. Dafforn, 2010
Published by the Royal Society of Chemistry, www.rsc.org

6.1   Introduction

A very sensitive tool for detecting orientation has existed for more than 100 years, namely birefringence (linear birefringence, *LB*) which typically measures light transmitted through a sample oriented at 45° between polarizers set at 0° and 90° (or the phase modulated equivalent which has improved signal:noise).

A huge drawback of *LB* over *LD* is the fact that *LB* cannot probe single transition moments by tuning to their wavelengths but is always an average over all transitions. As a result, it may be difficult with *LB* to tell anything about what is oriented and to what degree.

Ellipsometry, a technique frequently used for studying molecules adsorbed at surfaces by measuring how their refractive properties change the polarization of light, also has a similar drawback as *LB* relative to *LD*.

Most published *LD* data are for electronic transitions using visible or ultraviolet light, however, the analysis process is the same for vibrational transitions.

X is defined to be the propagation direction of the electromagnetic radiation throughout this book. *z*, the molecular orientation axis, may, in some cases, not be the usual symmetry defined *z*-axis. *x* is the molecular axis furthest from *Z*.

Consider the orientation of a molecule with a permanent dipole in an electric field. Only one structural parameter is needed. This is so because, for each and every value of the angle between the dipole moment and the electric field, all orientations around the field directions are equally probable. We say that the orientation distribution is uniaxial.

LD spectra are readily measured and usually easy to interpret at least qualitatively. However, not many laboratories use *LD* because the equipment required for orienting samples has only recently become commercially available. One aim of this book is to show that establishing *LD* facilities is fairly straightforward and that the effort is well-recompensed by the data that can be collected.

In Chapters 3 and 5 the focus is on applications of *LD* to biological macromolecular systems. When *LD* is used to study small molecules, a much more detailed analysis of the data is often possible, though sometimes a simple deduction of transition polarization directly from the sign of the observed *LD* is sufficient. The resulting structural and spectroscopic information may be of direct and immediate use or may be the input for analysis of other experiments such as analyzing the *LD* of the small molecule bound to DNA.

In this chapter the focus is on how to analyze an *LD* spectrum. We introduce simplifying assumptions and, depending on what we are after and what information we have, end up with a number of special cases, each appropriate to a specific class of applications. The analysis of this chapter is broadly based on the approach taken in references [8, 9, 10, 12]. Chapter 7 material builds on that of this chapter with a focus on molecular alignment.

6.2   Some *LD* definitions and proofs

General definition of *LD*

LD is defined in equation (1.1) as the difference in absorption of two linearly polarized light beams that are polarized at right-angles to one another and to the direction of propagation:

$$LD = A_{//} - A_\perp \qquad (6.1)$$

This notation is normally reserved for uniaxial samples with $//$ denoting the direction of the unique axis. So, although most *LD* experiments have (or are assumed to have) effective uniaxial orientation, in this chapter we use the more general definition

$$LD = A_Z - A_Y \qquad (6.2)$$

where A_Z is the absorbance of Z-polarized light, and similarly A_Y. $\{X, Y, Z\}$ form what is known as a laboratory-fixed axis system since these axes are used to identify macroscopic aspects of the experiment. $\{x, y, z\}$ is the notation used in this book for the molecule-fixed axis system; z is the molecular orientation axis. For any one molecule the two axis systems are related as illustrated in Figure 6.1. Unfortunately, each molecule in a sample probably has different values of the angles $\{\xi, \psi, \zeta\}$ defined in Figure 6.1.

Another key concept is that of the transition moment of a transition. In §1.3 *transition moment* was defined to be a vectorial property (*cf.* §11.1) having a well-defined direction (the transition polarization) within each molecule and a well-defined length (which is proportional to the square root of the absorbance). We denote the electric dipole transition moment by the vector $\boldsymbol{\mu}$. A_Z is then proportional to the square of the Z-component of $\boldsymbol{\mu}$, so the LD of a sample is:

$$LD = k\left\langle \mu_Z^2 - \mu_Y^2 \right\rangle \tag{6.3}$$

where $\langle\ \rangle$ denotes an average over the orientational distribution of the molecules in the sample, and k a positive constant.

With overlapping absorption bands the LD equations all have to be modified to be instead an average in which the contribution of each transition is weighted by the absorption coefficient of that absorption band.

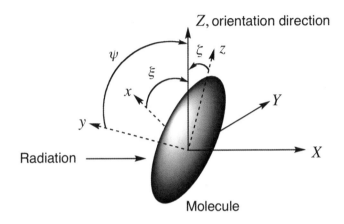

Figure 6.1 Macroscopic $\{X,Y,Z\}$ and molecular axes $\{x,y,z\}$ systems and the angles relating them. For uniaxial rod-like systems we write $\theta = \zeta$.

The main equation used in this book to interpret LD spectra is equation (3.3). Its derivation follows from Figure 3.3 and proceeds as follows. Assume first that the molecule of interest is perfectly oriented so that the molecular orientation axis lies along the macroscopic parallel direction, *i.e.* $z = Z$, $S = 1$. Since absorbance is proportional to the square of the component of the electric dipole transition moment, $\boldsymbol{\mu}$, which lies along the direction of the electric field of the light, we may write (*cf.* Figure 3.3)

$$\begin{aligned} A_Z &= k\mu^2 \cos^2\alpha \\ A_Y &= k\mu^2 \sin^2\alpha \sin^2\gamma \end{aligned} \tag{6.4}$$

Thus

$$LD = A_Z - A_Y = k\mu^2\left(\cos^2\alpha - \sin^2\alpha\sin^2\gamma\right) \tag{6.5}$$

Similarly, the isotropic absorbance is the average of the absorbance in the X, Y, and Z directions, so

$$\begin{aligned} A_{iso} &= \frac{1}{3}\left(A_Z + A_Y + A_X\right) = \frac{1}{3}\left(A_Z + 2A_Y\right) \\ &= \frac{1}{3}k\mu^2\left(\cos^2\alpha + 2\sin^2\alpha\sin^2\gamma\right) \end{aligned} \tag{6.6}$$

Since the experiment does not restrict the value of γ, we average over

$\sin^2 \gamma$. As $\left\langle \sin^2 \gamma \right\rangle = 1/2$ we have

$$LD = k\mu^2 \left(\cos^2 \alpha - \frac{1}{2} \sin^2 \alpha \right)$$

$$= \frac{k\mu^2}{2} \left(2\cos^2 \alpha - \sin^2 \alpha \right) \qquad (6.7)$$

$$= \frac{k\mu^2}{2} \left(3\cos^2 \alpha - 1 \right)$$

and

$$A_{iso} = \frac{k\mu^2}{3} \left(\cos^2 \alpha + \sin^2 \alpha \right) = \frac{k\mu^2}{3} \qquad (6.8)$$

A_{iso} is not necessarily the same as the absorbance of an oriented sample with unpolarized light. For a description of how to determine A_{iso} see §11.2.

Equation (3.3) follows upon acknowledging that actually the orientation is not perfect so including S we get

$$LD = \frac{3}{2} A_{iso} S \left(3\cos^2 \alpha - 1 \right) \qquad (6.9)$$

Reduced linear dichroism

The reduced linear dichroism, LD^r, as defined in equation (3.6) and illustrated in Figure 6.2 may be written

$$LD^r = 3 \left(\frac{A_Z - A_Y}{A_X + A_Y + A_Z} \right) = 3 \left(\left\langle \hat{\mu}_Z^2 \right\rangle - \left\langle \hat{\mu}_Y^2 \right\rangle \right) \qquad (6.10)$$

The factor of 3 in equation (6.10) arises because the isotropic absorbance

$$A = \frac{1}{3} \left(A_X + A_Y + A_Z \right)$$

is a rotational average in three dimensions.

The slight variations observed in Figure 6.2 may be due to small wavelength shifts in transition energies of (in this case solution *versus* film) resulting in the absorbance spectrum used to calculate LD^r not being quite synchronized with the LD.

where $\hat{\mu}_Z$ is the Z-component of the unit vector along $\boldsymbol{\mu}$. $\hat{\mu}_Z$ is the cosine of the angle between the transition moment polarization and the Z-axis, similarly $\hat{\mu}_Y$ is the cosine of the angle between $\boldsymbol{\mu}$ and the Y-axis, and $\hat{\mu}_X$ is the cosine of the angle between $\boldsymbol{\mu}$ and the X-axis. The advantage of using LD^r instead of LD is that dependence on the concentration of the sample, the path length and the dipole strength (intensity), μ^2, of the transition has been removed. Thus, across the absorption envelope of *a single transition* the LD^r should be constant. Thus, LD^r is a function only of the geometric arrangement in space of the transition moments relative to the orientation axis.

We could finish this chapter here since in principle equation (6.10) is the only relationship we need for interpreting an LD experiment. To use it we would first propose a model for how the chromophoric groups are arranged in space and how the transition moments are directed within each chromophore, then calculate the LD^r using equation (6.10) and finally average over all chromophores and all transitions. If the calculated LD^r spectrum did not agree with the experimental one, then the model would have to be revised. With many systems, by exploiting different parts of the spectrum and differently polarized transitions, our confidence in the structural conclusions from this approach would be high and it is in fact the preferred procedure for an object whose macroscopic *and* microscopic structure is well-defined, such as a crystal.

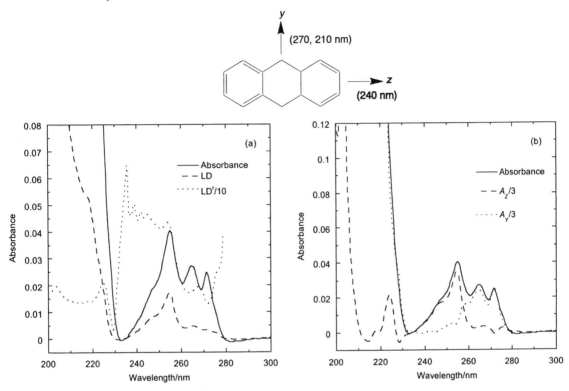

Figure 6.2 (a) Absorbance, *LD* and *LD'* spectra in a PE stretched film (5×) of dihydroanthracene. The approximately constant *LD'* signals in different parts of the spectrum indicate regions of single polarization. (b) The component spectra of dihydroanthracene (see below for methodology used to determine them) are also illustrated.[198]

6.3   Orientational and optical contributions to *LD'*

Unfortunately, it is usually the case that poor macroscopic alignment of the main structural entities composing the sample is superimposed on the uncertainty due to unknown transition polarizations. For example, in a polymer chain where our major concern is the structure of each link, we may only have information about the global orientation of the polymer. For many applications of the latter type, it is convenient to factorize the reduced *LD* into a product between an *optical tensor*, *O*, which depends on how the transition moment is oriented in a local coordinate system $\{x,y,z\}$ and an *orientation tensor*, *S*, which depends on how well the molecule is macroscopically oriented. In general we may write

$$\mathbf{S} = \begin{pmatrix} S_{xx} & S_{xy} & S_{xz} \\ S_{yx} & S_{yy} & S_{yz} \\ S_{zx} & S_{zy} & S_{zz} \end{pmatrix} \tag{6.11}$$

$$\mathbf{O} = \frac{1}{2} \begin{pmatrix} 2\mu_x\mu_x & \mu_x\mu_y & \mu_x\mu_z \\ \mu_y\mu_x & 2\mu_y\mu_y & \mu_y\mu_z \\ \mu_z\mu_x & \mu_z\mu_y & 2\mu_z\mu_z \end{pmatrix} \tag{6.12}$$

This product of two tensors is simply the sum of the products of the nine pair wise products of the tensor components, so

$$LD^r =$$

$$3S_{xx}\mu_x\mu_x + \frac{3}{2}S_{xy}\mu_x\mu_y + \dots$$

and

$$LD^r = 3O : S \qquad (6.13)$$

For molecules with all spectroscopic axes determined by symmetry, the identity of z, the molecular orientation axis, is usually apparent upon inspection. This simplifies the analysis required. For lower symmetry molecules (one or no axes determined by symmetry) the orientation method used for the LD always defines a unique z, but it may not be easily identified by an observer. In what follows we consider the most common special cases.

6.4 Uniaxial orientation

For the special (but in practice most common) case of uniaxial orientation, the off-diagonal terms of equation (6.11) vanish and

$$\mathbf{S} = \begin{pmatrix} S_{xx} & 0 & 0 \\ 0 & S_{yy} & 0 \\ 0 & 0 & S_{zz} \end{pmatrix} \qquad (6.14)$$

Since z is by definition the most oriented molecular axis and we take x to be the least oriented it follows that

$$S_{zz} \geq S_{yy} \geq S_{xx} \qquad (6.15)$$

Further, the trace of the orientation tensor is zero so

$$S_{xx} + S_{yy} + S_{zz} = 0 \qquad (6.16)$$

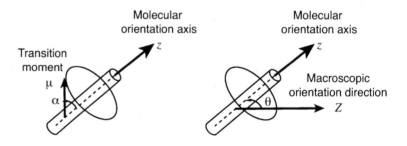

Figure 6.3 Geometry used for uniaxial orientation of rod-like molecules.

If we use ε_z instead of $\mu_z\mu_z$ *etc.* the reduced LD for a uniaxial system at a given wavelength λ is

$$LD^r(\lambda) = \frac{LD(\lambda)}{A_{iso}(\lambda)} = 3\left(\frac{S_{zz}\varepsilon_z(\lambda) + S_{yy}\varepsilon_y(\lambda) + S_{xx}\varepsilon_x(\lambda)}{\varepsilon_z(\lambda) + \varepsilon_y(\lambda) + \varepsilon_x(\lambda)}\right) \qquad (6.17)$$

where A_{iso} is the absorbance of an equivalent (same path length and same sample concentration) unoriented sample and $\varepsilon_z(\lambda)$ *etc.* are the extinction coefficients (at wavelength λ) corresponding to light polarized parallel to the respective coordinate axis. We shall return to equation (6.17) below as it is particularly useful for analyzing the *LD* spectra of small symmetric molecules, but emphasize that it is quite general, holding for complicated, low-symmetry molecules with overlapping transitions. For isolated transitions, equation (6.17) simplifies to

$$LD^r\left(\lambda, z\, polarized\right) = \frac{LD(\lambda)}{A(\lambda)} = 3S_{zz}$$

$$LD^r\left(\lambda, y\, polarized\right) = \frac{LD(\lambda)}{A(\lambda)} = 3S_{yy} \qquad (6.18)$$

$$LD^r\left(\lambda, x\, polarized\right) = \frac{LD(\lambda)}{A(\lambda)} = 3S_{xx}$$

Uniaxial rod-like orientation

For the further special case of unixial rod-like orientation (Figure 6.3), where $S_{xx} = S_{yy}$, we may write

$$S = S(\theta)\begin{pmatrix} -\frac{1}{2} & 0 & 0 \\ 0 & -\frac{1}{2} & 0 \\ 0 & 0 & 1 \end{pmatrix} = \frac{1}{2}\left(3\langle\cos^2\theta\rangle - 1\right)\begin{pmatrix} -\frac{1}{2} & 0 & 0 \\ 0 & -\frac{1}{2} & 0 \\ 0 & 0 & 1 \end{pmatrix} \qquad (6.19)$$

For uniaxial rod-like molecules $S_{xx} = S_{yy}$. It follows from equation (6.16) that

$$S_{xx} = S_{yy} = -\frac{1}{2}S_{zz}$$

where θ is the angle between the macroscopic orientation direction Z and the molecular orientation axis z (Figure 6.3). We use θ for this case instead of the ζ illustrated in Figure 6.1 to be consistent with the literature conventions for uniaxial rod-like systems. It conveniently reminds us that equation (6.19) strictly only holds for rod-like molecules in uniaxial samples. Since $\mu_z = \mu\cos\alpha$ (Figure 6.3), it follows that

$$\boldsymbol{\mu} = \mu(\sin\alpha\cos\beta, \sin\alpha\sin\beta, \cos\alpha)_{\{x,y,z\}} \qquad (6.20)$$

Note: $\sin^2\beta + \cos^2\beta = 1$

where $\{x,y,z\}$ denotes the coordinate system being used and β takes values between 0 and π. From equations (6.13) and (6.19) it follows that the reduced *LD* for a uniaxial rod-like system is

$$LD^r = 3O(\alpha)S(\theta)$$

$$= 3\left\{\frac{1}{2}\left(3\langle\cos^2\alpha\rangle - 1\right)\right\} \times \left\{\frac{1}{2}\left(3\langle\cos^2\theta\rangle - 1\right)\right\} \qquad (6.21)$$

Equation (6.17) may also be simplified for a uniaxial rod-like system

$$LD^r(\lambda) = \frac{LD(\lambda)}{A(\lambda)} = \frac{3}{2}S(\theta)\left(\frac{2\varepsilon_z(\lambda) - \left(\varepsilon_y(\lambda) + \varepsilon_x(\lambda)\right)}{\varepsilon_z(\lambda) + \varepsilon_y(\lambda) + \varepsilon_x(\lambda)}\right) \qquad (6.22)$$

In the Chapters 3 and 5 we used S as a simple scaling factor; it can, however, be used to provide more information about the system. In principle, S may be used to examine the mechanism of orientation of the sample once the optical factor (*i.e.* the transition moment direction α) is known. Conversely, if S can be assessed independently, the transition moment direction in the molecular frame may be determined. These two possibilities illustrate the two major classes of applications of *LD* spectroscopy: the structural and the spectroscopic applications.

Effective angle for uniaxal rod-like orientation

If we know α (Figure 6.3), we can determine the orientation parameter S from the LD^r. Unfortunately, even when we know S, this does not mean that we thereby have an exact value for the angle θ at which the molecules are oriented. First of all the cosine square is associated with a sign ambiguity (we do not know whether it is $+\cos\theta$ or $-\cos\theta$) and, secondly how to interpret averages such as $\langle \cos^2\theta \rangle$ in terms of structure is far from clear: $\langle \cos^2\theta \rangle \neq \cos^2\langle\theta\rangle$. The distribution of θ values, over which the $\cos^2$ function is averaged, may have features that are poorly characterized by a single angle value, θ_{eff} (denoting θ effective), defined by:

$$\cos^2\theta_{eff} = \langle \cos^2\theta \rangle \qquad (6.23)$$

For $\langle \cos^2\theta \rangle$ values close to 1 or 0 we can with some confidence say that θ is close to $0°$ or $90°$, respectively, but if, for example, $\langle \cos^2\theta \rangle = 1/3$ (*i.e.* $S=0$) we cannot say whether this is due to $\theta = 54.7°$ or to the orientational distribution being isotropic (*i.e.* all orientations equally probable).

54.7° is often called the 'magic angle' since in many experimental situations, including *LD* and solid state NMR, orientation at this angle leads to many of the features of a full rotational average.

Uniaxially oriented planar aromatic molecules with in-plane polarized transitions and no out-of-plane polarized transitions

Many chromophores of interest to chemists and biochemists are planar molecules whose only symmetry property is the reflection plane which contains all the atoms. When such a molecule is uniaxially oriented, it is generally true that the orienting forces align the molecules so that they present a minimum cross-sectional area to the orienting force, *i.e.* to Z. Thus, in this case, x is perpendicular to the molecular plane and y and z are somewhere within the plane of the molecule. We do not immediately know where y and z lie, but z is usually close to the 'longest' axis of the molecule (Figure 6.4). What follows is how we can determine z if we have a number of transitions whose polarizations in the molecular plane we know. The electric dipole transition moment for such a planar molecule in the molecular axis system is (Figure 6.4)

$$\boldsymbol{\mu} = (0, \sin\alpha, \cos\alpha)_{\{x,y,z\}} \qquad (6.24)$$

where α is the angle between $\boldsymbol{\mu}$ and z. Although our emphasis in this book is on electronic transitions, we note that infrared *LD* is particularly useful here since polarizations of some bond-stretch transitions are obvious upon inspection.

Our problem is that we do not know where y and z lie. So we set up a temporary axis system $\{x, y', z'\}$, where we choose y' and z' for our convenience. α' is the angle between $\boldsymbol{\mu}$ and z'. So

$$\boldsymbol{\mu} = (0, \sin\alpha', \cos\alpha')_{\{x,y',z'\}} \qquad (6.25)$$

The LD^r, expressed in terms of parameters in the $\{x, y', z'\}$ axis system is, from equation (6.13),

$$LD^r = 3\left(S_{y'y'}\sin^2\alpha' + S_{z'z'}\cos^2\alpha' + S_{y'z'}\sin\alpha'\cos\alpha'\right) \qquad (6.26)$$

The three orientation parameters are three unknowns in this equation. If our temporary molecular axes actually proved to be the same as the true molecular axes, then $S_{y'z'} = 0$. Otherwise we need three transitions to determine the unknowns and hence the true molecular orientation axis. Once we have done that we use

$$\tan(2\gamma) = \frac{S_{y'z'}}{S_{z'z'} - S_{y'y'}} \qquad (6.27)$$

(which is derived in §7.2) to determine γ, the angle between z' and z (Figure 6.4). The true molecular orientation axis system $\{x,y,z\}$ is then identified as illustrated in the example below.

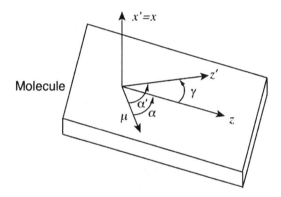

Figure 6.4 Orientation of a planar molecule illustrating true and temporary molecular axes systems.

The equation analogous to equation (6.26) for the true axis system, $\{x,y,z\}$, for in-plane polarized transitions of a planar molecule in a uniaxial sample is

$$LD^r = 3\left(S_{zz}\cos^2\alpha + S_{yy}\sin^2\alpha\right) \qquad (6.28)$$

Determination of a planar molecule's orientation axis of a molecule in a stretched film

To illustrate how equations (6.26) and (6.27) can be used in practice, consider methyllumiflavin which has three convenient infra-red bond-stretching vibrations whose polarizations have been determined.[199] As depicted in Figure 6.5, we choose the temporary axis system so that one of the carbonyl stretching transition moments has $\alpha' = 0$. The values of the LD^r for each transition are also given in the figure.

In this example we are ignoring experimental error. It must be considered in reality.

We begin by substituting the experimental data into equation (6.26) for each transition. It should be noted that the orientation parameters are molecular properties.

$$+0.070 = 3\left(S_{z'z'}\right)$$

$$+0.580 = 3\left(0.030 S_{z'z'} + 0.970 S_{y'y'} - 0.171 S_{y'z'}\right) \tag{6.29}$$

$$+0.110 = 3\left(0.250 S_{z'z'} + 0.750 S_{y'y'} - 0.433 S_{y'z'}\right)$$

From which it follows that

$$S_{z'z'} = 0.023$$

$$S_{y'y'} = 0.267 \tag{6.30}$$

$$S_{y'z'} = 0.392$$

Substituting the orientation parameters into equation (6.27) tells us that γ either equals 61° or 151°. By inspection of Figure 6.5 and guessing the orientation influence of the methyl groups, we deduce that the value of 61° depicted in Figure 6.5 is the correct one and the dashed line of Figure 6.5 is the molecular orientation axis z. Once z has been identified, since $\alpha' = \alpha + \gamma$, equation (6.28) can be used to determine the true orientation parameters: $S_{zz} = 0.38$; $S_{yy} = -0.08$. It then follows from equation (6.16) that $S_{xx} = -0.29$. Polarizations in the true axis system of electronic (and other vibrational) transitions can then be calculated as outlined in §6.6.

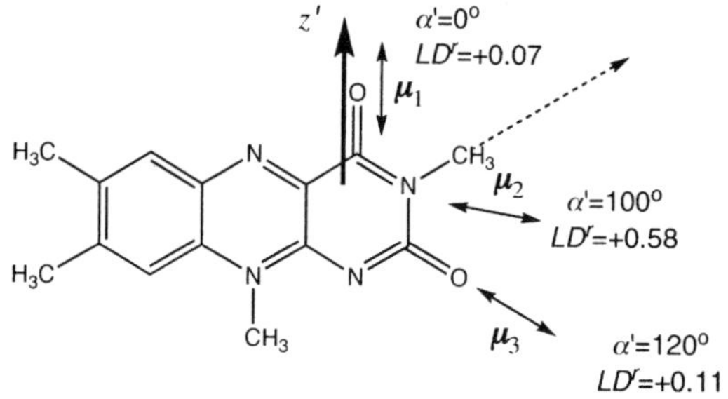

Figure 6.5 Methyllumiflavin indicating the temporary axis system and the polarizations and LD^r values for the three transitions used in the text. The dotted arrow indicates the calculated direction of z. Data are from reference [199].

6.5   Orientation triangle

If orientational behaviour is of primary interest, for example in order to characterize the interactions between a solute molecule and the surrounding anisotropic host, it is practical to display the S_{yy} and S_{zz} values in an orientation triangle as in Figure 6.6. Since $S_{xx} + S_{yy} + S_{zz} = 0$ (equation (6.16)), a plot of one of them *versus* another is enough to uniquely define a set of all three parameters. The borders and inside of the orientation triangle define all possible combinations of the two largest orientation parameters for any kind of situation and molecule. The position of molecules in the triangle depends on their shape and ability to orient.

- *Rod-shaped molecules* lie on the lower edge. For these molecules the z-axis is unique while the x- and y-axes are orientationally degenerate (*i.e.* $S_{xx} = S_{yy}$). Thus $S_{yy} = -0.5 S_{zz}$ so the line from (0,0) and ending at

(−0.5,1.0) represents rod-like molecules of increasing ability to align
in the medium.

- *Disc-shaped molecules*, for which y and z axes are degenerate (*i.e.* S_{yy} = S_{zz}) lie on the line starting at (0,0) and ending at (0.25,0.25) with molecules that orient better lying towards (0.25,0.25).

- *Molecules that are neither rods nor discs* are represented by points that may be anywhere inside the orientation triangle, with positions that may tell something about their orientational behaviour. Most planar molecules exhibit orientational properties that fall between the two extrema of rod-like and disk-like behaviour. For example, dihydroanthracene (Figure 6.2), whose S_{yy} and S_{zz} are both positive, lies towards the disc-like molecules compared with anthracene itself.

The minimum value of S_{xx} is −1 which corresponds to perfect orientation of the x-axis perpendicular to the orientation direction.

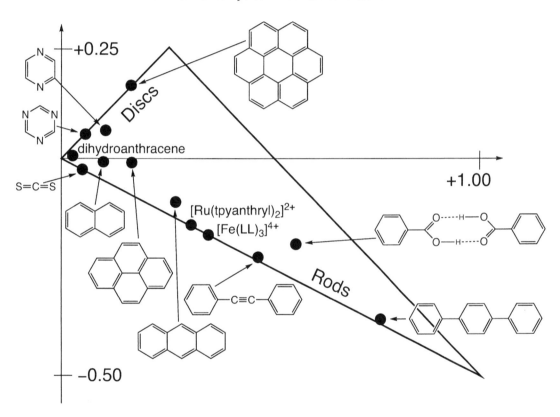

Figure 6.6 Orientation triangle for planar or rod-like aromatic molecules in stretched PE orientation matrices. Data are from reference [8] and calculated from data presented herein. The molecules are depicted with their z axes horizontal.

The orientations adopted by different molecules (which are mainly determined by steric forces) can be thought of in terms of how needles, coins, and rectangular objects become oriented when dropped into a densely packed hairbrush: the needles lie with their long axis parallel to the bristles, the circular coins align themselves with their planes parallel to the bristles of the brush, whereas a rectangular object's orientation will be driven by keeping its planar surface parallel to the bristles. The behaviour in Figure 6.6 refers to

Note that *s*-triazine, by symmetry, behaves like a disc, whereas pyrazine deviates somewhat from the disc-line.[197]

polyethylene; interestingly, many molecules that have an intermediate orientation behaviour, such as pyrene and naphthalene, show marked variations in where they lie in the orientation triangle in different polymer hosts.[200]

The position adopted by a given analyte in the orientation triangle is related to its *LD* in a very simple way: the *LD* for a transition along an axis in the molecule has the same sign as its *S*-parameter. For example, *n-π** transitions in pyrazine and *s*-triazine which are both polarized perpendicular to the molecular plane (*x*-polarized) give negative *LD* bands when oriented in films. The *y*-polarized transitions of pyrene oriented in films have almost no *LD* intensity consistent with its value of S_{yy} being nearly equal to 0 (it is neither rod not disc). Benzoic acid in the non-polar polyethylene host proved to be exclusively solubilized as a non-polar hydrogen-bonded dimer which results in an almost rod-like orientation (S_{zz}=+0.43; S_{yy}=–0.16; S_{xx}=–0.27) as shown in Figure 6.7. Nitrobenzene by way of contrast is a fairly discoid monomer that has much lower orientation parameters (S_{zz}=+0.023; S_{yy}=–0.001; S_{xx}=–0.022) which are very close to the origin in the orientation triangle of Figure 6.6.[201]

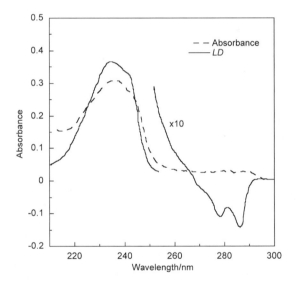

Figure 6.7 PE film *LD* spectra of the benzoic acid dimer (Figure 6.6) showing the long axis polarized positive *LD* band at 230 nm and the short axis negative *LD* band at 280 nm.[201]

6.6   Determination of transition polarizations

Polarizations of transition moments are often needed to determine structural information from *LD* spectra. For example, if transition moments of a molecule are known, when it binds to a macromolecule, *e.g.* DNA, its orientation on the macromolecule can be determined. Sometimes transition moments may be obtained from inspection of the molecular structure supported by a simple quantum-mechanical calculation, or may be determined experimentally as outlined below. Alternatively it may require a very sophisticated calculation of the kind undertaken in reference [202].

Many of the small molecules that are studied by *LD* have at least some symmetry elements. If a molecule has elements of symmetry, they restrict the possible polarizations of transition moments to being along rotation axes or parallel or perpendicular to reflection planes. In this section we consider the situation where the orienting force operates directly on the small molecule so its symmetry properties contribute to the orientation process. We are thus excluding the situations where the orientation is effected by another molecule to which the molecule is bound. The categories of molecular symmetry are outlined in Classes a–f below and illustrated in Figure 6.8. Point group symmetry labels are used to collect molecules into groups whose *LD* equations are the same. Some examples of molecules in the categories are also given.

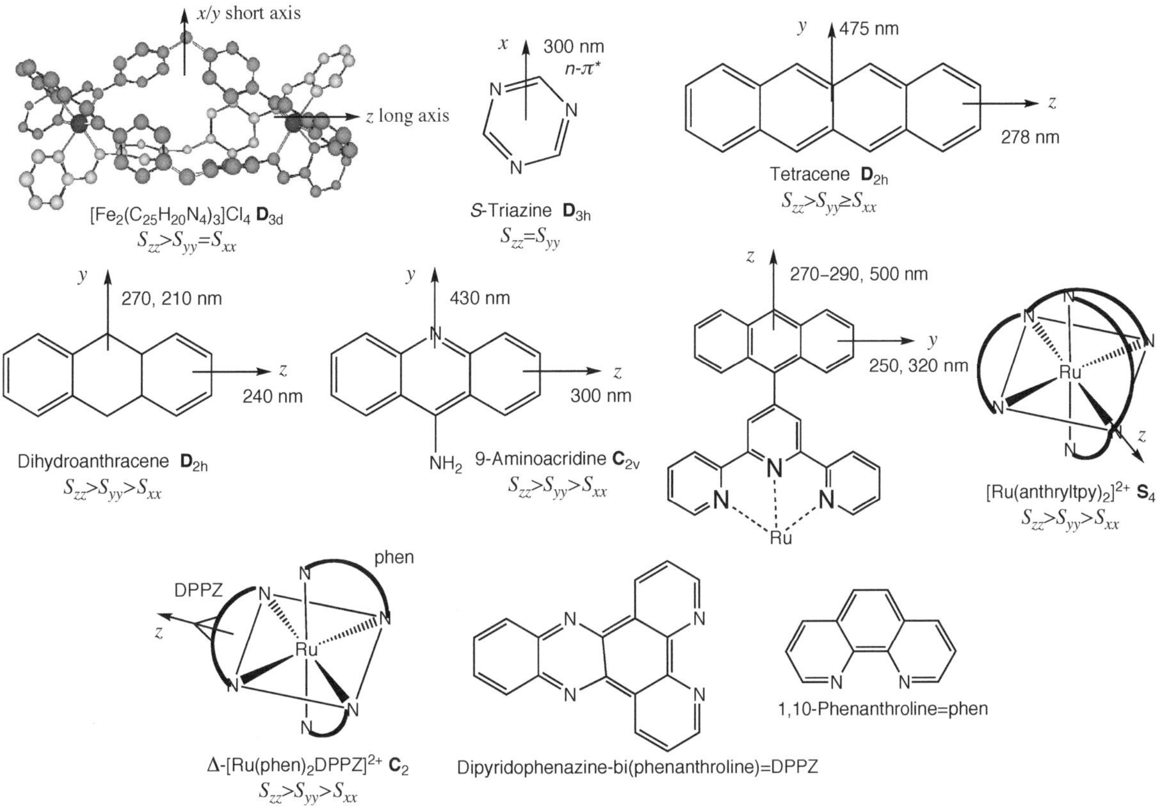

Figure 6.8 Examples of symmetry determined molecular orientation axes and transition polarization directions. The ligands of the metal complexes are denoted by thick curves whose molecular structure is given next to them.

Class a: $\mathbf{T}, \mathbf{T_d}, \mathbf{T_h}, \mathbf{O}, \mathbf{O_h}, \mathbf{I}, \mathbf{I_h}$, and the spherical point groups

For molecules with tetrahedral or higher symmetry the question of transition moment polarization is irrelevant since all polarizations occur for all transitions. Such molecules will not show *LD* unless they are distorted from their high symmetry geometry. Thus for this case

$$LD^r = 0 \qquad (6.31)$$

Class b: C_n, C_{nh}, C_{nv}, D_n, D_{nh}, $n \geq 3$, D_{nd}, $n \geq 2$, and S_n, $n \geq 4$

A molecule with one three-fold or higher rotation axis has transitions polarized either along this rotation axis or perpendicular to it. The polarization of any transition that is in the plane perpendicular to the major rotation axis is delocalized over all directions in that plane. z, the molecular orientation axis for such molecules, is either the unique axis (the n-fold rotation axis), in which case $S_{yy} = S_{xx} = -S_{zz}/2$, or (more unusually) it lies in the plane perpendicular to the unique axis of the molecule and is indistinguishable from y so $S_{zz} = S_{yy} = -S_{xx}/2$. In the former case, symmetry makes the orientation equivalent to the uniaxial rod case so for a z-polarized ($\alpha = 0°$) transition equation (6.18) gives

$$LD^r(z) = +3S_{zz} \qquad (6.32)$$

across the whole absorption band. Similarly, for an x/y polarized transition ($\alpha = 90°$)

$$LD^r(x/y) = -\frac{3}{2}S_{zz} \qquad (6.33)$$

It is thus easy to assign transition polarizations simply from the sign of the LD^r. LD^r signals of intermediate value (often sloping across an absorption envelope) indicate the presence of overlapping bands of different polarizations.

$[Fe_2(LL)_3]^{4+}$

This symmetry class is illustrated in Figure 6.9 for the film LD spectrum of a tetracationic iron triple helicate $[Fe_2(LL)_3]^{4+}$ of D_3 symmetry (*cf.* Figure 3.10).[86] If we assume (reasonably) for this case that the molecular orientation axis is the long axis of the triple helix and that the orientation is uniaxial (due to the x-y degeneracy of the metal complex itself), then the negative sign of the LD from 500–600 nm indicates that the transitions here are predominantly short axis (*i.e.* x/y or $\perp$) polarized transitions. By assuming they are pure $\perp$ polarized we get a lower bound on the magnitude of S_{yy} from equation (6.18).

$$\left|LD^r(600 \text{ nm})\right| \leq \left|LD^r(x/y)\right| = \frac{3}{2}S_{zz} \qquad (6.34)$$

Thus $S_{zz} \sim 0.35$ and $S_{yy} = S_{xx} \sim -0.18$. In this case the lower bound on the magnitude of S_{yy} is its value—so we conclude that the 500–600 nm region of the spectrum is indeed pure x/y polarized.

Component polarized spectra for $[Fe_2(LL)_3]^{4+}$ determined by incrementally increasing the orientation parameters (see below) that are given in Figure 6.9b. The in-ligand transitions below 400 nm are of mixed polarization since the LD^r magnitude never reaches twice that at 550 nm (equations (6.32) and (6.33)). More careful inspection shows that the 400 nm transitions have slightly more // than $\perp$ character.

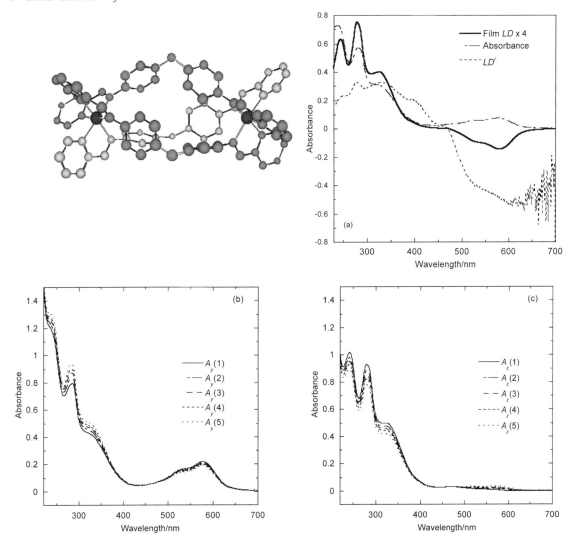

Figure 6.9 (a) Absorbance, *LD*, and *LD'* spectra of [Fe$_2$(LL)$_3$]$^{4+}$ (illustrated) in a PVA film made from a low-molecular weight PVA solution (10%(w/v)) that was allowed to dry then stretched 2×. (b) and (c) [Fe$_2$(LL)$_3$]$^{4+}$ component spectra.[86]

[Ru(tpyanth)$_2$]$^{2+}$

The film absorbance, *LD* and *LD'* for the **S**$_4$ ruthenium metal complex which has two terpyridine ligands each of which has an anthryl tail are illustrated in Figure 6.8. If *z* is as illustrated, the tpy transitions centred at ~320 nm should have negative film *LD* signals, which they indeed do (*LD'* = −0.45) as illustrated in Figure 6.10. The ~270–290 nm region should consequently be positive, as observed (*LD'* = 0.44). The anthracene chromophore's long axis polarized transition (Figure 2.2, Figure 2.3) at 250 nm is along the short axis of the metal complex (*y*-polarized) and is thus negative. We may then conclude that the metal ligand charge transfer transitions are dominated by z-polarized transitions, particularly at ~500 nm where the complex has a large positive LD^r signal ($LD^r = 0.9$). The largest positive LD^r is 0.925 suggesting $S_{zz} \sim 0.31$. The flatness of the 320 nm region of the LD^r suggests it is pure y-polarized, so $S_{yy} \sim -0.15$, resulting in $S_{xx} \sim -0.16$ since the three order parameters sum to zero. Thus $S_{xx} = S_{yy}$ within experimental error as expected for an S_4 system.

Class c: C_{2v}, D_2, and D_{2h}

Molecules with three two-fold rotation axes but no higher order rotation axes and those with one two-fold rotation axis and two reflection planes must have all transitions polarized either along a rotation axis or perpendicular to a reflection plane. z is one of the symmetry determined axes. In this case $S_{zz} > S_{yy} > S_{xx}$.

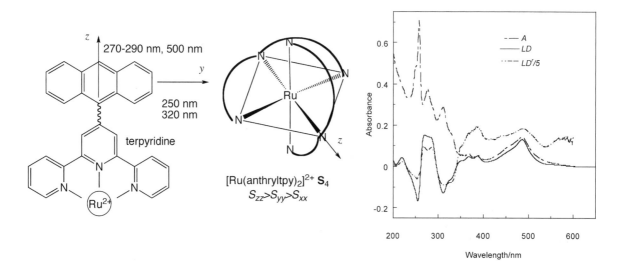

Figure 6.10 PVA film absorbance, LD and LD^r for $[Ru(tpyanth)_2]^{2+}$.[203]

Tetracene

Tetracene belongs to this class of molecules. Positive and negative peaks are apparent in the spectrum of Figure 6.11 showing the presence of in-plane $\pi \rightarrow \pi^*$ transitions polarized both parallel (z-polarized) and perpendicular (y-polarized) to the long symmetry axis of the molecule. Let us first assume that the strong positive peak at 278 nm is of pure z polarization. From equation (6.32), since the film absorbance at 278 nm is 0.13, we deduce

$$S_{zz} = \frac{LD^r}{3} \approx \frac{0.174}{3 \times 0.13} = 0.45 \tag{6.35}$$

The orientation process for this long molecule is thus very efficient. The most negative LD signal ($LD = -0.0032$, $A = 0.005$) at 477 nm leads us to conclude $S_{yy} \sim -0.22$. Since the orientation parameters sum to zero, we then deduce that $S_{xx} \approx S_{yy}$ (equation (6.16)) within experimental error. This means that tetracene actually approximates a uniaxial rod and behaves as if it is the higher symmetry of case (b).

The positive *LD* signal in the 350–400 nm region of tetracene's spectrum has an *LD*^{r} value only slightly lower than that of the long-axis polarized 278 nm transition. This reflects the significant extent of mixing of long-axis character into the short wavelength region of the short axis polarized transition. The situation for anthracene is much the same, as shown in Figure 2.3, though we initially ignored this coupling for tutorial purposes to produce the simplified spectra given in Figure 2.2.

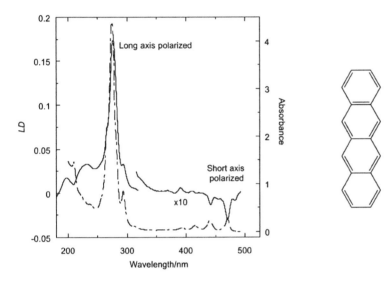

Figure 6.11 *LD* (solid lines) for the approximately uniaxial rod molecule tetracene oriented in a stretched polyethylene film (10× stretched pipette tip bag, solvent 5% CH$_2$Cl$_2$ in methylhexane) with absorbance (dashed lines) in methyl cyclohexane:diethyl ether (20:1).

Dihydroanthracene

The dihydroanthracene molecule of Figure 6.2 also belongs to the **C**$_{2v}$/**D**$_2$/**D**$_{2h}$ class of molecules. By symmetry we know it will align in a film so that its long axis is z. From its *LD*^{r} it is therefore apparent that the transitions between 235 nm and 255 nm are long-axis polarized and those below 225 nm and above 260 nm are short axis polarized. The optical tensors (equation (6.12)) for its long-axis and short-axis polarized transitions are respectively

$$O(z \text{ polarized transitions}) = \begin{pmatrix} 0 & 0 & 0 \\ 0 & 0 & 0 \\ 0 & 0 & 1 \end{pmatrix} \tag{6.36}$$

$$O(y \text{ polarized transitions}) = \begin{pmatrix} 0 & 0 & 0 \\ 0 & 1 & 0 \\ 0 & 0 & 0 \end{pmatrix} \tag{6.37}$$

So

$$LD^r \, (z \text{ polarized transitions}) = 3S_{zz} \qquad (6.38)$$

From Figure 6.2 it then follows that $S_{zz} = 0.045/3 = 0.015$. Similarly, $S_{yy}=0.015/3=0.005$, which means that $S_{xx} = -0.15-0.05 = -0.2$. From this we can deduce the component spectra that are illustrated in Figure 6.2b.

Class d: C_s and C_{2h} (planar molecules)

Molecules with only a reflection plane or with a reflection plane and a two-fold axis perpendicular to it, have transitions polarized either within or perpendicular to the plane. Each in-plane transition has a well-defined polarization, however, symmetry does not restrict the allowed directions.

To determine transition polarizations for such low symmetry planar molecules we first identify z (the molecular orientation direction) and then determine the S_{kk}, $k=x,y,z$. For a spectrum composed only of in-plane transitions (*e.g.* a DNA base with data collected down to 200 nm) equation (6.17) becomes

$$LD^r\left(\lambda\right) = 3\left(\frac{S_{zz}\varepsilon_z\left(\lambda\right)+S_{yy}\varepsilon_y\left(\lambda\right)}{\varepsilon_z\left(\lambda\right)+\varepsilon_y\left(\lambda\right)}\right) = 3\left(\frac{S_{zz}\dfrac{\varepsilon_z\left(\lambda\right)}{\varepsilon_y\left(\lambda\right)}+S_{yy}}{\dfrac{\varepsilon_z\left(\lambda\right)}{\varepsilon_y\left(\lambda\right)}+1}\right) \qquad (6.39)$$

from which we may write

$$\frac{\varepsilon_z\left(\lambda\right)}{\varepsilon_y\left(\lambda\right)} = -\frac{LD^r\left(\lambda\right)-3S_{yy}}{LD^r\left(\lambda\right)-3S_{zz}} \qquad (6.40)$$

For a single isolated transition (*i.e.* one with a well-defined but initially unknown transition polarization), the LD^r will be constant across the absorption envelope as will the ratio of z and y extinction coefficients. We may then determine the transition polarization for an in-plane transition of a planar molecule from

$$\frac{\varepsilon_z\left(\lambda\right)}{\varepsilon_y\left(\lambda\right)} = \frac{\cos^2\alpha}{\sin^2\alpha} = \cot^2\alpha \qquad (6.41)$$

Class e: C_2

Molecules with only a two-fold rotation axis have transitions polarized either along that axis or perpendicular to it. There is no restriction on the polarizations of transitions in the plane perpendicular to the rotation axis: z will be either along the symmetry axis or perpendicular to it. In some cases more can be deduced from the local symmetry of component parts.

[Ru(phen)₂DPPZ]²⁺

[Ru(phen)$_2$DPPZ]$^{2+}$

An interesting example of *LD* of a C_2 complex is given by the light-switch complex [Ru(phen)$_2$DPPZ]$^{2+}$ (Figure 6.8) oriented in a lamellar liquid crystal host (Figure 2.13).[144] The liquid crystal was deposited on a quartz surface; the analyte inserted into the liquid crystal perpendicular to the surface so that its C_2 axis (which runs through the DPPZ ligand) aligns with the orientation axis. All transitions are polarized either along z or perpendicular to it. The polarizations of the x/y polarized transitions are restricted by the nature of the chromophores in which they are localized. No *LD* signal was detected when the surface was aligned perpendicular to the light beam as there is no orientational force operative in the plane of the quartz surface. However, when the quartz was tilted with respect to the light beam, a spectrum resulted. Equation (6.17) is modified for this situation to become[8, 144, 204]

$$LD^r(\lambda) = \frac{LD(\lambda)}{A(\lambda)}$$

$$= 3\left(\frac{S_{zz}\varepsilon_z(\lambda) + S_{yy}\varepsilon_y(\lambda) + S_{xx}\varepsilon_x(\lambda)}{\varepsilon_z(\lambda) + \varepsilon_y(\lambda) + \varepsilon_x(\lambda)}\right) \frac{\left(\cos^2\omega\right)}{\left[n^2\left(1 - \cos^2\omega/n^2\right)^{1/2}\right]} \qquad (6.42)$$

where ω (in this case 75°) is the angle between the direction of propagation of the light beam and the plane of the quartz on which the sample is mounted and n is the refractive index. $n \sim 1.33$ for water and $n \sim 1.5$ for glycerol, so the final factor in equation (6.42) is ~ 0.030 for $\omega = 75°$. $LD = 0$ for normal incidence where $\omega = 90°$.

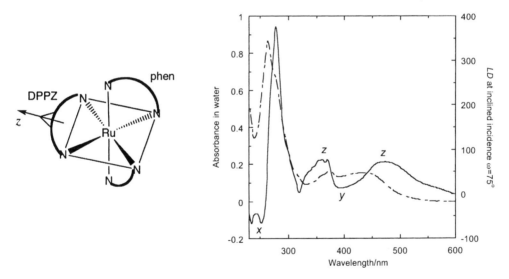

Figure 6.12 Absorbance (dashed line) and *LD* (solid line) of Λ-[Ru(phen)$_2$DPPZ]$^{2+}$ in a lamellar liquid crystal system when tilted at ω=75°. Transition polarizations are indicated on the spectrum.

In this example, the z-polarized transitions all have $LD^r \sim 0.015$, so

$$S_{zz} \sim \frac{0.015}{3} \frac{\left[n^2\left(1 - \cos^2\omega/n^2\right)^{1/2}\right]}{\left(\cos^2\omega\right)} \sim \frac{0.005}{0.03} \sim 0.17 \qquad (6.43)$$

The other transitions are localized within chromophores (parts) of the molecule, so their polarizations can also be assigned, even though they overlap. For example, the so-called $B(A_2)$ transition at 250 nm[176] is polarized out of the plane between the two phenanthroline ligands; it has a negative *LD* suggesting this direction is x.

Class f: C_1 and C_i

For molecules with no symmetry (except perhaps an inversion centre), symmetry is no help in determining transition moments or orientation axes.

6.7   Determination of polarizations of overlapping transitions

Trial and error methods

When we cannot isolate one transition from other transitions because their absorptions overlap, we need a way of separating out components. A method known as the 'Trial and Error Method' (TEM) to produce the component absorption spectra $A_z(\lambda)$ and $A_y(\lambda)$ for $\pi \rightarrow \pi^*$ transitions of planar molecules (where $A_x(\lambda) \approx 0$) was developed by Eggers and Thulstrup in the 1960s without any assumption of a uniaxial sample and developed further together with Michl.[7, 13, 16, 205] For high-symmetry molecules $A_z(\lambda)$ is obtained by identifying the linear combination

$$A_z(\lambda) \propto A_Z(\lambda) - \delta_z A_Y(\lambda) \tag{6.44}$$

> TEM can also conveniently be applied by forming linear combinations of $A(\lambda)$ and $LD(\lambda)$.[8]

which is a spectrum with a 'good' shape, that is everywhere non-negative, and has no feature that for some reason is known to belong to the orthogonal spectrum $A_y(\lambda)$. This is achieved by changing δ in small increments. The $A_y(\lambda)$ spectrum may then be determined in an analogous manner from

$$A_y(\lambda) \propto A_Y(\lambda) - \delta_y A_Z(\lambda) \tag{6.45}$$

This type of methodology was applied above to determine component spectra in Figure 6.9.

Using orientation parameters

An alternative approach used in Figure 6.2 for overlapping transitions of planar aromatic molecules, if one has determined the orientation parameters, is as follows. Since $\varepsilon_x(\lambda) = 0$ for π-π^* transitions of planar aromatic molecules, we may write

$$LD = S_{zz} A_z(\lambda) + S_{yy} A_y(\lambda) \tag{6.46}$$

where $A_z(\lambda)$ is the absorbance of the sample at wavelength λ if the radiation is polarized along the molecular axis z *etc.* Thus

$$S_{yy} A_z(\lambda) - S_{zz} A_z(\lambda) = S_{yy} A_y(\lambda) + S_{yy} A_z(\lambda) - LD \tag{6.47}$$

hence

$$A_z(\lambda) = \frac{S_{yy}A_y(\lambda) + S_{yy}A_z(\lambda) - LD}{S_{yy} - S_{zz}}$$

$$= \frac{3S_{yy}A_{iso}(\lambda) - LD}{S_{yy} - S_{zz}} \qquad (6.48)$$

where $A_{iso}(\lambda)$ is the isotropic absorbance at λ and similarly

$$A_y(\lambda) = \frac{S_{zz}A_z(\lambda) + S_{zz}A_y(\lambda) - LD}{S_{zz} - S_{yy}}$$

$$= \frac{3S_{zz}A_{iso}(\lambda) - LD}{S_{zz} - S_{yy}} \qquad (6.49)$$

Spectral decomposition

With broad, strongly overlapping bands such as in the nucleobases, purines, indoles, flavins, and many other biologically important chromophores, it is not so easy to determine 'good' component spectra. An alternative method for this situation[26, 206] has been used for a wide variety of planar chromophores and some macromolecules.[26, 207] It consists of the following steps:

(1) *Determine S_{zz}, S_{yy}, and the z-axis by an independent method.* For example, measure the infrared *LD* of three transitions with known directions and analyze the spectra as illustrated in §6.4.

(2) *Decompose $A_{iso}(\lambda)$ into (e.g. Gaussian) components* $A_{iso}(\lambda)_i$ each corresponding to a transition.

(3) *Fit the LD^r (or LD) spectrum* using these components and the two S parameters to determine the α value for each transition as described above.

(4) *Iteratively adjust* shapes and positions of the $A_{iso}(\lambda)_i$ and the angles α_i to optimise the fit of $A_{iso}(\lambda)_i$ and $LD(\lambda)$.

(5) *Resolve sign ambiguities* (is it $+\alpha_i$ or $-\alpha_i$?) either by comparing chromophores that are spectroscopically similar but have different orientations or by measuring the fluorescence anisotropy (see below).

6.8   Resolution of angle-sign ambiguities

As noted above, the use of *LD* to determine transition polarizations almost always results in two possible values for α. We shall therefore conclude this chapter with a brief look at how this ambiguity may be resolved.[26]

Fluorescence anisotropy

Fluorescence anisotropy (*FA*) is a fluorescence analogue of LD^r.[208] It requires fluorescence intensities to be measured with polarizers in both the excitation and emission beams of the fluorimeter. It can be a useful method for determining transition polarizations of fluorophores. If the instrument had perfect optics, then the experiment is performed by setting the excitation polarizer to be oriented so as to vertically polarize the light; the emission

In practice, to correct for inherent polarization effects of the instrument optics one should use[208]

$$FA = \frac{I_{VV} - GI_{VH}}{I_{VV} + 2GI_{VH}}$$

where

$$G = \frac{I_{HV}}{I_{HH}}$$

polarizer is then oriented first to transmit vertically polarized and then horizontally polarized light in the emission beam (Figure 6.13):

$$FA = \frac{I_{VV} - I_{VH}}{I_{VV} + 2I_{VH}} \tag{6.50}$$

where I_{VH} is the emerging intensity of light that has passed through a vertical excitation polarizer and a horizontal emission polarizer and I_{VV} the corresponding intensity for vertical emission polarizer setting

Let us assume for simplicity that there are no losses of polarization due to rotation of the chromophore during the lifetime of the emission process or to energy transfer between chromophores (this assumption is unlikely to be valid). For an isolated transition we may then write:[5]

$$FA = \tfrac{1}{5}\left(3\cos^2 \chi - 1\right) \tag{6.51}$$

where χ is the angle between the absorbing and emitting transition dipole moments. A derivation of equation (6.51) is given in §7.4. Whichever transition we choose to excite into, the light is (nearly) always emitted from the lowest excited state (of the same spin multiplicity as the initial excited state, assuming there is no singlet-triplet crossings).

FA is thus equal to +2/5 for transitions whose polarization is parallel to that of the lowest energy transition, since then $\chi = 0$. Its minimum value (corresponding to transitions polarized with absorbing and emitting moments pependicular to the lowest energy transition) is –1/5. To account for overlapping transitions a sum over transitions must be included.

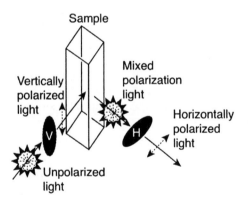

Figure 6.13 *FA* experiment with vertical excitation (**V**) and horizontal (**H**) emission polarizer settings (light source, detector, and monochromators omitted for clarity). Dotted arrows indicate polarizations of light present at each stage.

The potential of *FA* to aid in interpretation of an *LD* spectrum follows from the fact that it can be used to give the angle *between* transition moments. For example, consider a molecule with two transitions that have the same $\cos^2 \alpha$ value. The *FA* of the two transitions will give the angle between them. Depending on its magnitude it will eliminate some of the four possible sets of polarizations for the transition moments (unless $\alpha = 0$ or 90°).

- Both transitions are on the same side of z with $\alpha > 0$.
- Both transitions are on the same side of z with $\alpha < 0$.
- The transitions are on opposite sides of z, the first with $\alpha > 0$ and the second with $\alpha < 0$.
- The transitions are on opposite sides of z, the first with $\alpha < 0$ and the second with $\alpha > 0$.

Inspection may be sufficient to complete the task.

Substituent-perturbed orientation method

A second way to distinguish between the two mathematical solutions to the $\cos^2\alpha$ problem is the so-called 'substituent-perturbed orientation method'[206] which may be used to tell us on which side of the molecular orientation axis z the transition moment lies. Upon introducing into the chromophore a bulky substituent at a suitable position that does not coincide with the line of the orientation axis, the orientation axis will be rotated towards the substituent since it is in this direction that the solute molecule has been made longer. Thereby, angles between the orientation axis and such transition moments that lie on one side of the old orientation axis will be larger, whereas angles on the other side will be smaller, thus allowing the correct α value for the original molecule to be identified. It should be noted however, that some substituents change the transition polarizations as well as the orientation as illustrated in Figure 6.14. However, this approach can usually be used to choose between two options.

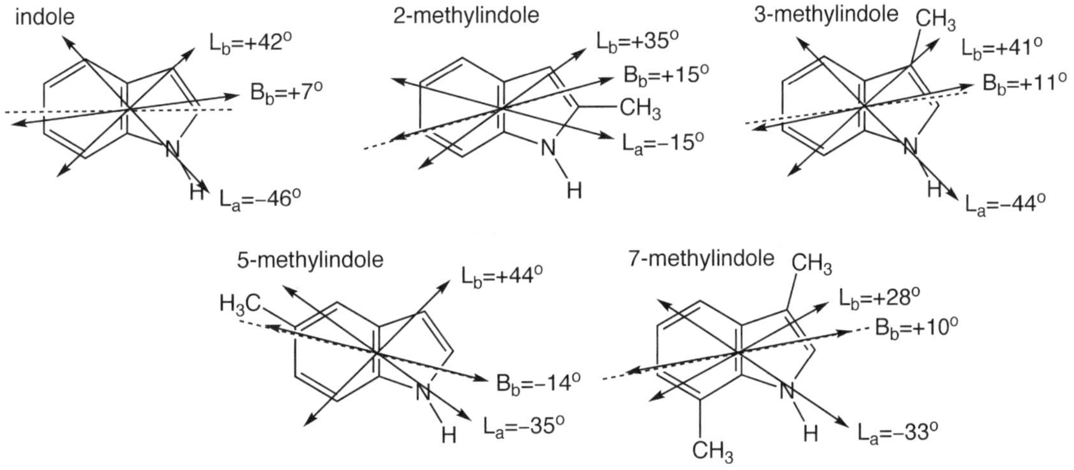

Figure 6.14 Illustration of the changes in transition moment polarizations of indoles upon substitution by methyl groups.[96, 206] Dotted lines indicate molecular orientation axes.

Calculations

An alternative to the experimental methods is to undertake a quantum mechanical calculation. Although the reliability of these varies, they are usually sufficiently good to choose between the two options identified from the *LD*.

6.9   *LD* of samples with a distribution of orientations

Usually the *LD* spectrum of a sample contains a number of more or less overlapping bands, each corresponding to a transition with a well-defined transition moment direction. For overlapping bands of different polarization this results in a wavelength-dependent LD^r since

$$LD^r(\lambda) = \frac{LD(\lambda)}{A_{iso}} = \left(\frac{LD_1(\lambda) + LD_2(\lambda)}{A_{iso1}(\lambda) + A_{iso2}(\lambda)} \right) \qquad (6.52)$$

Since for a pure transition, $LD_1(\lambda)$ has the same shape as $A_{iso1}(\lambda)$, and $LD_2(\lambda)$ has the same shape as A_{iso2} one could try to fit of the observed spectrum by determining k_1 and k_2 in the linear combination:

$$LD^r(\lambda) = \frac{LD(\lambda)}{A_{iso}} = \left(\frac{k_1 A_{iso1}(\lambda) + k_2 A_{iso2}(\lambda)}{A_{iso1}(\lambda) + A_{iso2}(\lambda)} \right) \qquad (6.53)$$

where the constants k_1 and k_2 depend only on geometric parameters. From comparison with equation (5.8) we may write

$$k_i = \frac{3}{2} S_i \left(3\cos^2 \alpha_i - 1 \right) \qquad (6.54)$$

for $k_i = k_1, k_2$. If the A_{iso} shapes have been determined (*e.g.* by fitting the observed absorbance spectrum to Gaussian or Lorentzian band shapes), and the orientation $S_1 = S_2 = S$ is known then the effective angles of the two transitions can be determined.

Now let us consider the situation with only one transition in the molecule of interest, but the molecules in the sample are distributed in two different environments which may induce a shift in the transition energies (and possibly also change the transition intensities and orientations). This situation is the same as that summarized in equation (6.52), but now 1 and 2 refer to populations not different transitions. If the intensity is actually the same in the two environments (no perturbation of transition moments in the molecules), so only the wavelengths are affected by environment, we have

The intensity of a transition of a planar π-electron system can be reduced by up to 50% different if it is moved into an environment where π-π stacking interactions occur. Thus, for example, a DNA binding ligand that is intercalated may have only half the absorbance intensity compared with when it is groove bound or free in solution.

$$LD^r(\lambda) = \frac{3}{2} S \left(\frac{x_1 \left(3\cos^2 \alpha_1 - 1 \right) A_{iso1}(\lambda) + x_2 \left(3\cos^2 \alpha_2 - 1 \right) A_{iso1}(\lambda + \delta)}{x_1 A_{iso1}(\lambda) + x_2 A_{iso1}(\lambda + \delta)} \right)$$

$$(6.55)$$

where we have assumed that the absorption spectrum of species 2 is shifted relative to that of species 1 by the wavelength increment δ. α_1 is the angle between some orientation axis and the transition moment for species 1 and x_1 is the mole fraction in population 1, similarly α_2 and x_1.

Figure 6.15 illustrates the kind of variation of LD^r with wavelength that we could expect for this two-environment example for a Gaussian absorption band assuming the environment causes a 5 nm shift of the band and the two species are oriented with $\alpha_1 = 0$ and $\alpha_2 = 30°$. The two angles may represent two distinct (static) conformations or the effect of structural dynamics that brings in the heterogeneity in orientation and environment. As we can see the net result is a variation of LD^r across the absorption envelope. Such variations are often seen as evidence for the presence of more than one transition (with different polarizations), but as we have shown here, this situation can also arise due to environmental and orientational heterogeneity. Since information about orientational distribution is generally difficult to obtain with any spectroscopic method (*cf.* Chapter 7) these environment-induced shifts can be considered rather an opportunity than a problem. Although a more common situation is a smooth orientation distribution, our two-state model is often very useful as a first decription of the orientation distribution.

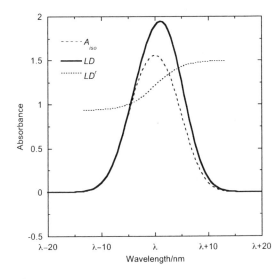

Figure 6.15 LD^r for a transition whose band shape is Gaussian but half of the molecules are oriented so their transition moments lie at 0° and half at 30° to the molecular orientation axis. $S = 1$. The intensity maximum of the first population is at wavelength λ/nm and that of the second population is $(\lambda + 5)$/nm.

An interesting case is where this type of analysis was applied was to determine the binding geometry of covalently DNA-bound benzopyrenediolepoxide (BPDE), whose unsaturated chromophore after the reaction with DNA becomes pyrene. Pyrene exhibits two orthogonal in-plane transition moments at 350 nm (long-axis) and 275 nm (short-axis). The covalent reaction leads to four isomers: opposite enantiomeric combinations of *syn* and *anti* forms. The most carcinogenic isomer of the four, (+)-*anti*-BPDE, was found to have these transitions oriented at angles near 30° and 70°, respectively, relative to the DNA helix axis.[209] This is a binding geometry consistent with its location in the minor groove—a location that has been speculated be the reason why (+)-*anti*-BPDE is not discovered by repair enzymes and therefore is such a strong carcinogen. The other three less toxic isomers have all negative LD for both the 350 and 275 nm transitions, consistent with an intercalative orientation in their DNA complexes.

For the carcinogenic (+)-*anti* isomer, a marked variation in the LD^r spectrum around 350 nm is observed suggesting an orientational and

Pyrene is the four ring molecule in Figure 6.6.

environmental heterogenity as a result of considerable conformational (static as well as dynamic) freedom—a negative *LD* component suggests possibly even a fraction of the molecules to be intercalated. Carrying out the analysis illustrated in Figure 6.15 for the 350 nm band by fitting the LD^r to equation (6.55), it was deduced that the pyrene may bind to DNA with a major fraction of the molecules oriented at 20° and one minor fraction at 70°.[210]

7 Molecular orientation principles

Most LD experiments can be undertaken and indeed analyzed without deeper understanding of how the molecules become oriented. However, in some cases, understanding of orientation mechanisms enables one to extract more information from the LD data.

Linear Dichroism and Circular Dichroism: A Textbook on Polarized-Light Spectroscopy
By Bengt Nordén, Alison Rodger and Timothy Dafforn
© B. Nordén, A. Rodger and T. Dafforn, 2010
Published by the Royal Society of Chemistry, www.rsc.org

7.1 Sample orientation for *LD* spectroscopy

The most easily derived distribution for an orientation distribution is the Boltzmann distribution

$$N(\theta) = N(\theta = 0)\exp\left[U(\theta)/(k_b T)\right]$$

where $N(\theta)$ is the number of molecules to be found in an interval centred at an angle θ from the macroscopic orientation direction, k_b is the Boltzmann constant, T is absolute temperature, and $U(\theta)$ is the potential energy of the molecule when subject to the orienting field.

In §2.3 a variety of different orientation methods are summarized. In practice, with the possible exception of a crystalline environment, the orientation is never perfect and we need to use the concept of an orientation distribution. Although the fact that S is less than 1 might be due to a segregated distribution between two distinct 'perfect' orientations (§6.9), this is seldom the case. Instead, it is generally the case that there is a smooth distribution with some molecules being better oriented and other worse depending on thermal fluctuations or their respective proximities to the polymer chains or other orienting force. Thus, the probability of finding a molecule at a certain orientation defines the orientation distribution function. The *LD* we measure follows directly from the orientation distribution, however, unfortunately, the reverse is not true: the orientation distribution cannot be deduced from *LD* data alone. To map an orientation distribution, for example in a crystal, a huge set of data is needed, such as a 3 dimensional diffraction picture—the task is formidable and in practice equal to solving the whole crystal structure.

The Scottish physicist Rev. John Kerr in 1875 discovered that if he applied a high voltage across parallel electrodes in a glass cell containing liquid nitrobenzene, and placed the cell between crossed polarizers, light was let through. When he switched off the field the image immediately went dark as the orientation birefringence vanished. This is also the principle of the very fast (nano second) camera Kerr shutters—fast because it takes only a few picoseconds for the nitrobenzene molecules to respond to the electric field and a few picoseconds to go back to random orientation when the field is switched off.[211]

Knowledge about the orientation mechanism sometimes permits us to derive a theoretical orientation distribution which may help the interpretation of *LD* data. For example, molecules that carry a permanent dipole moment or are polarizable may be oriented with the dipole or induced dipole parallel to a strong electric field (the birefringence then observed is called the Kerr effect) with a truly uniaxial (*cf.* Figure 3.4) orientational distribution. For most situations, the orientation parameters needed for structural applications are determined empirically (as in Chapters 1, 3, and 5) by, for example, adding an optical probe or by calibrating the macroscopic order parameter using another transition moment in the molecule. §7.2 and §7.4 provide the derivations of equations used in Chapter 6. §7.3 outlines how *LD* may be used to follow relaxation and induction of molecular orientation in flow. §7.5 contains a mainly theoretical treatise on orientation distributions. The chapter concludes with the derivation of *LD* equations for samples orientated on a cylindrical surface such as a flow distorted liposome or a carbon nanotube.

7.2 Equations for determination of molecular orientation

To describe the general situation of an *LD* experiment, we first define a laboratory-fixed (macroscopic) axis system, $\{X,Y,Z\}$, as illustrated in Figure 6.1. In this work we take X to be the direction of propagation of the light and Z to be the macroscopic direction of orientation. Y then completes the right-handed axis system. When studying *LD* it is also convenient to define the molecular axis system, $\{x,y,z\}$, using the 'molecular perspective' of the *LD* experiment. We take z to be the molecular orientation axis. This is the axis in the molecule which has maximum value of

$$\left\langle \cos^2 \zeta \right\rangle \tag{7.1}$$

where ζ is the angle between z and Z and $\langle \ \rangle$ denotes the average over all the molecules in the sample. x is the axis perpendicular to z that has the smallest value of

$$\left\langle \cos^2 \xi \right\rangle \tag{7.2}$$

where ξ is the angle between x and Z. ψ is similarly the angle between y and Z. In molecules of high symmetry, z may or may not be (but usually is) the highest order rotational axis. The two *LD* axis systems are illustrated in Figure 6.1.

Although few *LD* experiments involve perfect orientation, the orientation distributions generally have a degree of symmetry due both to the symmetry properties of the external orienting force and also to the structural symmetry of the molecules themselves. Consider first the external orienting force. Many *LD* experiments have *uniaxial orientation* which means that the probability of finding a given molecular axis at an angle θ from the macroscopic orientation direction, Z, is constant around a cone centred about Z (Figure 7.1). For some experiments, however, including rectangular squeezing of a gel (Figure 2.11) and flow alignment (Figure 2.5), the system behaves as if the orientation were *biaxial* (Figure 7.1) for which the constant probability contour for a given molecular axis traces out an ellipse. To visualize biaxial orientation, consider a fluid flowing between two parallel vertical plates. Z is the flow direction and X is perpendicular to the plates. The fluid is stationary at the walls, so the orienting force is a function of X but not of Y and Z. Thus it cannot be a uniform circle about Z. In some cases the shape of the molecules of interest or the averaging inherent in a combination of the method used to orient the molecules and the molecules themselves mean that we can frequently assume local uniaxial orientation even for a biaxial method. In such cases, we may use the simple orientation parameter S of equation. However, such simplicity cannot be assumed always to be operative.

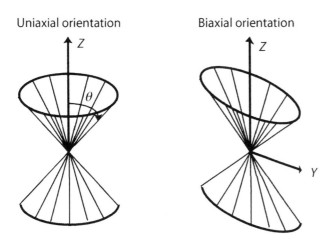

Uniaxial orientation   Biaxial orientation

Figure 7.1 Uniaxial and biaxial orientation.

How the external orienting force operates on the molecule depends on the nature of the molecule. In some experiments it is easy to determine the molecular orientation axis, z. Under electric field orientation, for example, z is the direction of the permanent or induced dipole moment (*e.g.* Figure 7.2). Orientation methods based on steric or shear forces usually have z corresponding with the 'longest' axis. However, this is not necessarily so. Imagine a planar, rectangular molecule. The orientational distribution function for this may be bimodal with the two diagonals having equal, maximal, tendency to be parallel to the orientation direction; however, the axis with maximum $\langle \cos^2 \zeta \rangle$ is the one bisecting the two diagonals; hence, the orientation axis is parallel to the long sides of the rectangle (Figure 7.2). In the case of flow-oriented DNA, although the orientation method is biaxial, the helical symmetry of the molecule renders the average local orientation to be uniaxial. For low symmetry molecules in anisotropic hosts such as a liquid crystal or a stretched polymer matrix, the identity of z may be far from clear and can only be determined by experiment as illustrated in §6.4.

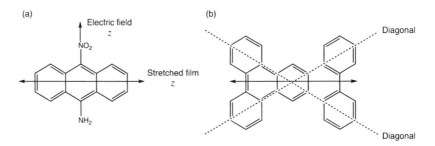

Figure 7.2 (a) Molecular orientation axes, z, under electric field orientation and stretched film orientation. (b) A planar molecule where z is not the longest axis.

General case

The orientation of a sample may be summarised by the orientation tensor which depends on how well the molecule is macroscopically oriented. In terms of the angles $\{\xi, \psi, \zeta\}$ (Figure 6.1) we may write

The form chosen for S leads to the simple final equations shown below.

$$S = \begin{pmatrix} S_{xx} & S_{xy} & S_{xz} \\ S_{yx} & S_{yy} & S_{yz} \\ S_{zx} & S_{zy} & S_{zz} \end{pmatrix}$$

$$= \begin{pmatrix} \frac{1}{2}\left(3\langle\cos^2\xi\rangle-1\right) & 3\langle\cos\xi\cos\psi\rangle & 3\langle\cos\xi\cos\zeta\rangle \\ 3\langle\cos\xi\cos\psi\rangle & \frac{1}{2}\left(3\langle\cos^2\psi\rangle-1\right) & 3\langle\cos\psi\cos\zeta\rangle \\ 3\langle\cos\xi\cos\zeta\rangle & 3\langle\cos\psi\cos\zeta\rangle & \frac{1}{2}\left(3\langle\cos^2\zeta\rangle-1\right) \end{pmatrix}$$

(7.3)

Uniaxial orientation

For the special case of uniaxial orientation of molecules whose x, y, z axes are chosen to fulfil equations (7.1) and equation (7.2), the off-diagonal terms of equation (7.3) vanish and

$$S = \begin{pmatrix} \frac{1}{2}\left(3\langle\cos^2\xi\rangle - 1\right) & 0 & 0 \\ 0 & \frac{1}{2}\left(3\langle\cos^2\psi\rangle - 1\right) & 0 \\ 0 & 0 & \frac{1}{2}\left(3\langle\cos^2\zeta\rangle - 1\right) \end{pmatrix} \qquad (7.4)$$

In this section the general equations that enable one to analyze how a molecule is oriented are given. An example of their application is given in §6.4. In solving any particular problem, we first choose a temporary molecular axis system $\{x',y',z'\}$, which may bear no relationship to the true orientation axis system but is convenient to use. Analogously to the definitions we used with the 'true' molecular axis system, we may express the direction of the macroscopic orientation axis Z by the vector

$$Z = \left(\cos\xi, \cos\psi, \cos\zeta\right)_{\{xyz\}} = \left(\cos\xi', \cos\psi', \cos\zeta'\right)_{\{x'y'z'\}} \qquad (7.5)$$

A convenient temporary axis system is usually one that ensures some terms in equation (6.17) are zero.

in respectively the $\{x,y,z\}$ and $\{x',y',z'\}$ coordinate systems. S' (usually) has no simplifying off-diagonal 0 elements so in general is written

$$\begin{aligned} S' &= \begin{pmatrix} S_{x'x'} & S_{x'y'} & S_{x'z'} \\ S_{y'x'} & S_{y'y'} & S_{y'z'} \\ S_{z'x'} & S_{z'y'} & S_{z'z'} \end{pmatrix} \\ &= \begin{pmatrix} \frac{1}{2}\left(3\langle\cos^2\xi'\rangle - 1\right) & 3\langle\cos\xi'\cos\psi'\rangle & 3\langle\cos\xi'\cos\zeta'\rangle \\ 3\langle\cos\xi'\cos\psi'\rangle & \frac{1}{2}\left(3\langle\cos^2\psi'\rangle - 1\right) & 3\langle\cos\psi'\cos\zeta'\rangle \\ 3\langle\cos\xi'\cos\zeta'\rangle & 3\langle\cos\psi'\cos\zeta'\rangle & \frac{1}{2}\left(3\langle\cos^2\zeta'\rangle - 1\right) \end{pmatrix} \end{aligned} \qquad (7.6)$$

We could now write general equations for determining $\{x,y,z\}$ from $\{x',y',z'\}$, however, they are complicated and not particularly helpful, so we consider instead the most common situations that arise.

A planar molecule in a uniaxially oriented system

In the temporary molecular axis system for a planar molecule we always choose $x' = x$ perpendicular to the molecular plane. The molecular orientation axis, z, for a planar uniaxially oriented molecule may then be written as the vector (Figure 6.4)

$$z = \left(0, \sin\gamma, \cos\gamma\right)_{\{x'y'z'\}} \qquad (7.7)$$

Thus, from equation (7.5) and the definition of ζ it follows that

$$\cos\zeta = Z \cdot z = \cos\psi'\sin\gamma + \cos\zeta'\cos\gamma \qquad (7.8)$$

Equation (7.8) is appropriate for a single molecule — hence no averaging.

By squaring equation (7.8) and comparing the result with equation (7.6) we may then write

$$3\langle\cos^2\zeta\rangle = 2\left\{S_{y'y'}\sin^2\gamma + S_{z'z'}\cos^2\gamma + \frac{1}{2}S_{y'z'}\sin2\gamma + \frac{1}{2}\right\} \qquad (7.9)$$

Since by definition (equation (7.2)) $\langle\cos^2\zeta\rangle$ is a maximum, its derivative with respect to γ must be zero, thus

$$3\frac{\partial\langle\cos^2\zeta\rangle}{\partial\beta} = 2\{(S_{y'y'} - S_{z'z'})\sin(2\gamma) + S_{y'z'}\cos(2\gamma)\} = 0 \qquad (7.10)$$

which upon rearranging gives:

$$\tan(2\gamma) = \frac{S_{y'z'}}{S_{z'z'} - S_{y'y'}} \qquad (7.11)$$

To derive an equation for the LD^r in terms of the temporary orientation parameters for the planar uniaxial case we proceed as follows and write

$\hat{\boldsymbol{\mu}}$ denotes the unit vector along $\boldsymbol{\mu}$.

$$\hat{\boldsymbol{\mu}} = (0, \sin\alpha', \cos\alpha')_{\{x'y'z'\}} \qquad (7.12)$$

where the final subscript indicates we are using the $\{x', y', z'\}$ axis system. To evaluate the LD^r we require $\langle\hat{\mu}_Z^2\rangle$ and $\langle\hat{\mu}_Y^2\rangle$ (*cf.* equation (6.3)). Now by definition

$$\hat{\mu}_Z = \mathbf{Z} \cdot \hat{\boldsymbol{\mu}} \qquad (7.13)$$

so

$$\langle\hat{\mu}_Z^2\rangle = \langle\{(\cos\xi', \cos\psi', \cos\zeta') \cdot (0, \sin\alpha', \cos\alpha')\}^2\rangle$$

$$= \langle\cos^2\psi'\rangle\sin^2\alpha' + \langle\cos^2\zeta'\rangle\cos^2\alpha' + 2\langle\cos\psi'\cos\zeta'\rangle\sin\alpha'\cos\alpha'$$

$$= \frac{1}{3}\{2(S_{y'y'}\sin^2\alpha' + S_{z'z'}\cos^2\alpha' + S_{y'z'}\sin\alpha'\cos\alpha') + 1\}$$

$$(7.14)$$

As the orientation is uniaxial, the component of the transition moment that is perpendicular to Z is equally likely to be oriented in any direction so

$$\langle\hat{\mu}_X^2\rangle = \langle\hat{\mu}_Y^2\rangle = \frac{1}{2}(1 - \langle\hat{\mu}_Z^2\rangle) \qquad (7.15)$$

The reduced LD then becomes

$$LD^r = 3(\langle\hat{\mu}_Z^2\rangle - \langle\hat{\mu}_Y^2\rangle)$$

$$= \frac{3}{2}(3\langle\hat{\mu}_Z^2\rangle - 1) \qquad (7.16)$$

$$= 3(S_{y'y'}\sin^2\alpha' + S_{z'z'}\cos^2\alpha' + S_{y'z'}\sin\alpha'\cos\alpha')$$

The relatively simple form of equation (7.16) is rewarding in view of the rather complicated definitions of the orientation tensor components used above.

7.3   Flow orientation of brain microtubules

Figure 3.26 illustrates the assembly of tubulin protein monomers into microtubule fibres. The resulting fibres are very stiff and can be modelled as a rigid rod. The main contribution to the observed LD signal is due to the anisotropic turbidity, LD^τ of the sample. If the sample is oriented in a

Couette flow cell and the LD^τ measured as a function of time after the cell stops rotating, it is possible to study the orientational relaxation of the microtubules. The relaxation becomes slower with increasing concentration and at concentrations above 1 mg/mL some orientation remains at 'infinite' time. Data such as those illustrated in Figure 7.3 may be fitted to a tri-exponential decay with a rapid time constant of 4 s, an intermediate one of 40 s, a slow one of 400 s, and a fraction of microtubules that remain oriented. The different rates reflect different reorientation processes in what is an inherently complex system.

Based on Rayleigh-Gans light scattering, the scattering dichroism of rod-shaped particles has been predicted to follow a simple linear relation[212]

$$LD^\tau = S\tau_{iso} \times 0.68$$

where τ_{iso} is the turbidity of the isotropic sample. This value is very near the upper limit of reduced scattering dichroism observed for flow-aligned microtubules. [213]

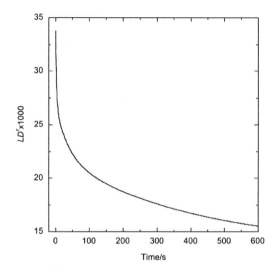

Figure 7.3 Decay of microtubules (0.37 mg/mL, 33 °C). *LD* as a function of time after the shear rate has been reduced from $G = 100 \text{ s}^{-1}$ to 0.[213]

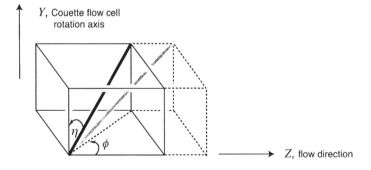

Figure 7.4 Schematic illustration of what happens to a microtubule rod (thick black line) when the flow of a cell is switched on (shear gradient is along X): it moves to be the thick grey line. η is the angle the microtubule makes with the Couette cell rotation axis, Y, and ϕ is the angle between the projection of microtubule (initially thin solid line across base of box, moving to be the dotted line) onto the X/Z plane and the X axis.[213]

Conversely, Couette flow *LD* can be used to follow the development of orientation when shear flow is applied. For an outer rotating Couette cell, the flow rate, v, changes along the X axis going from zero at the inner wall of the Couette cell (front of Figure 7.4) to maximum at the outer edge. By way of contrast, the shear gradient, $G = dv/dX$, is constant. Thus, if $\phi \neq 90°$ in Figure 7.4, then the ends of any microtubule rod will be in stream lines moving with different velocities:

$$v = v_{mid} \pm \frac{1}{2} GL \cos\phi \qquad (7.17)$$

where v_{mid} is the velocity of the middle of the rod and L is the length of the rod.[213] If for simplicity we take the front end of the microtubule in Figure 7.4 to be the origin, then the back end lies at:

$$L\left(\sin\eta\cos\varphi, \cos\eta, \sin\eta\sin\phi\right)_{\{X,Y,Z\}} \qquad (7.18)$$

At time δt (which is very small) after the flow is switched on, the back-end of the microtubule lies at

$$\frac{L\left(\sin\eta\cos\varphi, \cos\eta, \sin\eta\sin\phi + \delta t G \sin\eta\cos\phi\right)}{\sqrt{A}}$$

A Taylor's series expansion of
$$\left(1 + 2w\right)^{-\frac{1}{2}} = 1 - w$$
for small w.

$$\cong \frac{L\left(\sin\eta\cos\varphi, \cos\eta, \sin\eta\sin\phi + \delta t G \sin\eta\cos\phi\right)}{\sqrt{1 + 2\delta t \sin\eta\sin\phi G \sin\eta\cos\phi}} \qquad (7.19)$$

$$\cong L\left(\sin\eta\cos\varphi, \cos\eta, \sin\eta\sin\phi + \delta t G \sin\eta\cos\phi\right)$$
$$\times \left(1 - \delta t \sin\eta\sin\phi G \sin\eta\cos\phi\right)$$

where

$$A = \sin^2\eta\cos^2\varphi + \cos^2\eta + \sin^2\eta\sin^2\phi$$
$$+ 2\delta t \sin\eta\sin\phi G \sin\eta\cos\phi + (\delta t)^2 G^2 \sin^2\eta\cos^2\phi \qquad (7.20)$$

Thus, during the time interval δt, the back end moves by

$$\frac{\left(\delta X, \delta Y, \delta Z\right)}{L}$$
$$\cong -\left(\sin\eta\cos\varphi, \cos\eta, \sin\eta\sin\phi\right)$$
$$+ \left(\sin\eta\cos\varphi, \cos\eta, \sin\eta\sin\phi + \delta t G \sin\eta\cos\phi\right)$$
$$\times \left(1 - \delta t G \sin\eta\cos\phi\sin\eta\sin\phi\right) \qquad (7.21)$$
$$\cong \delta t G \sin\eta\cos\phi \times$$
$$\left(-\sin^2\eta\cos\phi\sin\phi, -\cos\eta\sin\eta\sin\phi, -\sin^2\eta\sin^2\phi + 1\right)$$

Since

$$\delta Y = L(\cos(\eta + \delta\eta) - \cos\eta) \qquad (7.22)$$

where $\delta\eta$ is the change in η during δt, it follows (using another Taylor series expansion) that

$$\frac{\delta Y}{L} = -\delta t G \sin^2\eta\cos\eta\cos\phi\sin\phi$$
$$= \left(\cos(\eta + \delta\eta) - \cos\eta\right) \cong -\sin\eta(\delta\eta) \qquad (7.23)$$

So

$$\frac{d\eta}{dt} = G \sin\eta \cos\eta \cos\phi \sin\phi \tag{7.24}$$

Similarly

$$\frac{(\delta X)}{L} \cong \delta t G \cos\phi(-\cos\phi\sin\phi)$$
$$= \left(\cos(\phi + \delta\phi) - \cos\phi\right) \cong -\sin\phi(\delta\phi) \tag{7.25}$$

So

$$\frac{d\phi}{dt} = G \cos^2\phi \tag{7.26}$$

By solving equations (7.24) and (7.26) numerically for a large ensemble of 20,000 randomly generated rods distributed initially isotropically, the time development of the orientation parameter S when flow is switched on was determined.[213]

Conversely, since the orientation parameter for this system

$$S = \left\langle \sin^2\phi\sin^2\eta \right\rangle - \left\langle \cos^2\eta \right\rangle \tag{7.27}$$

the time evolution of S and hence of the *LD* follows from the time dependence of ϕ and η.

7.4   Derivation of the relationship between fluorescence anisotropy and transition polarization

In Chapter 6 the use of FA (also called fluorescence polarization anisotropy, FPA) to help resolve sign ambiguities in determining $\cos^2\alpha$ was outlined. Equation (6.51) is derived below (assuming a perfect instrument)

$$FA = \frac{I_{VV} - I_{VH}}{I_{VV} + 2I_{VH}} \tag{7.28}$$

where, *e.g.*, I_{VH} is the emerging intensity of the light beam that has passed through a vertical excitation polarizer and a horizontal emission polarizer, assuming that there is no loss of polarization due to rotation of the chromophore during the lifetime of the emission process or to energy transfer between chromophores.

χ is defined to be the angle between $\boldsymbol{\mu}^{ex}$ (the transition dipole moment *via* which the excitation occurs) and $\boldsymbol{\mu}^{em}$ (the electric dipole transition moment *via* which emission occurs). For this derivation we define the molecule fixed axis system such that the two transition moments define the y/z plane of the molecular axis system $\{x, y, z\}$:

$$\boldsymbol{\mu}^{em} = \mu^{em}(0,0,1)_{\{x,y,z\}} \tag{7.29}$$

and

$$\boldsymbol{\mu}^{ex} = \mu^{ex}(0,\cos\chi,\sin\chi)_{\{x,y,z\}} \tag{7.30}$$

where the parenthetic subscripts denote the coordinate system being used. As usual in the laboratory fixed axis system, X is the direction of propagation of light and Z is the principal photo-orientation direction; in this case Z is the vertical direction and Y is the horizontal direction in the *FA* definition of equation (7.28). Also let θ be the angle between z and Z and ϕ the angle between the projection of z onto the X/Y plane and Y. Thus

$$(0,0,1)_{\{x,y,z\}} = \left(\sin\phi\sin\theta,\cos\phi\sin\theta,\cos\theta\right)_{\{X,Y,Z\}} \tag{7.31}$$

Similarly

$$(0,1,0)_{\{x,y,z\}} = \cos\gamma\left(\sin\phi\cos\theta,\cos\phi\cos\theta,-\sin\theta\right)_{\{X,Y,Z\}}$$
$$+ \sin\gamma\left(\cos\phi,-\sin\phi,0\right)_{\{X,Y,Z\}} \tag{7.32}$$

where γ takes any value between 0 and π. We therefore may write

$$\boldsymbol{\mu}^{em} = \mu^{em}\left(\sin\phi\sin\theta,\cos\phi\sin\theta,\cos\theta\right)_{\{X,Y,Z\}} \tag{7.33}$$

and

$$\boldsymbol{\mu}^{ex} = \mu^{ex}\left\{ \begin{array}{l} \sin\chi\begin{pmatrix}\cos\gamma\sin\phi\cos\theta+\sin\gamma\cos\phi,\\ \cos\gamma\cos\phi\cos\theta-\sin\gamma\sin\phi,-\cos\gamma\sin\theta\end{pmatrix} \\ +\cos\chi\left(\sin\phi\sin\theta,\cos\phi\sin\theta,\cos\theta\right) \end{array} \right\}_{\{X,Y,Z\}} \tag{7.34}$$

Since

$$I^{VV} = I^{ZZ} = \left(\mu_Z^{ex}\right)^2\left(\mu_Z^{em}\right)^2$$
$$I^{VH} = I^{ZY} = \left(\mu_Z^{ex}\right)^2\left(\mu_Y^{em}\right)^2 \tag{7.35}$$

Thus

$$I_{VV} = \left(\mu^{em}\right)^2\left(\mu^{ex}\right)^2\left(-\sin\chi\cos\gamma\sin\theta+\cos\chi\cos\theta\right)^2\cos^2\theta$$
$$I_{VH} = \left(\mu^{em}\right)^2\left(\mu^{ex}\right)^2\left(-\sin\chi\cos\gamma\sin\theta+\cos\chi\cos\theta\right)^2\cos^2\phi\sin^2\theta \tag{7.36}$$

where γ: $0\to\pi$, ϕ: $0\to2\pi$, and θ: $0\to2\pi$. Upon averaging over these angles (integrating) it follows that

$$I_{VV} = \left(\mu^{em}\right)^2\left(\mu^{ex}\right)^2\left(\frac{4}{15}\cos^2\chi+\frac{2}{15}\right)$$
$$I_{VH} = \left(\mu^{em}\right)^2\left(\mu^{ex}\right)^2\left(-\frac{2}{15}\cos^2\chi+\frac{4}{15}\right) \tag{7.37}$$

For an isolated transition we may then write:

$$FA = \frac{I_{VV}-I_{VH}}{I_{VV}+2I_{VH}} = \frac{2}{15}\left(\frac{3\cos^2\chi-1}{5}\right) = \frac{1}{5}\left(3\cos^2\chi-1\right) \tag{7.38}$$

7.5   Orientation distributions

In the preceding text we have several times touched upon the problem of determining the orientation factor, S, and the related issue of determining information about the orientation distribution. While S can often be determined experimentally using transitions of known polarization, the distribution of orientations over different angles is harder to determine.

When determining an angle α from $<\cos^2\alpha>$, not only is there an ambiguity about the sign of α (§6.8), since both $+\alpha$ and $-\alpha$ give the same value of $<\cos^2\alpha>$ (and hence same LD), but there is also an ambiguity related to the averaging. Sharp and broad distributions may give the same value of $<\cos^2\alpha>$. For example, $(3<\cos^2\alpha>-1)=0$ both for $\alpha=54.7°$, the magic angle, and for a completely randomly oriented sample (*i.e.* one where the orientation distribution is uniform over a sphere). Below we shall work ourselves through a case of non-empirical determination of the orientation parameter S based on the orientation modeled for a stretched polymer. Thereafter the use of 'moments' as source of information about the orientation distribution will be considered.

In the limits of orientation close to $\alpha=0$ or 90°, so $<\cos^2\alpha>$ is close to 1 and 0, respectively, the distribution question is less of an issue.

Amorphous deformation model for stretched polymer matrices

A simple model for the orientation of molecules in a uniaxially stretched polymer may be described by considering a large ensemble of unit vectors (the molecules). Before the stretching force is applied, the vectors are uniformly distributed so they are pointing in all directions from the centre to the surface of a sphere (imagine toothpicks put into an orange). This isotropic distribution may be described by the function

$$f(\theta) = \sin\theta \qquad (7.39)$$

where θ is the angle between a unit vector and some arbitrary axis along which we shall align the sample (Z). The $\sin\theta$ dependence means that the number of toothpicks is vanishingly small close to the pole ($\theta=0$) and correspondingly bigger at the equator ($\theta=90°$). $f(\theta)$ can be used to express the probability of finding a unit vector in the infinitesimal angular interval (θ, $\theta+d\theta$).

Now assume that the sphere with unit vectors evenly spread over its surface is deformed into a spheroid by compression around the equator and elongation at the poles, and that each unit vector follows its original surface element. The deformation can be characterized by the ratio of the long (a) and short (b) axes of the ellipsoid.

$$R_S = a/b \qquad (7.40)$$

In the case of a stretched polymer film R_S is the ratio between the major and minor axes of the ellipse into which a circle drawn on the surface of the film is changed upon stretching the film. If the degree of stretch instead is characterized in terms of how much the film is extended

$$R_S = \left(a/a_o\right)^{3/2} \qquad (7.41)$$

where a_o refers to the unstretched film.

It has been shown that the distribution $f(\theta)$ within this deformation model

Equation (7.41) has been shown to hold for PE for $a/a_o = 2$–3 (*i.e.* a 200–300% stretch); an observation supporting the assumptions of uniaxial deformation and constant volume.[214]

may be written in a simple analytical form as a function of stretch:[215]

$$f(\theta) = R_S^2 \sin\theta \left[1 + \left(R_S^2 - 1\right)\sin^2\theta\right]^{-3/2} \tag{7.42}$$

This is illustrated in Figure 7.5. We now define the polymer orientation parameter

$$S_P = \frac{1}{2}\left(3\left\langle\cos^2\theta\right\rangle - 1\right) \tag{7.43}$$

where the average is over the orientation distribution $f(\theta)$. Integration[214] leads to

$$S_P = \frac{1}{2}\left\{3\left(\frac{R_S^2}{R_S^2 - 1}\right)\left[1 - \frac{\arcsin\left(\left(R_S^2 - 1\right)^{1/2}/R_S\right)}{\left(R_S^2 - 1\right)^{1/2}}\right] - 1\right\} \tag{7.44}$$

The orientations of elongated molecules in stretched films have been found to be well described by the amorphous deformation, both in the limit of very small stretch ratios as well as at greater degrees of stretch up to practically perfect orientation adopted at stretch ratios above 10.

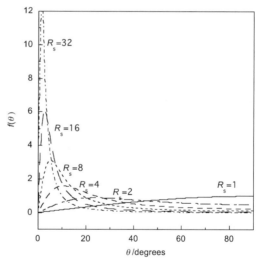

Figure 7.5 $f(\theta)$ depicted for increasing stretching ratios $R_S = 1$ (no stretch), 2 (200% stretch), 4, ... The distribution of the unit vectors (molecules) becomes more and more aligned parallel with the orientation direction ($\theta = 0°$) as the stretch increases.

It has been suggested that the uniaxial orientation of solute molecules in polymer hosts can be considered as built up of microscopic anisotropic systems that gradually get more aligned as the polymer host is stretched.[214] The total *LD* may then be described as a product between the polymer orientation factor S_P and the *LD* of equation (6.13):

$$LD^r = 3S_P \frac{\left(S_{xx}\varepsilon_x + S_{yy}\varepsilon_y + S_{zz}\varepsilon_z\right)}{\left(\varepsilon_x + \varepsilon_y + \varepsilon_z\right)} \tag{7.45}$$

By and large this is generally a good approximation, if one may assume that local domains (*e.g.* crystallites) that are aligned upon stretching are forming microscopic hosts for the solute molecules. A critical test, using the perpendicularly polarized 275 nm and 238 nm transitions of pyrene, indicated that the orientation on the molecular level is dependent on the macroscopic aligning forces—presumably through the packing of the polymer chains. Thus equation (7.45) is useful, but an over-simplification.

Moments for describing molecular orientation distribution

Some extra information about the shape of the orientation distribution may be obtained from the study of higher moments of the distribution function:

$$\left\langle \cos^n \theta \right\rangle = \int \cos^n f(\theta)d\theta \tag{7.46}$$

The orientation distribution function is expanded as a sum of Legendre polynomials (Wigner rotation matrices) with the moments appearing as the coefficients:

$$f(\theta) = \sum_0^\infty c_n(\theta)P_n(\theta) \tag{7.47}$$

where

$$c_n(\theta) = \frac{2n+1}{2} \int_0^{\pi/2} f(\theta)P_n(\theta)d\theta \tag{7.48}$$

and the first five Legendre polynomials for $\cos\theta$ are

$$P_0(\theta) = 1$$

$$P_1(\theta) = \cos\theta$$

$$P_2(\theta) = \frac{1}{2}\left(3\cos^2\theta - 1\right) \tag{7.49}$$

$$P_3(\theta) = \frac{1}{2}\left(5\cos^3\theta - 3\cos\theta\right)$$

$$P_4(\theta) = \frac{1}{8}\left(35\cos^4\theta - 30\cos^2\theta + 3\right)$$

It is not a coincidence that the shapes of P_0, $P_1(\theta)$, $P_2(\theta)$, *etc.* may be identified with those of the hydrogen atom atomic orbitals.

Table 7.1 R_s values for the amorphous deformation model. Note that $R_s = 4$ and $\langle\cos^2\theta\rangle$ corresponds to $S = 0.56$. Error is the percentage error if only the first four terms in the expansion of equation (7.46) are used.

R_s	$\langle\cos\theta\rangle$	$\langle\cos^2\theta\rangle$	$\langle\cos^3\theta\rangle$	$\langle\cos^4\theta\rangle$	Error (%)
1	0	0.333	0	0.200	0
1.44	0	0.429	0	0.288	1
2	0	0.527	0	0.388	9
4	0	0.704	0	0.592	57

The possibility of describing the orientation of molecules in a stretched film by a truncated series containing only the first four terms of the expansion has been found to work quite well for electric field orientation, where both odd and even moments are nonzero.[216] However, for the amorphous deformation model described above the deviations become serious even at stretching ratio $R_S = 2$ as shown in Table 7.1.

7.6   Orientation of molecules bound to cylindrical bodies in shear flow

We considered shear-deformed liposomes (§3.7 and §5.3) and carbon nanotubes (§5.11) as cylindrical systems where the analytes are oriented uniformly about the orientation axis of the cylinder. The *LD* under such circumstances is not described by equations (3.3) and (3.6) but by equation (3.7) because there is an additional element of rotational averaging required. We assume the orientation is locally uniaxial. Let z be along the long-axis of the cylinder as in Figure 7.6. z is uniformly distributed about Z the macroscopic orientation axis in a manner accounted for by S.

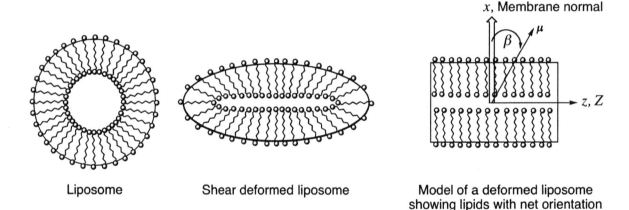

Liposome   Shear deformed liposome   Model of a deformed liposome showing lipids with net orientation

Figure 7.6 Schematic of liposomes distorted in shear flow.

This approach is analogous to that used with DNA where we take care of the biaxility of the Couette flow orientation by factorizing it into an orientation term, *S*, defined by equation (7.27), and an optical term (*cf.* §6.3).

In this case it is convenient to define the orientation of a transition moment, $\boldsymbol{\mu}$, of an analyte by sitting on the analyte and defining x to be the normal to the surface of the cylinder that goes through the analyte. Let β be the angle between x and $\boldsymbol{\mu}$. The analyte orientation will not be affected by the shear flow (the forces are too small), so on average any analyte transition moment, $\boldsymbol{\mu}$, will be uniformly distributed about the x axis. Let ψ be the angle between the projection of $\boldsymbol{\mu}$ onto the y/z plane and z. Thus in the local cylinder coordinate system

$$\boldsymbol{\mu} = \mu\big(\cos\beta, \sin\beta\sin\psi, \sin\beta\cos\psi\big)_{\{x,y,z\}} \tag{7.50}$$

where μ is the magnitude of $\boldsymbol{\mu}$. The reduced linear dichroism is by definition

$$LD^r = S\frac{A_Z - A_Y}{A_{iso}} = 3S\frac{\left(\mu_Z^2 - \mu_Y^2\right)}{\mu^2} \tag{7.51}$$

Now, $\mu_z = \mu_Z$ and μ_Y may be written as the dot product of the transition moment vector and the vector for the Y axis in the $\{x,y,z\}$ coordinate system. Thus

$$\mu_Y = \boldsymbol{\mu} \cdot \boldsymbol{Y}$$

$$= \mu\left(\cos\beta, \sin\beta\sin\psi, \sin\beta\cos\psi\right)_{\{x,y,z\}} \cdot \left(\sin\gamma, \cos\gamma, 0\right)_{\{x,y,z\}} \tag{7.52}$$

$$= \mu\left(\cos\beta\sin\gamma + \sin\beta\sin\psi\cos\gamma\right)$$

where γ take values from 0 to 2π. Thus,

$$LD^r = 3S\begin{pmatrix} \sin^2\beta\cos^2\psi - \cos^2\beta\sin^2\gamma - \sin^2\beta\sin^2\psi\cos^2\gamma \\ -2\cos\beta\sin\gamma\sin\beta\sin\psi\cos\gamma \end{pmatrix} \tag{7.53}$$

Both ψ and γ take values from 0 to 2π, so upon averaging over them:

$$LD^r = 3S\frac{2\sin^2\beta - 2\cos^2\beta - \sin^2\beta}{4} \tag{7.54}$$

$$= \frac{3}{4}S\left(1 - 3\cos^2\beta\right)$$

which is equation (3.7).

For the special case of $\boldsymbol{\mu}$ lying parallel to the surface normal, *i.e.* inserted into the cylinder, $\beta = 0°$, and $LD^r/S = -3/2$. If, however, the molecule is oriented so that $\boldsymbol{\mu}$ lies on the surface of the cylinder without any preferred orientation then $\beta = 0$ giving $LD^r/S = +3/4$. So, in general, for this orientation geometry, the maximum positive LD^r signal is half that of the maximum negative LD^r signal. This is in contrast to the situation for the standard geometry, such as for flow oriented DNA, where lower and upper limits are $-1/2$ and $+3$.

8 Analysis of circular dichroism: electric dipole allowed transitions

Most transitions that we study are electric dipole allowed and gain much of their CD intensity by coupling with other electric dipole allowed transitions. Understanding the underlying mechanisms for this type of CD, and how CD can be non-empirically interpreted in structural terms, forms the subject matter of this chapter.

Linear Dichroism and Circular Dichroism: A Textbook on Polarized-Light Spectroscopy
By Bengt Nordén, Alison Rodger and Timothy Dafforn
© B. Nordén, A. Rodger and T. Dafforn, 2010
Published by the Royal Society of Chemistry, www.rsc.org

8.1 Introduction

CD probes optical activity of electronic transitions if we use UV/visible light. If infra-red radiation is used then we measure vibrational optical activity.

The absorption spectrum of a solution of chiral molecules measured with left circularly polarized radiation would differ by at the very most 2% from that measured with right circularly polarized light. The difference between the absorbances measured with left and right circularly polarized light rarely exceed ±0.1%. A *CD* spectrum is thus the result of a small difference between two large absorbances. It gives only the helical or asymmetric part of the change that occurs when radiation is absorbed. The interaction between light and a molecule depends on a combination of the electric and magnetic fields of the radiation, whereas absorbance, to a first approximation, depends only on the electric field. The smallness of the *CD* signal is due to the fact that, at a given field strength, magnetic field interaction energies are some thousand times weaker than the corresponding electric field interaction energies.

At least in principle, a *CD* spectrum contains all the information we might wish to know about the asymmetry of the system we are studying. In reality, how *CD* is used and interpreted varies considerably depending on the systems studied. Whilst by no means always the case, it is often true that large molecular systems are analyzed qualitatively or empirically (*cf.* Chapter 4) whereas small systems are analyzed in more quantitative ways with a more secure theoretical foundation.

In order to interpret a *CD* spectrum in geometric terms it is important that we understand how the signal arises. One reason *CD* was only gradually adopted as a structural tool is that there is a 50% chance of correctly guessing which enantiomer is in the solution, and it has not always been obvious that the success rate has been improved by using *CD*. The aim of this chapter and the next is to provide a sound basis for relating *CD* and geometry for electric dipole allowed transitions. Magnetic dipole allowed transitions form the subject matter of the next chapter. The equations used in both chapters are derived in Chapter 10.

8.2 Ways of analyzing *CD*

Rotatory strength is proportional to the area under a plot of $\Delta\varepsilon$ *versus* frequency.

Our conclusion from the previous chapters is that we would expect to see a *CD* signal for a collection of chiral molecules if the photons incident upon the sample have the correct energy to cause a transition. In a *CD* experiment the molecules are typically randomly oriented, otherwise *LD* (see previous chapters) effects often dominate the *CD* signal due to imperfect optics in the spectrometer (§2.6). The equation that summarizes the requirement of parallel $\boldsymbol{\mu}$ and $\boldsymbol{m}$ for the helical electron displacement required for *CD* of a sample of randomly oriented molecules is the Rosenfeld equation (*cf.* §10.5):

$$R = \mathrm{Im}\{\boldsymbol{\mu} \cdot \boldsymbol{m}\} \tag{8.1}$$

The magnetic dipole moment operator contains that factor $i = \sqrt{-1}$ and hence can give rise to complex functions (*cf.* Chapter 10).

where R is the *CD* intensity or *CD* strength or rotational strength or rotatory strength, 'Im' denotes 'imaginary part of', $\boldsymbol{\mu}$ is the electric dipole transition moment for the transition from the final to the initial state, and $\boldsymbol{m}$ is the magnetic dipole transition moment for the reverse transition.

Methods of analysis of *CD* spectra may be categorized as follows:

* empirical—based on experience with related systems
* *ab initio*—calculation directly from equation (8.1) using complete molecular wave functions
* chromophoric—where a molecule is divided into separate chromophores and some level of calculation is performed.

The aim is usually to get maximum structural information for minimum effort.

The strict definition of a *chromophore* is a sub-unit of a system whose wave functions have no overlap with the rest of the system; electronic wave functions in a chromophore therefore have no electron exchange with the rest of the system. In practice for *CD*, a chromophore is usually identified as a moiety within a molecule whose normal absorption is more or less independent of the rest of the molecule. Chromophores of a molecule can often be identified with functional groups.

Empirical *CD* analysis

Most applications of *CD* spectroscopy in the scientific literature involve qualitative and/or empirical analyses of spectra. By qualitative we simply mean an observation such as 'when … was done, the *CD* spectrum changed, therefore we concluded that … had happened'. For example, the loss of structure in an RNA molecule upon heating may be followed using *CD* (Figure 4.4). Alternatively, the proof of a chiral (or at least enantioselective) synthesis or separation may be acquired using *CD*. Convergence of *CD* amplitude upon repeated purification is then a criterion of pure enantiomers.

If the analysis of the *CD* data is quantified by measuring the change in the *CD* as a function of concentration, temperature, ionic strength, *etc.*, and/or by comparison with 'related' systems (as is the case for protein structure determination, §4.5) then we describe the analysis as empirical. When using 'related' systems to interpret a new *CD* spectrum it is vital that the systems are related spectroscopically as well as by molecular and electronic structure.

Ab initio CD analysis

Ab initio calculations are useful for individual molecules. However, a great deal of work is required to ensure the calculation is reliable, and the results usually only relate to the single system for which the calculation has been performed. In some instances *ab initio* calculations on a particular class of compounds have proved useful in a wider context, for example, where Lightner *et al.* used the calculation results for β-adamantanones (Figure 9.4b) to relate back to underlying *CD* mechanisms. In this case a particular calculation was used to advance our understanding of the theory of *CD*.[217]

Chromophoric analysis of *CD* spectra

The key feature of a chromophore is that it can be considered as a spectroscopically well-defined sub-unit of a molecule that is only slightly perturbed by the rest of the system. We shall limit our consideration to achiral chromophores. Thus the isolated chromophores have no intrinsic *CD* and we shall concentrate on deducing the *CD* induced into a particular transition by the chromophore's environment.

Given that the chromophores are achiral in the analysis below, then for every transition either $m = 0$, or $\mu = 0$, or they are perpendicular to one another. The $m = 0$, $\mu \neq 0$ transitions are called *electric dipole allowed* (eda), *magnetic dipole forbidden* (mdf); these form the subject matter of the remainder of this chapter. The $m \neq 0$, $\mu = 0$ transitions are called *electric*

The same analysis may be applied to chiral chromophores, but for these the observed signal will be dominated by the intrinsic *CD*, so to see the induced *CD*, the intrinsic signal must be subtracted.

dipole forbidden (edf), *magnetic dipole allowed* (mda); they are analyzed in the next chapter.

The significance of this division is due to the way *CD* intensity is induced into the two different types of transitions. Eda transitions require the induction of a magnetic component, whereas mda transitions require an induced electric component. The dependence of the different kinds of induced moments on the geometry of the system is very different, so when we wish to extract geometric information from *CD* we must be aware of which situation we have. The *CD* spectrum induced into eda transitions is also different if it arises from the coupling of identical chromophores rather than from the coupling of non-identical chromophores. In this chapter we quote the results for eda transitions and discuss their physical significance. The derivations are given in Chapter 10.

8.3   Degenerate coupled-oscillator *CD*: general case

We use the labels **A** and **C** to denote the chromophore of the coupling system. **A** is an achiral chromophore where the transition of interest is located and **C** is the chromophore that provides the chiral perturbation to **A** transitions. In the degenerate coupled-oscillator case **A** also provides an equivalent chiral perturbation to **C**.

If we can measure a *CD* spectrum for a transition that is eda but (within its own chromophore, **A**) mdf then it must have acquired some magnetic character (equation (8.1)). Most commonly, the required helix of the electron rearrangement during the transition arises because an eda transition in another part of the molecule, μ^c in chromophore **C**, causes the electron motion in the transition of interest, μ^a, to be helically 'deflected'. The induced magnetic character arises since the linear motion of electrons in **C** makes a tangent to a circle about **A** (Figure 8.1). Conversely, μ^a gives magnetic character to μ^c.

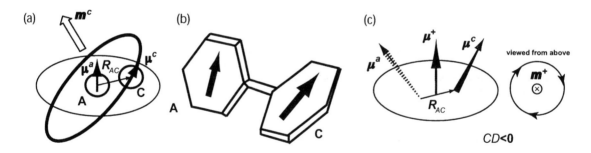

Figure 8.1 (a) Schematic illustration of the coupling of two electric dipole transition moments in a chiral system where the coupling of two dipoles in different chromophores produces a net helical motion of electrons in each transition. (b) Biphenyl. (c) Illustration of μ^a and μ^c combining to give a net electric, μ^+, and net magnetic dipole moment, m^+. As illustrated, by the right-hand rule (*cf.* Figure 1.2, Figure 4.1, and Figure 4.24), m^+ points into the page so the *CD* is negative.

If **A** and **C** are identical then the most significant coupling takes place between the transitions occurring at the same energy, ε. The transition moments of degenerate transitions have the same length but different origins and orientations. We refer to *CD* arising from their coupling as *degenerate coupled-oscillator CD*. The two electric dipole transition moments couple together to make one of two helices depending upon whether they are in-phase or out-of-phase—so we expect to see a *CD* spectrum with two peaks as some molecules will be in-phase and some out-of-phase. The two resulting helices of electron motion have equal magnitude and opposite sign *CD* bands at energies very close to ε. As a result, what is measured is illustrated in Figure 8.2, where significant cancellation of component bands leads to the observed spectrum. In this section the equations that describe the magnitude and sign of the *CD* from the coupling of eda transitions in two distinct identical achiral chromophores, **A** and **C**, are given and applied for a number of different cases and compared with experimental data. §8.4 contains a qualitative approach; §8.5 contains a more quantitative one. The full derivation is given in §10.6.

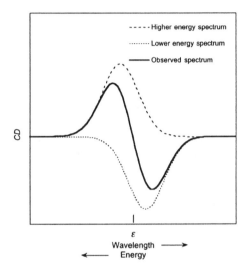

Figure 8.2 *CD* resulting from the coupling of one transition in each of two chromophores where both transitions are eda and degenerate (occurring at energy ε). The characteristic form of an excitonic spectrum (bold line) results from cancellation of overlapping positive and negative bands. In reality the energy splitting is smaller and the cancellation larger than depicted here.

8.4   Qualitative approach to exciton *CD*

The coupled-oscillator (or exciton) theory is an extremely useful tool for determining the absolute configuration of a chiral compound. It is in fact, apart from X-ray crystallographic determination, the most reliable non-empirical method for this task. For the practical purpose of assignment of *CD* signs to a given enantiomeric form, a full quantitative calculation is often unnecessary and a qualitative analysis along the lines given below will suffice. In addition, this qualitative approach will provide us with a way to understand the physical significance of the equations used and derived below. A brief dicsussion of coupled-oscillator *CD* was given at the end of Chapter 4. Here we expand on that discussion.

See §11.1 for a discussion of vector products. The cross product of two vectors is another vector perpendicular to both of the original ones. A dot product gives the amount one vector aligns with another.

Both the case of two non-degenerate transitions and that of two degenerate transitions (the true exciton case) provide *CD* signals given by a scalar triple product (§10.6):

$$\boldsymbol{\mu}^a \cdot \boldsymbol{R}_{AC} \times \boldsymbol{\mu}^c \qquad (8.2)$$

If $\boldsymbol{\mu}^a$ and $\boldsymbol{\mu}^c$ are perpendicular, then they do not interact and no CD signal is observed.

where $\boldsymbol{\mu}^a$ and $\boldsymbol{\mu}^c$ are transition moments in the two chromophores that are chirally arranged relative to one another in space and $\boldsymbol{R}_{AC}$ is the distance vector connecting them as illustrated in Figure 8.1c.

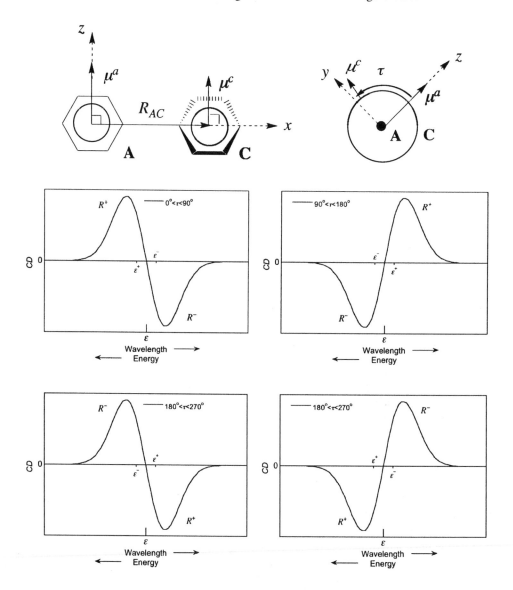

Figure 8.3 Geometry of a biphenyl **A/C** system and its exciton CD spectra for 'short-axis' polarized transitions as a function of τ, the torsional angle between the chromophores (assuming Gaussian band shapes for the two transitions). The spectra illustrated are the net CD spectra that would be observed after partial cancellation of the bands occurring at ε^+ and ε^-.

We may here see an analogy with the CD as theoretically provided by the Rosenfeld equation:

$$\boldsymbol{\mu} \cdot \boldsymbol{m} \qquad (8.3)$$

The second vector $R_{AC} \times \mu^c$ in equation (8.2) may be regarded as the magnetic moment that the electric moment μ^c exerts on μ^a by orbiting at a radius of R_{AC} around the μ^a centre.

For degenerate transitions we first consider the effects that the electric moments have on each other. From coupling between the transition moments of the two degenerate transitions, the following net exciton transition moments result

$$\mu^+ = \frac{1}{\sqrt{2}}\left(\mu^a + \mu^c\right)$$

$$\mu^- = \frac{1}{\sqrt{2}}\left(\mu^a - \mu^c\right)$$

(8.4)

where the + and − signs correspond, respectively, to in-phase and out-of-phase combinations of the transition moments occurring at respectively ε^+ and ε^-. To determine the sign of the in-phase *CD*, view the system from above both μ^a and μ^c and project them onto the plane that is perpendicular to μ^+ and contains R_{AC} as illustrated in Figure 8.3. Trace a circle through the projections. If the rotation of charge is anti-clockwise, m^+ by the right hand rule is parallel to μ^+ and the *CD* is positive. Figure 8.1 illustrates the situation where $CD^+ > 0$ and $CD^- < 0$. The energy ordering of the two transitions follows by placing positive charges at the positive tips of μ^a and μ^c and determining whether this arrangement or μ^a and $-\mu^c$ has the least repulsion. In practice if the μ^a / μ^c torsion angle is less than 90° then ε^+ is at a higher transition energy than ε^- as illustrated in Figure 8.3 for biphenyl.

8.5   Quantitative approach to exciton *CD*

If we took the mirror image of the **A**/**C** system illustrated in Figure 8.1, then the *CD* spectrum would be the inverse of the one illustrated in Figure 8.2. It is therefore crucially important that we have a protocol for defining the geometry of the **A**/**C** system when we wish to determine *CD* arising from the coupling of the electric dipole transition moments on **A** and **C**, μ^a and μ^c, respectively.

Wherever possible throughout this book we shall illustrate the **A** and **C** chromophores as in Figure 8.3 with **A** in front and **C** behind. The vector from the **A** origin to the **C** origin, R_{AC}, then goes back into the page from **A** to **C**. In most of the examples we shall look at, we define the *x*-axis to lie along R_{AC}, and *z* to lie along the *y-z* projection of μ^a (*i.e.* μ^a has *y* = 0). It is also convenient to define three angles: α ($0 \le \alpha \le 180°$) the angle between R_{AC} and μ^a, γ ($0 \le \gamma \le 180°$) the angle between $-R_{AC}$ and μ^c, and τ the torsion angle passed through in going from the *y-z* projection of μ^a (*i.e.* *z*) in an *anticlockwise* direction to the *y-z* projection of μ^c. Thus

> If $0° < \tau < 180°$, the three vectors μ^c, μ^a, and R_{AC} (CAR) form a right-handed coordinate system. This requires that you can put your thumb along μ^c, your fore-finger along μ^a and your middle finger along R_{AC} without breaking any bones (*cf.* Figure 1.2, Figure 8.4).

$$R_{AC} = R_{AC}(1,0,0)$$

(8.5)

$$\mu^a = \mu^a\left(\cos\alpha,0,\sin\alpha\right)$$

(8.6)

$$\mu^c = \mu^c\left(-\cos\gamma,\sin\gamma\sin\tau,\sin\gamma\cos\tau\right)$$

(8.7)

and

$$\cos \tau = \frac{\hat{\boldsymbol{\mu}}^a \cdot \hat{\boldsymbol{\mu}}^c - \hat{\boldsymbol{\mu}}^a \cdot \hat{R}_{AC} \hat{R}_{AC} \cdot \hat{\boldsymbol{\mu}}^c}{\sqrt{\left[1 - \left(\hat{\boldsymbol{\mu}}^a \cdot \hat{R}_{AC}\right)^2\right]\left[1 - \left(\hat{\boldsymbol{\mu}}^c \cdot \hat{R}_{AC}\right)^2\right]}}$$

$$= \frac{\hat{\boldsymbol{\mu}}^a \cdot \hat{\boldsymbol{\mu}}^c + \cos \alpha \cos \gamma}{\sqrt{\left[1 - \left(\cos \alpha\right)^2\right]\left[1 - \left(\cos \gamma\right)^2\right]}} \qquad (8.8)$$

$$= \frac{\hat{\boldsymbol{\mu}}^a \cdot \hat{\boldsymbol{\mu}}^c + \cos \alpha \cos \gamma}{\sin \alpha \sin \gamma}$$

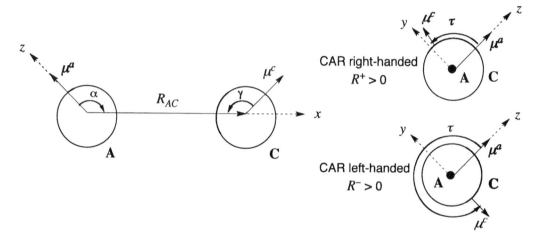

Figure 8.4 Diagram illustrating the geometry and coordinates for the **A/C** system described in the text. Note that τ is the angle taken in the anticlockwise direction between the *projections* of the electric dipole transition moments onto the *y-z* plane when the observer is looking down the *x*-axis from A to C. $0° < \tau < 180°$ if $\boldsymbol{\mu}^c \times \boldsymbol{\mu}^a \cdot R_{AC} > 0$ (*i.e.* if the three vectors form a right-handed parallelepiped) and $180° < \tau < 360°$ if $\boldsymbol{\mu}^c \times \boldsymbol{\mu}^a \cdot R_{AC} < 0$.

Degenerate coupled-oscillator *CD* equations

In order to analyze *CD* spectra in geometric terms, we need to have equations relating the spectra to the geometry of the system. These are derived in Chapter 10 and the results summarized here. Two alternative representations of the equations are given: one using vectors and vector products, the other using components of vectors and the angles defined in Figure 8.4. As illustrated in Figure 8.2, when degenerate transitions on two chromophores couple we find two *CD* bands of opposite sign that occur at energies slightly above and slightly below the transition energies of the isolated chromophores at energies

When we use the $\varepsilon^{\pm} = \varepsilon^{+} \pm V$ it means that ε^{+} follows the upper of the two signs when there is a choice. So *e.g.* from equation (8.9)
$$\varepsilon^{+} = \varepsilon + V.$$

$$\varepsilon^{\pm} = \varepsilon \pm V \qquad (8.9)$$

where

$$V = \frac{\boldsymbol{\mu}^a \cdot \boldsymbol{\mu}^c - 3R_{AC} \cdot \boldsymbol{\mu}^a \boldsymbol{\mu}^c \cdot R_{AC}}{R_{AC}^3} \qquad (8.10)$$

The two *CD* bands have signs and magnitudes given by the following vectorial equation (*cf.* §11.1)

$$R^{\pm} = \pm \frac{\varepsilon}{4\hbar} \left\{ \boldsymbol{\mu}^c \times \boldsymbol{\mu}^a \cdot \boldsymbol{R}_{AC} \right\}$$ (8.11)

where R^+ refers to the transition that occurs at energy ε^+ and R^- to the one that occurs at energy ε^- (*cf.* Figure 8.2).

Note: whether ε^+ is larger or smaller than ε^- depends on the sign of *V*.

Equation (8.11) describes the magnitude of the area under each *CD* band resulting from the coupling of $\boldsymbol{\mu}^a$ and $\boldsymbol{\mu}^c$; it is also the volume of the parallelepiped the vectors define. The actual appearance of the spectrum is dependent on the band shape, which is usually Gaussian or Lorentzian, rather than single sharp lines at a precise energies (as implied by the the the equations).

From equations (8.9) and (8.11) we can describe the *CD* spectrum we expect to see for the coupling of two degenerate transitions on two chromophores.

- There are two bands of equal magnitude and opposite sign, R^+ centred at ε^+ and R^- centred at ε^-.

Note that either ε^+ or ε^- might be higher in energy.

- The *CD* strength *R* depends on the strength of the transitions in the isolated chromophores and on their orientations relative to one another and to the vector connecting the two chromophores.
- The sign of the *CD* of the '+' state is positive if $\boldsymbol{\mu}^c$, $\boldsymbol{\mu}^a$, and R_{AC} form a right handed axis system and negative if they form a left-handed axis system.
- *V* is small, so the two bands are close to one another and, since they have opposite signs, largely cancel each other's *CD* where they overlap.

In some applications it is easier to have equations (8.9)–(8.11) written in terms of angles using the right-handed coordinate system $\{x, y, z\}$ illustrated in Figure 8.4 and given by equations (8.5)–(8.8). The degenerate coupled-oscillator *CD* strengths may then be written:

$$R^{\pm} = \pm \frac{\varepsilon \mu^2 R_{AC}}{4\hbar} \sin\alpha \sin\gamma \sin\tau$$ (8.12)

Here and subsequently, energies and *CD* magnitudes will be given in unspecified units. The units required for quantitative calculations are given in §11.5.

and the energies of the two bands are

$$\varepsilon^{\pm} = \varepsilon \pm \frac{\mu^2 (\sin\alpha \sin\gamma \cos\tau + 2\cos\alpha \cos\gamma)}{R_{AC}^3}$$ (8.13)

A little care is needed in applying equations (8.12)–(8.13) as they are not independent.

- Although the *CD* strength, *R*, is a maximum when τ, α, γ are all $90°$, under these circumstances $V = 0$, so both the positive and negative *CD* signals are centred at ε and exactly cancel, resulting in no *CD* signal.

When $\tau = \alpha = \gamma = 90°$ the **A/C** system is achiral as it has an S_4 improper rotation axis.

- According to equation (8.12) the *CD* strength should get larger as the distance between the chromophores increases. However, the energy gap concomitantly decreases so more and more cancellation occurs, thus avoiding the nonsense situation of two chromophores too far apart to interact giving infinitely large *CD* signals. The extent of cancellation depends upon the shapes of the bands but is approximately linear in V. From equations (8.12) and (8.13) we thus conclude that the *CD* signal effectively decreases as R_{AC}^{-2}.

- The *CD* of two enantiomers are equal in magnitude and opposite in sign. This may be seen by reflecting the system about the plane containing μ^a and R_{AC}, *i.e.* the *x-z* plane, as then $\tau \to 360° - \tau$. Thus, $\cos\tau$ remains unchanged and $\sin\tau \to -\sin\tau$ leaving the energy splitting the same but the *CD* bands of opposite sign. Hence the net inversion of the *CD* spectrum.

8.6   Degenerate coupled-oscillator *CD*: some examples

Biphenyl: $\alpha = \gamma = 90°$

When $\alpha = \gamma = 0°$, μ^a and μ^c are perpendicular to R_{AC} so they are themselves the projections in the *y-z* plane.

Equations (8.12) and (8.13) are simplest when $\alpha = \gamma = 0$ or $\alpha = \gamma = 90°$, *i.e.* when the transitions are polarized either parallel or perpendicular to the line connecting the chromophore origins. If we may ignore electron exchange (conjugation) between the rings, biphenyl (Figure 8.1) in a twisted conformation provides us with such an example with each ring of biphenyl having two possible transition polarizations.

Consider first two degenerate transitions with $\alpha = \gamma = 0$. The electric dipole transition moments for these transitions are co-planar (in fact co-linear), so they form an achiral system and have no *CD* strength. Consistent with this, upon substituting $\alpha = \gamma = 0$ into equation (8.12) a value of 0 results.

In practice, unless the biphenyl has bulky substituents or a bridge between the rings, the energetic barrier for converting one enantiomer to the other is so small that biphenyl can never be resolved into samples containing only one enantiomer.

If, however, two degenerate $\alpha = \gamma = 90°$ transitions couple then the resulting *CD* strengths and transition energies are

$$R^\pm = \pm \frac{\varepsilon \mu^2 R_{AC} \sin\tau}{4\hbar} \tag{8.14}$$

$$\varepsilon^\pm = \varepsilon \pm \frac{\mu^2 \cos\tau}{R_{AC}^3} \tag{8.15}$$

where τ is the dihedral angle traced in going in an anticlockwise direction when viewed down R_{AC} from μ^a to μ^c as illustrated in Figure 8.3. When $\tau = 0$ or $180°$ the *CD* strength is zero and when $\tau = 90°$ or $270°$ there is no energy splitting between the two bands, so no *CD* spectrum is observed. These are the achiral arrangements of the biphenyl system. Maximum magnitude *CD* intensities are expected for τ in the region of $\pm 45°$ and $\pm 135°$. The gas-phase experimental value for τ is $44°$,[218] meaning each biphenyl molecule has close to the maximum possible *CD* signal. However, in any real sample there are equal proportions of both enantiomers, so we observe no experimental *CD* spectrum for biphenyl.

The precise value of maximum *CD* signal in an experiment will depend upon the shapes of the bands and how the two components cancel.

Put your right thumb along μ^a; your right hand fingers may then be curled to lie along μ^c.

To understand the origin of the Figure 8.3 spectra, consider a short-axis polarized transition moment in each phenyl oriented so that $\tau \sim 45°$. The two moments may couple in-phase or out-of-phase. As may be seen from Figure 8.3, μ^c circles about μ^a in a right-handed direction as does μ^a about μ^c. As the in-phase coupling of the two moments has net electric dipole transition moment $\mu = \mu^a + \mu^c$, and the net magnetic dipole transition moment of the in-phase coupled moments is a circling of charge about $\boldsymbol{\mu}$, the net effect of the in-phase coupling of the two moments is a right-handed helix of electron motion in space. Thus we expect a positive *CD* signal for $0 < \tau < 90°$ as predicted by equation (8.12) and illustrated in Figure 8.1 and Figure 8.3.

Before leaving our first example we note a few cautions in using *CD* to analyze molecular geometry.

- Despite the opposite handedness of the molecules giving the first and third spectra in Figure 8.3, these spectra *look* the same because *both* the *CD* signs and the energy order are inverted in going from one to the other.
- In order experimentally to test Figure 8.4, we would need to consider substituted biphenyl systems and it may not be appropriate to assume that the phenyl transitions are unchanged by the substitution. We shall therefore leave this example as an illustration of the simplest case of degenerate coupled-oscillator *CD*.
- The transitions used in the analysis must be *independent, i.e.* there must not be significant *conjugation* between the sub-units into which we divide the molecule. The π system of, for example, butadiene, cannot satisfactorily be treated as two independent ethylene groups.
- Although we have stated that the coupling between transitions of identical chromophores is dominated by the degenerate coupling of one transition with the same one on the other chromophore if other transitions are close in energy they may interfere significantly. For example, the coupling of $\pi \rightarrow \pi^*$ transitions with $\sigma \rightarrow \pi^*$ transitions seems to be the reason for the mixed success of assignments of the geometry of non-conjugated chiral dienes.[219]

The greater the extent of conjugation, the more planar a molecule becomes, so if $\tau \sim 45°$ in biphenyl then we can assume there is little conjugation.

6,15-dihydro-6,15-ethanonaphthol[2,3-c]pentaphene: $\alpha = \gamma \neq 90°$

The next level of complexity is provided by the two anthracenes of the molecule illustrated in Figure 8.5. Each anthracene chromophore has a weak short axis polarized transition at ~360 nm and a strong long axis one at 252 nm (Figure 1.8). In contrast to the biphenyl case, the transition polarizations are neither parallel nor perpendicular to $\boldsymbol{R}_{AC}$ and τ is defined by the rigid geometry of the molecule (Figure 8.5). For the long axis polarized transitions $\alpha^{//} = \gamma^{//} = 137°$ and $\tau^{//} = 259°$ for the enantiomer illustrated in Figure 8.5. Equations (8.12) and (8.13) thus become

$$R^{\pm} = \pm \frac{\varepsilon \mu^2 R_{AC}}{4\hbar} \sin\alpha \sin\gamma \sin\tau$$

$$= \mp 0.46 \frac{\varepsilon \mu^2 R_{AC}}{4\hbar} \tag{8.16}$$

and

$$\varepsilon^{\pm} = \varepsilon \pm \mu^2 \left(\sin\alpha \sin\gamma \cos\tau + 2\cos\alpha \cos\gamma\right) R_{AC}^{-3}$$

$$= \varepsilon \pm 0.98 \mu^2 R_{AC}^{-3} \tag{8.17}$$

So we expect to see a symmetric exciton couplet at ~250 nm with the lower energy (longer wavelength) band having the positive sign and the higher energy band the negative sign.

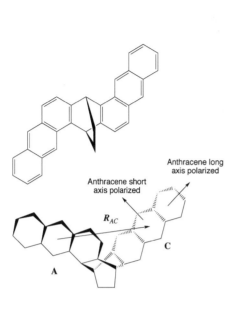

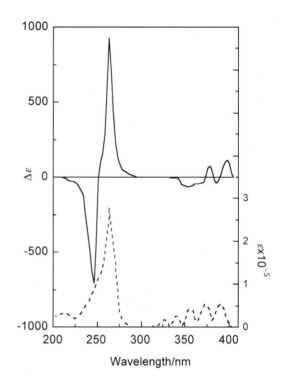

Figure 8.5 Geometry, absorption, and *CD* of 6*R*,15*R*-(+)-dihydro-6,15-ethanonaphthol[2,3-*c*]pentaphene. Spectra sketched from reference [220].

Similarly, as the angles for the short-axis polarized transition are $\alpha^\perp = \gamma^\perp = 73°$ and $\tau^\perp = 112°$,

$$R^\pm = \pm 0.85 \frac{\varepsilon \mu^2 R_{AC}}{4\hbar} \tag{8.18}$$

and

$$\varepsilon^\pm = \varepsilon \mp 0.17 \mu^2 R_{AC}^{-3} \tag{8.19}$$

As might be expected, the signs of the in-phase and out-of-phase components are reversed for the short-axis polarized transitions relative to those of the long-axis polarized transitions, however, the energy order of the bands is also reversed. Therefore the *CD* spectra of the two geometries have the same sign exciton effects. The *CD* strength in the 380 nm region is significantly smaller than that at 260 nm since coupled-oscillator *CD*, as seen from equation (8.12), scales with oscillator strength μ^2 (normal absorption intensity) as well as the geometry factor. In the 380 nm region, the pair-wise couplings of the vibronic components of the transition are apparent resulting in the oscillating signs observed in the *CD*.

Rather than simply show that we can write equations to reproduce the experimental *CD* spectra, the usual application would be to determine which enantiomer was present in a sample by finding the value of τ that gives the experimentally observed *CD* sign.

In-ligand transition of a *bis*-chelate metal complex: $\alpha \neq \gamma \neq 90°$

Although metal complexes[221] are chemically different from the previous example, the *CD* arising from the coupling of long-axis polarized transitions of the ligands in *bis*-chelate transition metal complexes can be understood in the same way as for the *bis*-anthracene molecule. In this application the metal is not part of any chromophore, but can be considered as a structural unit placing the ligands in a chiral orientation in space relative to one another. For simplicity, let us assume the ligating atoms define a perfect octahedron, then a Δ enantiomer has the coordinates and dipoles shown in Figure 8.6.

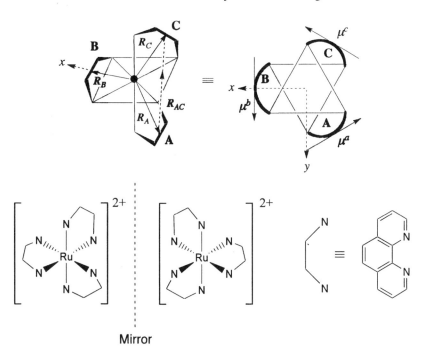

Figure 8.6 Geometry described in the text for Δ *bis*- and *tris*-chelate complexes and illustration of Δ (left) and Λ (right) enantiomers of [Ru(1,10-phenanthroline)$_3$]$^{2+}$.

The $\{x,y,z\}$ axis system used in Figure 8.6 is that defined by the $\mathbf{D}_3$ symmetry of the next example rather than that of Figure 8.4. The vectors from the metal to the centres of the chelates **A** and **C** in Figure 8.6 axes are:

$$R_A = \rho\left(-\frac{1}{2}, \frac{\sqrt{3}}{2}, 0\right) \qquad (8.20)$$

and

$$R_C = \rho\left(-\frac{1}{2}, -\frac{\sqrt{3}}{2}, 0\right) \tag{8.21}$$

so

$$R_{AC} = \sqrt{3}\rho(0, -1, 0) \tag{8.22}$$

where ρ is the distance from the metal to the centre of a chelate. The long-axis polarized chelate electric dipole transition moments as illustrated are:

$$\boldsymbol{\mu}^a = \mu\left(-\frac{1}{2}, -\frac{1}{2\sqrt{3}}, \frac{\sqrt{2}}{\sqrt{3}}\right) \tag{8.23}$$

and

$$\boldsymbol{\mu}^c = \mu\left(\frac{1}{2}, -\frac{1}{2\sqrt{3}}, \frac{\sqrt{2}}{\sqrt{3}}\right) \tag{8.24}$$

where μ is the magnitude of the electric dipole transition moments. Thus (Figure 8.4) $\cos\alpha = -\cos\gamma = (2\sqrt{3})^{-1}$, $\sin\alpha = \sin\gamma = \sqrt{(11/12)}$ and $\cos\tau = 5/11$. Further, CAR is right-handed so $0 < \tau < 180°$ and thus $\sin\tau = 4\sqrt{6}/11$.

The *CD* strengths and transition energies from the coupling of long-axis polarized transition moments on two chelates is thus (from equations (8.9) and (8.11) or equations (8.12) and (8.13))

$$R^{\pm} = \pm\frac{\varepsilon\mu^2\rho}{2\sqrt{2}\hbar}$$

$$\varepsilon^{\pm} = \varepsilon \pm \frac{\mu^2}{12\sqrt{3}\rho^3} \tag{8.25}$$

where μ^2 is the dipole strength of the isolated ligand transition. With the directions chosen for the electric dipole transition moments in Figure 8.6, the in-phase R^+ band has a positive *CD* strength, occurs at the higher energy, and is polarized perpendicular to x, the two-fold axis of the system. The out-of-phase R^- band has a negative *CD* and is polarized along x.

In reality, the ligating atoms seldom define a regular octahedron. Imagine the distortion from octahedral being such as to twist the chelates more towards being coplanar. In this case, τ increases. A *very* large twist will actually cause the second term in the energy expression to change sign and the R^+ state to swap to the lower energy position making it appear that the *CD* sign has inverted—although in fact it is the energy ordering that has inverted.

In-ligand transition of a *tris*-chelate metal complex: $\alpha_1 = \alpha_2 = \alpha_3 \neq 90°$

The previous example leads directly to the in-ligand *CD* for Δ *tris*-chelate transition metal complexes.[221] As there are now three chelates, to evaluate the *CD* we also require vectors for the chelate **B** in Figure 8.6:

$$\boldsymbol{R}_B = \rho(1,0,0)$$

$$\boldsymbol{\mu}^b = \mu\left(0, \frac{1}{\sqrt{3}}, \frac{\sqrt{2}}{\sqrt{3}}\right) \qquad\qquad (8.26)$$

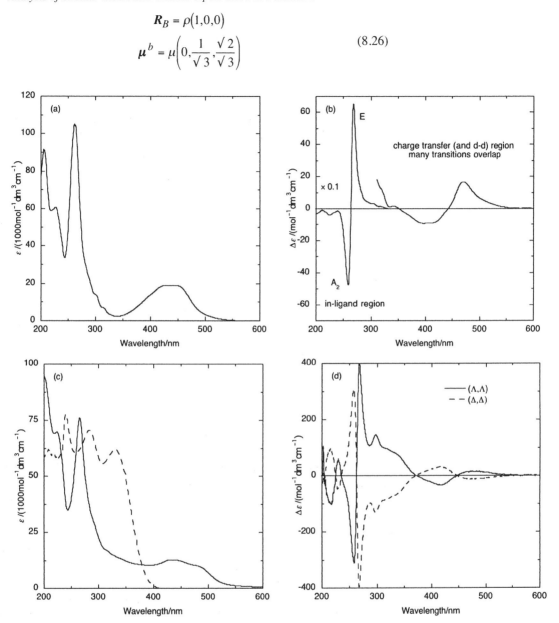

Figure 8.7 (a) Absorbance and (b) *CD* spectrum of (a) Λ-[Ru(1,10-phenanthroline)$_3$]$^{2+}$ in water.[222] (c) Absorbance spectra of [Ru$_2$(phen)$_4$LL]$^{4+}$ (solid line) and of LL (dashed line) in water. (d) *CD* spectra of (Δ,Δ) and (Λ,Λ)-[Ru$_2$(phen)$_4$(LL)]$^{4+}$ in water.[117] LL is illustrated in Figure 3.10.

However, since the system is symmetric, the coupling of each pair of ligands leads to the same contribution, so we can simple multiply equation (8.25) by a factor of 3 and divide by 3/2 to account for the rotational averaging of the three-fold rotational axis (see Chapter 10) due the higher symmetry of the system. The *CD* strengths and transition energies for the in-ligand *CD* of a Δ configuration *tris*-chelate are thus:[221]

$$R\left(\frac{z}{x/y}\right) = R\left(\frac{A_2}{E}\right) = \pm \frac{\varepsilon \mu^2 \rho}{\sqrt{2}\hbar}$$

$$\varepsilon\left(\frac{z}{x/y}\right) = \varepsilon\left(\frac{A_2}{E}\right) = \varepsilon \pm \frac{\mu^2}{12\sqrt{3}\rho^3}$$

$$(8.27)$$

$R(z)$ is the z–polarized *CD* band (A_2) that results from the in-phase coupling of the three dipoles. It occurs at the higher energy (lower wavelength) and is positive. The out-of-phase couplings give the second *CD* band which has the same magnitude but opposite sign from the first band. Its polarization is x/y or E, with its transition density spread out over the whole x/y plane. An experimental spectrum is illustrated in Figure 8.7b. It should be noted that an experimental spectrum also includes the non-degenerate coupling contributions between all transitions so is not quite as symmetric as the equations suggest.

Helical peptides

A somewhat simplistic model for determining the backbone $\pi \rightarrow \pi^*$ *CD* of α-helical peptides is to assume we may represent the peptide by a right-handed helix of transition moments oriented along each C=O bond (Figure 3.17) as illustrated in Figure 8.8. Each neighbouring pair of dipoles has $\tau \sim 300°$ if they are both directed towards the oxygen, and $\tau = 120°$ when they have opposite orientations. Although N residues would lead to N transitions when they couple, we need only consider the highest and lowest energy ones; the others will essentially cancel each other out between the two extrema.

The lowest energy excited state (and hence longer wavelength component of the transition) will result from the head-to-tail couplings of neighbouring transition moments. For an α-helix, head-to-tail coupling of neighbouring $\pi \rightarrow \pi^*$ electric dipole transition moments occurs when they are both pointing from C to O along the C=O bond since $\tau > 180°$. As Figure 8.8 illustrates, the net electric dipole transition moment of this state is parallel to the helix axis, and the net magnetic dipole transition moment is anti-parallel to it, making a left-handed helix of electron displacement and therefore a negative *CD*. The head-to-head, tail-to-tail coupling gives the highest energy transition whose polarization is perpendicular to the helix axis and has a net right-handed helical charge displacement and a positive *CD*. The two components (208 nm and 190 nm) of the $\pi \rightarrow \pi^*$ α-helix *CD* (§3.7) illustrate this discussion.

Combined *CD* and *LD* spectroscopy

The coupled $\pi \rightarrow \pi^*$ transitions of the amide chromophores in polypeptides provide an excellent example of how *CD* and *LD* spectroscopy can provide complementary structural information. As noticed above, the two coupling modes that give rise to the high energy positive *CD* and the low energy negative *CD* bands are associated with net electron transition moments that lie, respectively, perpendicular and parallel to the helix axis and thus give rise to respectively, negative and positive *LD* bands in a flow oriented α-helix. Since the net $\boldsymbol{\mu}$ is simply a vector sum of the $\boldsymbol{\mu}$s of the individual chromophores (with appropriate signs to match the phase of the coupling mode) the *LD* signal directly relates to the structure of the polypeptide and may be analyzed as outlined in Chapters 3 and 5.

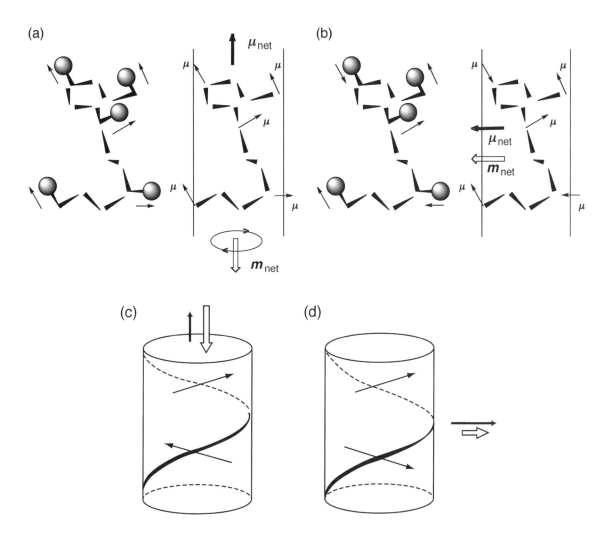

Figure 8.8 Schematic illustration of the backbone and carbonyl groups (oxygens indicated by balls) of an α-helical peptide approximately indicating (a) head-to-tail and (b) head-to-head couplings of the peptide electric dipole transition moments and the resulting net electric and magnetic dipole transition moments. An alternative somewhat simplified model is to consider the π-π^* transitions of only two amide chromophores on a right-handed α-helix as shown in (c) and (d). By applying the six steps outlined in §4.7, the qualitative features of the exciton *CD* and *LD* spectra immediately emerge.

8.7   Non-degenerate coupled-oscillator *CD*: general case

In the final sections of this chapter the *CD* spectra expected for two non-identical achiral chromophores **A** and **C** whose eda transitions are arranged in a chiral geometry and couple are described. The required induced magnetic dipole transition moments arise in the same way as when degenerate oscillators couple—a linear motion of charge at a distance (in another chromophore) has a circular (magnetic) component locally. The main differences between the non-degenerate and degenerate cases are:

- the two *CD* bands resulting from the coupling of two non-degenerate transitions occur at (strictly very close to) the energies of the two non-degenerate transitions, rather than as an exciton couplet where most of the *CD* intensity is cancelled, and
- the coupling between non-degenerate transitions is not as strong as that between degenerate transitions and it decreases when either the energy difference between the transitions or the distance between the chromophores increases.

If a transition is eda and mda but they are perpendicular, then the coupled-oscillator mechanism generally dominates the observed *CD*.

The Figure 8.4 coordinate system used for the degenerate coupled-oscillator analysis can be used for the non-degenerate situation. The only change in notation required is that we need to distinguish the transition energies and transition moment magnitudes of the transitions in the two chromophores since they are now different. As shown in Chapter 10, the *CD* induced at ε_a into the transition on **A** by its coupling with an electric dipole transition moment of different energy in **C**

$$R(\varepsilon_a) = \frac{-\varepsilon_a \varepsilon_c V}{\hbar\left(\varepsilon_c^2 - \varepsilon_a^2\right)}\left\{\boldsymbol{\mu}^c \times \boldsymbol{\mu}^a \cdot \boldsymbol{R}_{AC}\right\}$$

$$= \frac{-\varepsilon_a \varepsilon_c}{\hbar\left(\varepsilon_c^2 - \varepsilon_a^2\right)}\left\{\frac{\boldsymbol{\mu}^c \cdot \boldsymbol{\mu}^a - 3\hat{\boldsymbol{R}}_{AC} \cdot \boldsymbol{\mu}^a \boldsymbol{\mu}^c \cdot \hat{\boldsymbol{R}}_{AC}}{R_{AC}^2}\right\}\left\{\boldsymbol{\mu}^c \times \boldsymbol{\mu}^a \cdot \hat{\boldsymbol{R}}_{AC}\right\}$$

$$(8.28)$$

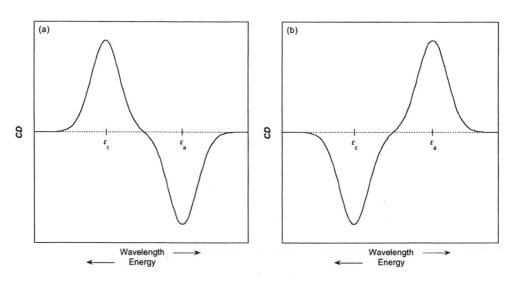

Figure 8.9 Schematic illustration of non-degenerate coupled-oscillator *CD* spectra for $\alpha = \gamma = 90°$: (a) $0 \le 2\tau < 180°$ and (b) $180° \le 2\tau < 360°$ (see Figure 8.4 for definition of τ).

The alternative version of equation (8.28) in terms of the angles defined in Figure 8.4 is given below. As in the degenerate coupled-oscillator case, it is very important to determine the sense of τ correctly as it defines the handedness of the system.

$$R(\varepsilon_a) = \frac{-\varepsilon_a \varepsilon_c (\mu^a \mu^c)^2}{\hbar(\varepsilon_c^2 - \varepsilon_a^2)R_{AC}^2}\{\sin\alpha\sin\gamma\cos\tau + 2\cos\alpha\cos\gamma\}\sin\alpha\sin\gamma\sin\tau$$

$$(8.29)$$

A number of general points about non-degenerate coupled-oscillator *CD* spectra follow from equations (8.28) and (8.29).

- The *CD* strength *decreases* with *increasing* **A** to **C** distance according to R_{AC}^2. This is the smallest distance-dependence for any *CD* mechanism; coupled-oscillator *CD*s are therefore generally larger than ones arising from any of the chromophore coupling mechanisms discussed in the next chapter.
- $R^{\pm} \to -R^{\pm}$, *i.e.* $CD \to -CD$, for the mirror image system.
- It is often the case that $\alpha = \gamma = 90°$. When this situation arises the structural geometry factor in equation (8.29) is

$$\cos\tau\sin\tau = \frac{1}{2}\sin(2\tau) \qquad (8.30)$$

so the *CD* is a maximum when $\boldsymbol{\mu}^a$ and $\boldsymbol{\mu}^c$ are oriented at 45° to one another.

- If we wish to know the combined effect of many transitions in **C** on an **A** transition, we simply introduce a sum over *c* into equations (8.28) and (8.29).
- The *CD* induced into the ε_c transition of **C** by $\boldsymbol{\mu}^a$ follows upon permuting the '*a*' and '*c*' labels and the '*A*' and '*C*' labels in equation (8.28) and (8.29) (note $\boldsymbol{R}_{AC} = -\boldsymbol{R}_{CA}$). Thus, the *CD* induced by an **A** transition into a **C** transition is equal in sign and opposite in magnitude from the one induced by the **C** transition into the **A** transition as illustrated in Figure 8.9.

8.8   Non-degenerate coupled-oscillator *CD*: some examples

DNA dinucleotide: $\alpha = \gamma = 90°$

Consider two non-identical DNA nucleo-bases (*e.g.* guanine and adenine) that are a small fragment of B-DNA, so they are vertically stacked parallel (but skew) to one another with their planes perpendicular to the vector between their origins (Figure 8.10). Thus, $\alpha = \gamma = 90°$. With ten bases per turn of DNA, the twist between the long-axes of adjacent base pairs is 36°. For simplicity we assume that each base has only one transition moment polarized either parallel to its long axis or perpendicular to it (parallel to the short axis). We make the base that has the lower energy transition **A**, and look at the dinucleotide from above with **A** on top. Figure 8.10 shows the value of τ for the possible transition moment polarization combinations of this simple model.

Note that if the two stacked bases are identical, the only thing that matters is the helical twist (here 36°) which will give be the angle between identical, degenerate transition moments (*cf.* Figure 4.24).

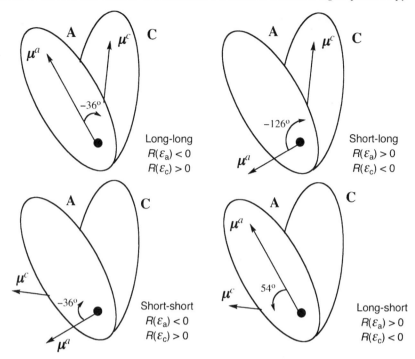

Figure 8.10 Geometry and *CD* signs (assuming $\varepsilon_a < \varepsilon_c$) for a 'B-DNA' dinucleotide with two different bases. Each nucleotide is assumed to have only one electric dipole transition moment, either long axis or short axis polarized. The τ angles for each situation are indicated.

In this geometry, equation (8.29) for the *CD* of the transition of the top base becomes

$$R(\varepsilon_a) = \frac{-\varepsilon_a \varepsilon_c (\mu^a \mu^c)^2}{2\hbar\left(\varepsilon_c^2 - \varepsilon_a^2\right)R_{AC}^2} \sin 2\tau \qquad (8.31)$$

Thus the 'long-long' and 'short-short' combinations both give a negative *CD* at ε_a, and the mixed combinations both give a positive *CD* at ε_a. The *CD* signals at ε_c are equal in magnitude but opposite in sign from those at ε_a (Figure 8.10).

It seems immediately reasonable from the above discussion that left-handed Z-DNA should have a *CD* signal more-or-less a mirror image of that found for B-DNA (*cf.* Figure 4.5) on the basis of equation (8.31). While this might be approximately true, we must realize that in reality the base transition moments are not only long or short axis polarized. In addition, the Z-DNA structure is not a simple mirror image of the B-DNA structure since the presence of the chiral ribose moieties makes it a diastereomeric system. A further warning about the real situation follows from considering that each base has several transitions occurring at similar transition energies and DNA has four different bases. Experimental DNA spectra are thus the result of complicated coupling patterns between many overlapping transitions.

For the case of two identical bases in a B-DNA conformation (a degenerate coupled-oscillator example), $\tau = (360 - 36)°$. R^+ therefore occurs at the higher energy with negative sign giving rise to the common exciton-like pattern of the long wavelength region of B-DNA spectra (*e.g.* Figure 4.3).

(a)   (b)

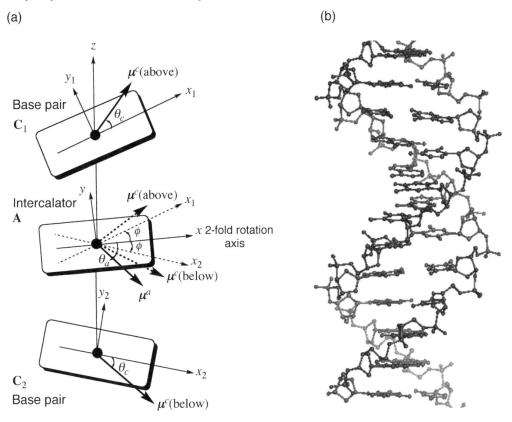

Figure 8.11 (a) Geometry of a dinucleotide/intercalator system where the dinucleotide has a two-fold rotation axis about x. (b) B-DNA.

Intercalator in a random sequence DNA polynucleotide

In this example we wish to determine the *CD* induced into a transition of an achiral planar aromatic molecule that is intercalated (sandwiched) between two base pairs of a B-DNA molecule. The plane of the intercalator lies parallel to those of the bases, so, as in the previous example, $\alpha = \gamma = 90°$. Thus equation (8.31), where μ^a is now the intercalator transition moment, describes the *CD* for this situation. A sum over at least the neighbouring base pairs and all their transitions should also be included. The geometry of the dinucleotide/adduct situation is illustrated in Figure 8.11.

Although this is a fairly unrealistic description of DNA it forms a starting point for fuller theoretical treatments.[119, 124, 125, 225]

In a random sequence DNA with random intercalative binding a given base pair will be found above and below an intercalator equally often. So we consider the situation of Figure 8.11 where $\tau(\text{above}) = (360-\theta_c-\theta_a-\phi)°$ and $\tau(\text{below}) = (360-\theta_c+\theta_a-\phi)°$. The *CD* from the coupling of the intercalator transition with transitions on each base is, in terms of the angles defined in Figure 8.11, therefore (*cf.* equation (8.29))[118, 119]

$$R(\varepsilon_a) = \sum_c \frac{\varepsilon_a \varepsilon_c \left(\mu^a \mu^c\right)^2}{2\hbar\left(\varepsilon_c^2 - \varepsilon_a^2\right)R_{AC}^2} \left\{\sin 2(\theta_c - \theta_a + \phi) + \sin 2(\theta_c + \theta_a + \phi)\right\}$$

$$= \sum_c \varepsilon_a \left(\mu^a\right)^2 \cos(2\theta_a) \left\{\frac{\varepsilon_c \left(\mu^c\right)^2 \sin(2\theta_c + 2\phi)}{2\hbar\left(\varepsilon_c^2 - \varepsilon_a^2\right)R_{AC}^2}\right\} \qquad (8.32)$$

The sign of $\sin(2\theta_c + 2\phi)$ in equation (8.32) is determined by the DNA helix pitch (2ϕ in Figure 8.11) and the polarizations of transition moments within the base pairs (recall a summation over all base transitions is implied in equation (8.32)). The sign of $\cos(2\theta_a)$ is determined by the polarization of the intercalator transition and the orientation of the intercalator in the pocket.

Schipper *et al.*[118, 119] deduced that the average sign of the DNA factor (the term in parentheses) is negative. Therefore, long-axis polarized transitions of intercalators oriented with their long axis 'parallel' to the DNA bases (thus having $\theta_2 \sim 0$) will have $CD < 0$ in a random sequence DNA. Long-axis polarized transitions of intercalators whose long axis pokes out into the DNA grooves will have $CD > 0$. The converse signs are to be expected for short axis polarized transitions. It is essential to note that it is the transition moment polarizations that determine the *CD* not the direction of the long axis of the molecule as has sometimes been assumed (unless of course the transition is long axis polarized).

As must be the case, the same geometry/*CD* correlations are derived for a random sequence DNA if *y* is used as the two-fold rotation axis rather than *x*. The symmetry related base pairs would then be *e.g.* 5'-T–A-3' and 3'-A–T-5'. Care must be taken in applying equation (8.32) to alternating homopolymers such as poly[d(G–C)]$_2$ since the value of θ_c is not the same for the same transition in 5'-G–C-3' and 5'-C–G-3'. Thus, the *CD* induced into an intercalator in a purine–pyrimidine site is expected to be quite different (even of different sign)[124, 125] from when it is in a pyrimidine–purine site.

Charge transfer transitions of *tris*-chelate metal complexes: $\alpha = \gamma = 90°$

In addition to eda in-ligand transitions, *tris*-chelate transition metal complexes also have eda charge transfer transitions, where electron density is transferred from metal to ligand orbitals or *vice versa*. We define the charge transfer chromophore, **A**, to be the metal and directly ligating atoms and take it to have D_{3d} symmetry so it may be treated as an achiral guest in the middle of the chiral host composed of three chelates C_1, C_2, and C_3 (Figure 8.12). **A** transitions are polarized either along the three-fold axis, *z*, for which $\alpha = 90°$, or anywhere in the plane perpendicular to *z*. The two-fold degenerate *x/y* polarized transitions may be considered as one polarized along *x* (*i.e.* pointing towards a chelate) with $\alpha = 0°$, so no *CD*, and one polarized along *y* with $\alpha = 90°$. In calculating the *CD* of planar aromatic chelates we need only consider the chelate long-axis polarized transitions as short-axis polarized ones, which are co-planar, give no *CD*. Thus $\gamma = 90°$ for each C_j transition.[119]

Equation (8.32), though not necessarily the sign of the DNA factor, also holds for B-DNA non-alternating homopolymers such as [poly(dG)].[poly(dC)].

We return to the question of how to describe the MLCT transitions in a metal complex with aromatic chelates in §8.9.

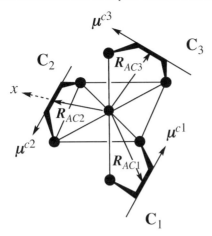

Figure 8.12 Geometry of the *tris*-chelate charge transfer system for a Δ complex. Components of the **A** chromophore are indicated by solid circles and the three **C** chromophores by thick lines. The axis system is as in Figure 8.6.

The charge transfer transition of a Δ complex as drawn in Figure 8.12 has $\cos\tau = \sqrt{2}/\sqrt{3}$ and $\sin\tau = 1/\sqrt{3}$, so the z-polarized (along the three-fold axis of the molecule) *CD* is[119]

$$R(z) = R(A_2)$$

$$= -\frac{\sqrt{2}\varepsilon_a\varepsilon_c\left(\mu^a\mu^c\right)^2}{\hbar\left(\varepsilon_c^2 - \varepsilon_a^2\right)R_{AC}^2} \qquad (8.33)$$

The factor of three (which cancels the $(1/\sqrt{3})^2$) comes from the three chelates. The A_2 label is the symmetry label for the transition. Note, this is opposite in sign from the same polarization in-ligand *CD* band and also equal in magnitude but opposite in sign from the x/y charge transfer band since $\tau(E) = 360° - \tau(A_2)$ so

$$R(A_2) = -R(E) \qquad (8.34)$$

The resulting *CD* spectrum (Figure 8.7b) at the charge transfer transition energy looks similar to (but more spread out than) the in-ligand region spectrum. In the in-ligand case, the overlapping bands are the result of two degenerate transitions on *different* chelates coupling together to give two transitions that are close in energy with opposite sign *CD*s. In the charge transfer case, however, two nearly degenerate (the A_2 and E) charge transfer transitions on the *same* chromophore each non-degenerately couple with the chelate transitions. The two *CD* bands have opposite signs because the A_2 and E transitions are perpendicularly polarized.

Note that $\mu^a\mu^a$ is the oscillator strength for the A_2 charge transfer band. The intensity of the E band is twice that of the A_2 band. $(\mu^c)^2$ is the oscillator strength of the in-ligand transition.

The energy ordering of the charge transfer transitions is difficult to determine, in contrast to the in-ligand case. As the charge transfer and in-ligand *CD* have *opposite* signs for a given polarization, the polarizations of the charge transfer bands can thus be determined from the *CD* spectrum.

In-ligand transitions of unsymmetric *bis*-chelate metal complexes: $\alpha = \gamma \neq 90°$

In the previous section we determined the *CD* resulting from the coupling of in-ligand long-axis polarized transition moments on two identical chelates in a *bis*-chelate transition metal complex. If we now consider two distinct ligands (such as 1,10-phenanthroline and 2,2'-bipyridine) we can use the same geometry as in the above degenerate *bis*-chelate example, but must use the non-degenerate coupled-oscillator equation, equation (8.29). Thus, the CD expected at ε_a is:

$$R(\varepsilon_a) = \frac{-\varepsilon_a \varepsilon_c (\mu_a \mu_c)^2}{2\sqrt{6}\hbar\left(\varepsilon_c^2 - \varepsilon_a^2\right)R_{AC}^2}$$

$$= \frac{-\varepsilon_a \varepsilon_c (\mu_a \mu_c)^2}{6\sqrt{6}\hbar\left(\varepsilon_c^2 - \varepsilon_a^2\right)\rho^2} \tag{8.35}$$

The ε_c transition has CD equal in magnitude but opposite in sign from that of the transition at energy ε_a. The exact 'equal and opposite' is in fact seldom observed because in practice more than one transition on each chromophore takes part in the coupling. Experimental data may be found in *e.g.* reference [226].

Cyclodextrin inclusion compounds.

The CD of large formally rather complicated systems is sometimes as simple as that of much smaller ones, especially if the system has high symmetry. One such system is a cyclodextrin guest/host complex (Figure 8.13). Cyclodextrins are α-1,4-linked D-glucose oligomers. α-cyclodextrin has six glucose units, β-cyclodextrin seven, and so on.

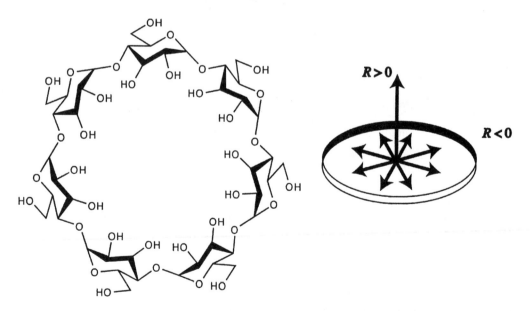

Figure 8.13 β-Cyclodextrin and the polarizations (indicated by the arrows) of the transitions of some guests inserted into its cavity and the signs of the resulting guest *ICD* signals.

Since, cyclodextrins have been found able to catalyse reactions involving their guests, it is of interest to know the orientation of the guest inside the cavity. The *CD* induced into the transitions of known polarization of the guest may tell us that immediately. Most *CD* experiments have been done with β-cyclodextrin since its cavity has just the right size to accommodate a single aromatic molecule such as benzene or naphthalene.

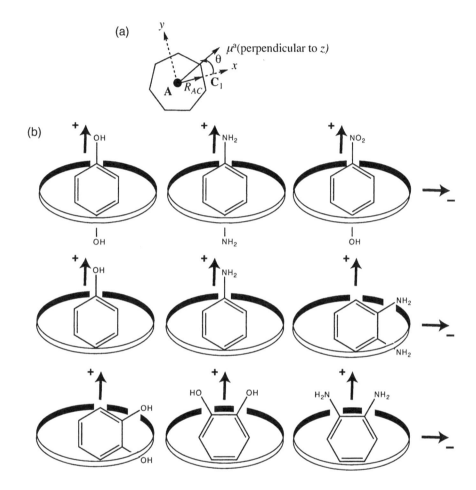

Figure 8.14 (a) Geometric parameters for a guest transition polarized perpendicular to z. (b) Some cyclodextrin-guest complexes showing the guest transition polarizations (thick arrow), the sign of the induced *CD*, and the orientation of the guest consistent with these data.

We shall make two physically reasonable assumptions about the geometry of the cyclodextrin host/guest system.

(1) One of the symmetry-determined axes of the guest molecule (chromophore **A**) aligns on average with the seven-fold rotation axis of β-cyclodextrin. This means that any guest transition is either polarized along z, the symmetry axis of the cyclodextrin (so $\alpha = 90°$), or is polarized perpendicular to z. In the latter case (see Figure 8.14), if $\alpha = \theta$ for the guest/glucose number 1 pair, then

The reduction of the *CD* intensity to 1/2 that of the z-polarized value if a guest transition is polarized in the x/y plane is the result of the averaging due to the symmetry of the cyclodextrin host.

$\alpha = \theta + (360/7)°$ for the second, and so on. A number of different transitions on each glucose will be important and their γ and τ values will all be different.

(2) The guest lies at the centre of the cyclodextrin, so

$$R_{AC} = R_{AC}\left(\cos\left(\frac{(n-1)360°}{7}\right), \sin\cos\left(\frac{(n-1)360°}{7}\right), 0\right) \qquad (8.36)$$

where n indexes the glucose units.

Let us consider the case of guests whose transitions are not degenerate with those of the cyclodextrins (the usual case since the cyclodextrin transitions start below 200 nm). The *CD* induced into a z-polarized transition by each glucose unit is the same, so the net *CD* induced by a β-cyclodextrin is seven times that of glucose number one, which is located with its origin along the x-axis. For a z-polarized guest transition equation (8.29) thus becomes

$$R(\varepsilon_a, z) = \frac{-7\varepsilon_a\left(\mu^a\right)^2}{\hbar R_{AC}^2}\left\{\sum_c \frac{\varepsilon_c\left(\mu^c\right)^2 \sin^2\gamma\sin\tau\cos\tau}{\left(\varepsilon_c^2 - \varepsilon_a^2\right)}\right\} \qquad (8.37)$$

where the factor of '7' accounts for the seven glucose units (*cf.* Table 7.1) and the summation is over all transitions on glucose unit number one.

If the guest lies oriented in the cavity so that its transition moment is in the x/y plane then the induced *CD* is half the magnitude of equation (8.37) (due to the symmetry-induced averaging, *cf.* §10.7) and of opposite sign (*cf.* Table 7.1).

The term in large brackets in equation (8.37) may be treated as a parameter of a specific glucose unit since for typical guest molecules (which are aromatic) $\varepsilon_c \gg \varepsilon_a$, so $\varepsilon_c^2 - \varepsilon_a^2 \approx \varepsilon_c^2$. Therefore to determine the orientation of a guest within the cyclodextrin cavity we need only to know the sign of the glucose parameter. This can be deduced empirically from the sign of the dextrin–induced *CD* of para disubstituted benzene molecules which will have their long axis lying along z for steric reasons. Some results are summarized in Figure 8.12 and Figure 8.14.[119]

8.9   *CD* of dimetallo helicates

In the final section of this chapter we consider the *CD* spectrum of the dimetallo helicate shown in Figure 8.15. As the conjugation of this class of chelate is broken at the central bridge, we can spectroscopically model the helicates as two *tris*-chelate metal complexes stacked vertically.

In-ligand *CD*

The *CD* arising from the in-ligand couplings of single metal *tris*-chelate complexes is given by equation (8.27). The equations for the bimetallo *tris*-chelate *CD* is therefore approximately twice that of equation (8.27).[227]

$$R\left(\frac{z}{x/y} \right) = R\left(\frac{A_2}{E} \right) = \pm \frac{\sqrt{2}\varepsilon\mu^2\rho}{\hbar}$$

$$\varepsilon\left(\frac{z}{x/y} \right) = \varepsilon\left(\frac{A_2}{E} \right) = \varepsilon \pm \frac{\mu^2}{12\sqrt{3}\rho^3}$$

(8.38)

Thus the A_2 band occurs at the higher energy (shorter wavelength) and is positive in sign for the P-enantiomer. The energy ordering is the same for the M-enantiomer, but the sign pattern is inverted. The exciton band found between 250 nm and 300 nm arises from the coupling of in-ligand bands with absorbance maximum at 275 nm. Equation (8.38) and the spectra of Figure 8.15 let us identify the handedness of enantiomers eluting from the cellulose column to be as indicated in the figure.

MLCT *CD*

The longer wavelength bands in the *CD* spectra of Figure 8.15 are less clearly excitonic in nature than the shorter wavelength in-ligand ones, arising as they do from a complicated overlay of different metal-ligand charge transfer (MLCT) and in-ligand bands. However, it is convenient to have an empirical rule based only on the longest wavelength *CD* band. To this end it is necessary to understand the longest wavelength MLCT band *CD*. The dilemma here is that the available data on transition polarizations of long wavelength charge transfer transitions for low spin iron(II) and ruthenium(II) complexes (including our own stretched film *LD* assignments of transition polarizations in Fe_2L_3)[228] show significantly more (>90%) E than A_2 intensity at the long wavelength end of the absorbance spectrum, but the *CD* spectra show similar magnitude positive and negative bands—which, assuming the *CD* signals arises from a similar mechanism such as that summarized in equations (8.33) and (8.34), requires both A_2 and E absorption intensities to be of similar magnitudes.

The dilemma is resolved by realising that the E-polarized MLCT transitions that occur from a metal *d* orbital to the short-axis directed ligand π^* orbitals will have significant absorbance intensity that involves only planar electron movement. Thus much of the absorption intensity comes from transitions that have zero *CD*. By way of contrast, MLCT transitions from the metal into long-axis directed π^* states will give rise to both E and A_2 MLCT absorbance intensity, both of which will have a helical twist and give rise to *CD* signals. The net effect of these factors will be that the absorbance spectrum is dominated by E-polarized transitions in regions where there are short-axis polarized MLCT bands, whereas the *CD* spectrum will have similar magnitude contributions from the two polarizations—in accord with experimental observation. Using the enantiomer assignment made from the 275 nm in-ligand band (for which we also know the energy ordering of components, see above), we conclude that the longest wavelength P (Δ,Δ) enantiomer MLCT *CD*, which is positive in sign, is due to an A_2 transition (equation (8.33)). The longest wavelength *CD* band for all such helicates is thus A_2 and can be used to assign the handedness of related helicates.[227]

By convention the right-handed helical form of a *tris*-chelate is denoted Δ, but the right-handed helical form of a di-metallo triple helicate is denoted P. To confuse matters, note that the helicity of *tris*-chelates is apparently reversed when one views the complex looking aong a two-fold (C_2) axis. Thus

$$P(C_3) = M(C_2).$$

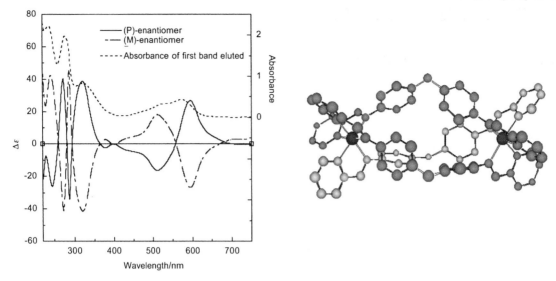

Figure 8.15 Absorbance of $[Fe_2(LL)_3]^{4+}$ (shown) and the *CD* spectra of the two compounds eluted from a cellulose column (mobile phase: 0.02 M NaCl) converted to $\Delta\varepsilon$ using concentrations determined from the absorbance magnitude and the extinction coefficient ($\varepsilon_{574\ nm}$ = 16,900 mol^{-1} cm^{-1} dm^3).[27, 88, 229] The spectrum labelled (M) is the enantiomer eluted first. Labels M and P were assigned on the basis of the *CD*.

9 Analysis of circular dichroism: magnetic dipole allowed transitions and magnetic *CD*

The CD equations for magnetic dipole allowed transitions form the major subject matter of this chapter. They are important for the analysis of optical activity of chiral ketones and transition metal complexes. The chapter concludes with a presentation of MCD, the CD induced into achiral molecules by applied magnetic fields.

Linear Dichroism and Circular Dichroism: A Textbook on Polarized-Light Spectroscopy
By Bengt Nordén, Alison Rodger and Timothy Dafforn
© B. Nordén, A. Rodger and T. Dafforn, 2010
Published by the Royal Society of Chemistry, www.rsc.org

9.1   Introduction

Most of this chapter is devoted to the *CD* of magnetic dipole allowed (mda)/electric dipole forbidden (edf) transitions. The final section includes a brief consideration of the *CD* induced into electric dipole allowed (eda)/magnetic dipole forbidden (mdf) transitions by a permanent magnetic field: so-called magnetic *CD* or *MCD*. The formalism used is consistent with that used throughout this book, though the origin of most of the theory presented is much older. The mechanisms giving rise to the *CD* of mda transitions and to *MCD* are quite different, but they naturally fit together since, in addition to their similar titles, both techniques often enable transitions whose existence is undetected or merely suspected in a normal absorption spectrum to be identified.

9.2   Magnetic dipole allowed transitions

The non-bonding *n* orbital of Figure 9.1 has been adapted to the C_{2v} symmetry of the carbonyl chromophore.

In §4.2 a magnetic dipole transition moment was defined to be an axial vector, $\boldsymbol{m}$, which interacts with the magnetic field of the electromagnetic radiation to cause a circling of electron density about the transition polarization direction. Examples of a transition which, to a first approximation, has an magnetic dipole transition moment but no electric dipole transition moment are the $n \rightarrow \pi^*$ transition of carbonyls and *d-d* transitions of metal complexes which are illustrated in Figure 9.1.

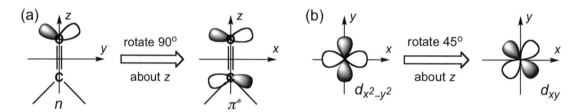

Figure 9.1 Schematic illustrations of magnetic dipole allowed (mda) transitions corresponding to rotation of charge. (a) $n \rightarrow \pi^*$ transition of the carbonyl chromophore. Note rotation of axis system. (b) A $d \rightarrow d$ transition.

The *n* orbital is essentially a combination of two *p*-orbitals.

There is no net linear motion of charge because any linear motion of charge along the symmetry defined axes of the system results in equal magnitude overlap of, for example, the positive lobe of the *n*-orbital with positive and negative lobes of the π^* orbital. So there is no net effect.

To understand mda transitions, consider what happens during the carbonyl $n \rightarrow \pi^*$ transition: electron density that is initially in the *n* orbital is *rotated* about the *z*-axis into the π^* orbital. As can be seen in Figure 9.1a, after rotation there is a net overlap of the positive lobe (white) of the *n* orbital with the positive lobe of the π^* orbital, and similarly for the negative (black) lobes. Thus the transition is mda. However, the transition is edf because there is no net linear motion of charge.

The *CD* of the $n \rightarrow \pi^*$ transition of carbonyl compounds has received more attention than that of any other type of transition. The biological importance of many molecules (including steroids) containing the carbonyl chromophore has justified this attention. However, the real reasons are probably more pragmatic. (i) Steroids were readily available in enantiomerically pure form; (ii) steroids have a *CD* signal that is comparatively easy to measure since the factor (the dissymmetry factor, equation (11.71))

$$\frac{\varepsilon_\ell - \varepsilon_r}{\varepsilon_\ell + \varepsilon_r} \qquad (9.1)$$

is comparatively large, making the *CD* an easily detectable percentage of the total absorbance; and (iii) fairly reliable empirical rules concerning the *CD* sign were deduced at an early stage for the carbonyl $n \to \pi^*$ transition making *CD* very useful. Other examples of mda transitions are some of the *d-d* transitions of metal complexes for much the same reasons, though producing enantiomerically pure compounds is more challenging. These two systems form the basis of this chapter.

9.3 $n \to \pi^*$ carbonyl transition and the octant rule

Long before any theoretical understanding of the carbonyl $n \to \pi^*$ transition *CD* had been achieved, it was found empirically that the *CD* of this transition has a sign determined by the factor[230]

$$-xyz \qquad (9.2)$$

where (x, y, z) (Figure 9.2) are the coordinates of the net perturbers in the rest of the carbonyl molecule. Each atom (or group of atoms) of the molecule induces a contribution to the observed *CD* whose sign is determined by the octant in which the atom lies. Any atom that has an equivalent 'partner' in a neighbouring octant has its contribution to the *CD* cancelled so it is not a net perturber and may be ignored.

A relationship such as equation (9.2) where the *CD* sign is related to whether a substituent is in one region of space or another is often referred to as a *sector rule*.

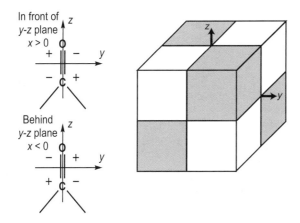

Figure 9.2 The octant rule.[230] Grey shading denotes negative octant-rule *ICD* and white denotes positive octant-rule *ICD*.

Hydrocarbon carbonyls and the octant rule

Let us first consider examples of carbonyl molecules where, apart from the carbonyl oxygen, all the atoms are carbons or hydrogens (*e.g.* Figure 9.3). By convention we ignore the hydrogens (which have small polarizabilities so can be considered as very weak perturbers) and consider only the carbons.

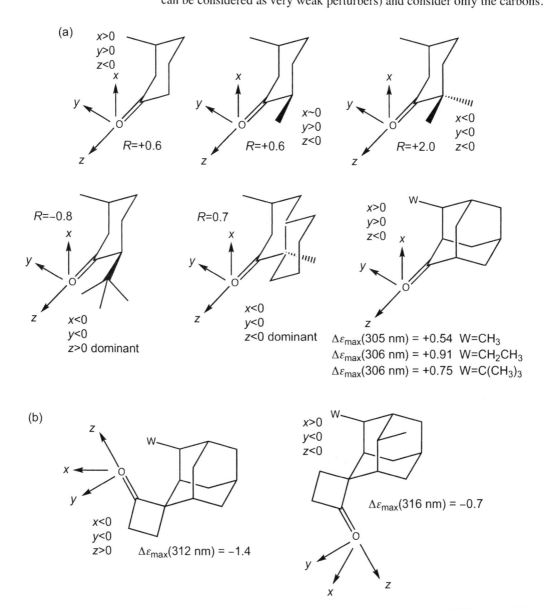

Figure 9.3 (a) Mainly 'rear octant' net-perturber carbonyls. Signs of x, y, and z are indicated for net perturbers. *CD* strengths (R) or $\Delta\varepsilon_{max}$ are given in each case. (b) Related front and rear octant net-perturber molecules.[217, 231, 232]

Building a molecular model or representing the molecule by its projection onto the plane perpendicular to the carbonyl bond (*i.e.* the x-y plane) enables the octant sign of any part of a molecule to be determined. Most molecules have all their atoms 'behind' the carbonyl with negative values for z (the so-called 'rear' octants). The molecules of Figure 9.3a are such rear octant molecules. An example of a 'front' octant molecule is illustrated in Figure

9.3b. The observed *CD* intensities, either as rotatory strength, R, or $\Delta\varepsilon_{max}$ are given in the Figure 9.3.

Heteroatomic carbonyls and the octant rule

A simple series of molecules that illustrates the applicability of the octant rule to heteroatomic carbonyls is provided by the equatorially substituted adamantanones of Figure 9.4. In this case, for W = CH_3, Cl, Br, and I as perturbers, the rotatory strength scales with the polarizability of the substituent, giving rotatory strengths of +2, +13, +26, and +43 respectively for the 4(*S*) isomers.[233] This indicates that the *CD* is related to an electronic interaction with the substituent atom rather than to some property related to the motions of the nuclei or steric effects of bulkiness of the substituent groups.

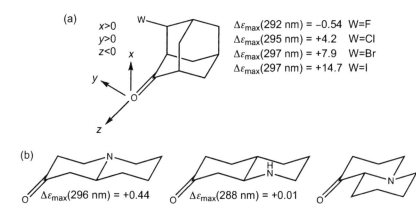

(a)

$x>0$
$y>0$
$z<0$

$\Delta\varepsilon_{max}$(292 nm) = −0.54 W=F
$\Delta\varepsilon_{max}$(295 nm) = +4.2   W=Cl
$\Delta\varepsilon_{max}$(297 nm) = +7.9   W=Br
$\Delta\varepsilon_{max}$(297 nm) = +14.7 W=I

(b)

$\Delta\varepsilon_{max}$(296 nm) = +0.44 $\Delta\varepsilon_{max}$(288 nm) = +0.01

Figure 9.4 (a) Equatorial adamantanones: carbonyl $n\rightarrow\pi^*$ CD intensities for (1*S*,3*R*)-4(*S*)(e)-haloadamadamantanones.[233] (b) Nitrogen containing 'anti-octant' carbonyl compounds. The *CD* maxima of the analogous compounds with N replaced by C are +1.35, +1.35 and +1.00 respectively.[234]

For W = F, the rotatory strength for the 4(*S*) equatorial adamantanone isomer is negative, suggesting, at first sight, a contravention of the octant rule. The same effect is noted when the substituent is deuterium.[235] A number of apparently 'anti-octant' nitrogen-containing compounds have been found (Figure 9.4b).[234] The key to this apparent break-down of the octant rule is provided by the fact that the magnitude of induced *CD* signal that a particular perturber provides scales with its polarizability. When there is a deuterium or fluorine substituent, there is usually a hydrogen in a mirror image position in a neighbouring octant. Since the polarizability of the C–H bond is larger than that of both the C–D and C–F bonds, the H contribution dominates and determines the *CD* sign rather than the D or F. Similarly, a nitrogen atom, although referred to as the substituent, contributes less to the *ICD* than a corresponding C in a mirror image position since the polarizability of C is larger than that of N. Careful inspection of the full molecular structure is thus required before application of the octant rule.

D is heavier than H. This causes C–D vibrational energy levels to be lower than those for C–H. This in turn means C–H is more affected by the anharmonicity of the potential energy surface, so its average bond length is longer than that of C–D, so its polarizability is greater.

Flexible molecules and the octant rule

In flexible carbonyls, such as substituted cyclohexanones, there is generally a distribution over many different conformations. If one conformation strongly predominates this may dominate the *CD*, although one should recall the possibility that there may be some less abundant conformer that, by having a very intense *CD* signal, could be important for the net result. A strongly temperature dependent *CD* is often due to slight shifts in equilibria between conformers having widely different *CD*s. Another source of temperature effects is rearrangement of solvent molecules (see below).

Solvent effects and the octant rule

β-axially substituted adamantanones (Figure 9.5) have sometimes been observed to give 'anti-octant' effects, especially if the substituent is bulky. Computer simulations have been used to show that in such cases the solvent molecules may be arranged so as to provide a net chiral effect from the solvent.[236-238] For example, a large β-axial substituent can cause a 'solvent deficit' in its octant relative to another octant. Such effects are generally temperature dependent, whereas the normal *CD* of conformationally rigid molecules is essentially temperature independent. Thus the octant contributions of the solvent molecules must be included in any rigorous treatment aiming to account for the observed *CD* spectrum.

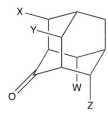

Figure 9.5 β-axial adamantanones indicating all axial substituent positions.

Perturbers near nodal planes of octants

If all the net perturber(s) lie on or very near one of the planes defining the octants, then *x* or *y* or *z* is small, so $-xyz$ is small. It has been observed in such a case that the sign of the *CD* does not always correlate with the octant rule. A number of published papers have endeavoured to show that such a situation requires the central nodal surface through the middle of the C=O bond to be curved. Opposite directions of curvature have been proposed.[239, 240] A simpler rationale for inconsistency between experiment and the octant rule follows naturally from the equations that show why the octant rule works so well most of the time. In this case, the assumptions required to derive the octant rule break down if $-xyz$ is small.

9.4 $n \rightarrow \pi^*$ carbonyl transition and the dynamic coupling model: justification of the octant rule

In Chapter 8 we saw that eda/mdf transitions required induced magnetic dipole transition moments in order to have a non-zero *CD* spectrum. An induced magnetic dipole transition moment can be gained by coupling of the electric dipole transition moment of the transition with that of another eda transition in a different chromophore. Similarly, in order for the $n{\to}\pi^*$ (or any other mda) transition of an achiral chromophore, **A**, to have a *CD* spectrum, it must acquire electric character in a direction parallel to that of its magnetic dipole transition moment. Since magnetic effects are orders of magnitude weaker than electric effects, the mechanism analogous to that adopted by eda transitions, namely where magnetic dipole transition moments on different chromophores couple, does not give rise to any detectable *CD*. Instead we need to look for magnetic–electric couplings.

The greatest magnitude electric dipole transition moment induced into an mda transition of a carbonyl chromophore results from the chromophore vibrating and becoming non-planar. Formally this is expressed as the coupling of the electronic magnetic dipole transition moment with vibrational motions: as **A** vibrates in a mode that removes a reflection plane of the carbonyl chromophore, the electron redistribution of the $n{\to}\pi^*$ transition gains some linear character. The normal absorption intensity of an mda transition is almost wholly due to this mechanism. However, on average, a vibration causes both left-handed and right-handed helices of electron motion, so no *net CD* signal arises due to vibronic coupling.

> The magnetic dipole transition moment is an integral over a product of two factors: the wave function describing the final state and the wave function describing the ground state after the magnetic dipole operator has acted on it. As the effect of the magnetic dipole operator is to rotate the ground state wave function, **m** is non-zero if the ground state can be rotated into a function that has net overlap with the excited state (*cf*. Figure 9.1).

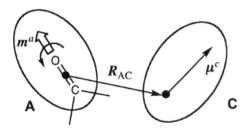

Figure 9.6 Geometry of a carbonyl **A/C** system (cf. Figure 8.3).

The largest induced electric dipole transition moment that has net chirality comes from the *dynamic coupling* of the magnetic dipole transition moment of the $n{\to}\pi^*$ transition, $\boldsymbol{m}^a$ in **A**, with electric dipole transition moments, $\boldsymbol{\mu}^c$, in chromophores, $\mathbf{C}_j$, that make up the chiral environment of **A**. The full equations are derived in Chapter 10. Here we take the final equations and express them with a slightly simpler notation. We express the *CD* strength for any mda transition in terms of an **A** optical factor, $f(\mathbf{A})$, and a **C** optical/geometry factor, $g(\mathbf{C})$, (which includes the energy terms):

$$R = f(\mathbf{A})g(\mathbf{C}) \tag{9.3}$$

> If there is more than one **C** chromophore then we introduce a sum over $\mathbf{C}_j$
> $$R = \sum_j f(\mathbf{A})g(\mathbf{C}_j)$$

Since the carbonyl chromophore has $\mathbf{C}_{2v}$ symmetry, the first non-zero term of the *CD* strength for the carbonyl $n{\to}\pi^*$ transition has (see §10.8)

$$f(\mathbf{A}) = \mathrm{Im}\!\left[Q^a_{xy} m^a_z\right] \tag{9.4}$$

> The quadrupole transition moment $\boldsymbol{Q}^a$ is a second rank tensor, it thus has two subscripts.

and a **C** optical/geometry factor:

$$g(\mathbf{C}) = \frac{-6\varepsilon_c}{\left(\varepsilon_c^2 - \varepsilon_a^2\right)R_{AC}^4}\,\mu_z^c\left[\mu_x^c\hat{\mathbf{R}}_{ACy} + \mu_y^c\hat{\mathbf{R}}_{ACx} - \sum_{k=x,y,z}5\mu_k^c\hat{\mathbf{R}}_{ACx}\hat{\mathbf{R}}_{ACy}\hat{\mathbf{R}}_{ACk}\right]$$

$$(9.5)$$

where ε_c is the transition energy for $\mathbf{C}$ transition moments $\boldsymbol{\mu}^c$, ε_a is the transition energy for the $n \rightarrow \pi^*$ transition which has magnetic dipole transition moment $\boldsymbol{m}^a$ and electric quadrupole transition moment $\boldsymbol{Q}^a$. Q_{xy}^a is the x-y component of $\boldsymbol{Q}^a$. $\boldsymbol{R}_{AC}$ is the vector from the origin of chromophore $\mathbf{A}$ to the origin of chromophore $\mathbf{C}$ as in Figure 8.3, and $\hat{\boldsymbol{R}}_{AC}$ is the unit vector along that line (Figure 9.6).

The definition of the chromophores of a carbonyl molecule

The achiral chromophore $\mathbf{A}$ for carbonyls is composed of the C=O bond and parts of the bonds to the two carbon atoms to which it is joined. The rest of the molecule is divided into the chromophores, $\mathbf{C}_j$, which are subunits of the molecule that contain electronic transitions (represented by transition moments or polarizabilities) and which do not exchange electrons with the rest of the molecule. For a molecule where (apart from the carbonyl group) all the valence electrons are in sigma bonds, the valence electron density of the molecule is, to a good approximation, localized in two-centre two-electron bonds between pairs of atoms. For such molecules taking the $\mathbf{C}_j$ to be bonds is a good approximation. Alternatively, the $\mathbf{C}_j$ are often defined to be atoms such as carbons and nitrogens. Although the bond approximation is the better one, in most cases the difference in xyz for the middle of a bond and one of the terminal atoms of the bond is not great.

If there are π-bonds or non-bonding electrons in a molecule, then the unit where the electrons are localized is often somewhat larger than a single bond between two atoms, *e.g.* if the molecule has a $-NH_2$, $-COOH$, or $-C_6H_5$ group, these will need to be treated as units. The correct identification of spectroscopic chromophores can usually be achieved by dividing the molecule into chemical functional groups.

The octant rule

To derive the octant rule from equations (9.3)–(9.5), we take the $\mathbf{C}_j$ to be the atoms of the molecule, take the origin to be in the middle of C=O, and replace R_{ACx} by x for each atom, R_{ACy} by y *etc.* The final term of equation (9.5) with $k=z$ is the isotropic polarizability component; it immediately gives the octant rule. The assumptions made in this derivation are discussed below and in §10.8.[241] The main one is that isotropic polarizabilities are much larger than anisotropic polarizabilities. Thus,

$$R \approx \frac{30\varepsilon_c}{\left(\varepsilon_c^2 - \varepsilon_a^2\right)R_{AC}^4}\,f(\mathbf{A})\mu_z^c\mu_z^c\left[\hat{\boldsymbol{R}}_{ACx}\hat{\boldsymbol{R}}_{ACy}\hat{\boldsymbol{R}}_{ACz}\right]$$

$$(9.6)$$

$$\approx \frac{30\varepsilon_c}{\left(\varepsilon_c^2 - \varepsilon_a^2\right)R_{AC}^4}\,f(\mathbf{A})\mu_z^c\mu_z^c\left[xyz\right]$$

or more simply

$$R \approx \frac{30\alpha_c(\varepsilon_a)}{R_{AC}^4} f(\mathbf{A})[xyz] \qquad (9.7)$$

where the dynamic polarizability of the $\mathbf{C}_j$ chromophores at energy ε_a is

$$\alpha_c(\varepsilon_a) = \frac{1}{3}\sum_c \frac{\varepsilon_c \boldsymbol{\mu}^c \cdot \boldsymbol{\mu}^c}{\varepsilon_c^2 - \varepsilon_a^2} \qquad (9.8)$$

As the carbonyl $n \rightarrow \pi^*$ transition has $\varepsilon_c > \varepsilon_a$, the octant rule requires $f(\mathbf{A})$ to be negative for carbonyl.

In applications of the octant rule, the groups for which the product xyz is determined are sometimes taken to be atoms, sometimes bonds, and sometimes functional groups as discussed above. The main assumptions made in going from equation (9.5) to equation (9.6) are that the diagonal components (*e.g.* $\mu_z^c \mu_z^c$) of the dynamic polarizability of each $\mathbf{C}_j$ are much larger than the off-diagonal components (*e.g.* $\mu_x^c \mu_z^c$) and that xyz is not close to zero. If xyz is close to zero, then one of the first two terms in the square brackets of equation (9.5) may be larger than the last term. This is one of the reasons why the octant rule breaks down when the net perturber lies close to one of the planes defining the octants (see above). We have further simplified equation (9.5) by using an average of the three diagonal polarizability terms of the $\mathbf{C}_j$ by assuming that $\mu^c \mu^c \approx \mu_z^c \mu_z^c$ etc.

> The *uv* component of the polarizability describes how the electrons move in the *v* direction if they are pushed in the *u* direction)

Two features of the full octant rule that are omitted from equation (9.2) follow from equation (9.7). The first is the already noted dependence on net perturber polarizability. The second is the R_{AC}^{-4} distance dependence of the CD magnitude. This enables us to assess the relative contributions of two or more net perturbers should they be present in a carbonyl molecule: the further away they are from the carbonyl the less effect they have.

Thus, to apply the octant rule in the revised form indicated by equation (9.7), first identify the chromophores, $\mathbf{C}_j$, into which the non-carbonyl part of the molecule is to be divided (usually sigma bonds and functional groups). Second, determine for each $\mathbf{C}_j$

- its isotropic polarizability[242]
- its distance from the carbonyl chromophore origin (which we take to be the centre of the C=O bond)
- the unit vector from the carbonyl origin to the $\mathbf{C}_j$ origin (this is (x,y,z)).

Equation (9.7) may then be evaluated for each $\mathbf{C}_j$. We assume the parameter $f(\mathbf{A})$ is negative and constant for all ketone carbonyls.

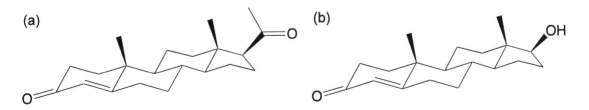

Figure 9.7 (a) Progesterone and (b) testosterone.

Electronic causes for failures of the octant rule

The octant rule may also break down when the carbonyl group cannot be regarded as an **A**-chromophore that is spectroscopically isolated from the rest of the molecule. Some steroids, such as progesterone and testosterone (Figure 9.7) are cases in point, since the carbonyl is conjugated with the carbon framework of the molecule.

Another example where the octant rule is not applicable is for molecules where substituents are linked to a carbonyl chromophore by a 'planar zig-zag' chain. Such substitutents often induce a larger *CD* signal than expected from evaluating equation (9.7). They also cause the transition energy to be red-shifted relative to the same substituent in other positions. The reason probably lies in the overlap of the *n* orbital with orbitals of the bonds of the zig-zag. The interaction between the *n* orbital and the σ antibonding orbitals of the zig-zag is illustrated in Figure 9.8a.[243] The zig-zag may be thought of as providing an enhanced communication of the substituent's polarizability-induced helical twisting of the $n{\to}\pi^*$ transition. Similarly, σ orbitals on Z_2, Z_4 (Figure 9.8a) interact with the π^* orbital of the carbonyl.

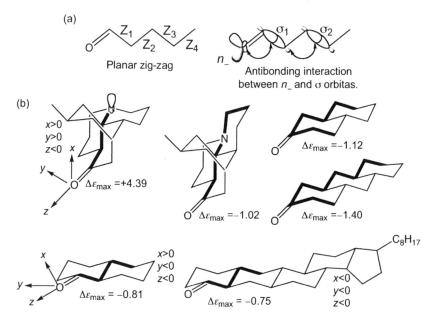

Figure 9.8 (a) Antibonding interactions of n_- and σ orbitals in planar zig-zag carbonyls.[243] (b) Planar zig-zag molecules (zig-zags shown by thick lines) where an enhanced intensity and red-shifted *CD* is observed. Non-zig-zag contributions are in some cases still dominant perturbers.[234,244]

9.5   *d-d* transitions of transition metal complexes: symmetry dependence of the *CD* of mda transitions

Another class of mda/edf transitions that have been extensively studied are the *d-d* transitions (Figure 9.1b) of transition metal complexes. These transitions usually occur at lower energies (longer wavelengths) than the charge transfer and in-ligand transitions considered in Chapter 8. They are therefore usually clearly identifiable in the spectrum even though the extinction coefficients of edf transitions are generally orders of magnitude smaller than those of eda transitions. In contrast to the situation for the metal complex eda transitions studied in Chapter 8, where we could describe the CD for any molecule in terms of degenerate and non-degenerate coupled-oscillator equations, with the mda/edf *d-d* transitions the form of the CD equation depends both on the symmetry of **A** and on the polarization of the intrinsic magnetic dipole transition moment of the mda transition.

One can take the *d-d* chromophore **A** to be the metal and directly ligating atoms. For *z*-polarized mda (the magnetic dipole transition moment lies along *z*) *d-d* transitions of metal complexes with C_{2v} **A** symmetry, the equations derived for the CD of the $n \rightarrow \pi^*$ carbonyl transition are appropriate, so such transitions follow the octant rule as summarized in equation (9.7). For *x*- and *y*- polarized mda *d-d* transitions of such C_{2v} systems, the first non-vanishing term in the CD expressions are respectively:

> We are considering mda/edf *d-d* transitions here. Some *d-d* transitions are mdf/edf. The latter are derived from T_{2g} symmetry transitions in octahedral metal complexes, whereas the mda transitions come from T_{1g} transitions.

$$R(x) = \frac{6\alpha_c(\varepsilon_a)}{R_{AC}^3} \operatorname{Im}\left[\mu_y^a m_x^a\right]\left[xy\right] \qquad (9.9)$$

$$R(x) = \frac{6\alpha_c(\varepsilon_a)}{R_{AC}^3} \operatorname{Im}\left[\mu_x^a m_y^a\right]\left[xy\right] \qquad (9.10)$$

The geometry dependence of equations (9.9) and (9.10) is a quadrant, *xy*, instead of an octant rule and the dependence on the **A**-**C** distance is inverse third rather than inverse fourth power (Figure 9.10). Thus the CD of an *x*- or *y*-polarized *d-d* transition in a C_{2v} metal complex is expected to be larger than that for a *z*-polarized transition.

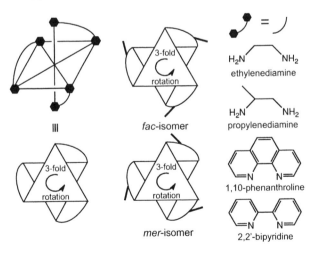

Figure 9.9 Δ-enantiomer *tris*-chelate transition metal complexes illustrating the geometry of the *fac* and *mer* isomers adopted by [Co(propylenediamine)₃]³⁺. The *lel* conformer of [Co(ethylenediamine)₃]³⁺ has the C–C bond of each ethylenediamine approximately parallel to the three-fold axis, whereas the *ob* conformers has them skewed to the three-fold axis.

Unfortunately, it is seldom possible to use the quadrant *versus* octant *CD* strength magnitude difference to help assign *d-d* transitions for C_{2v} systems. The reason for this lies in the nature of the transition and the high template symmetry of the molecules. The fact that we can usefully label these transitions as *d-d* transitions as such indicates that the transitions are between fairly pure *d* orbitals. Thus, as in the higher symmetry octahedral complexes such as $[Co(NH_3)_6]^{3+}$, in C_{2v} complexes *x-*, *y-*, and *z-* polarized transitions occur very close together in energy (they are degenerate for octahedral complexes) and so significant amount of cancellation occur between overlapping *CD* bands of opposite signs.

When we wish to consider higher symmetry molecules, such as the *tris*-chelate complexes of Figure 9.9 it proves necessary to go to higher order **A** moment products and much more complicated geometry dependences. Some possibilities are summarized in Figure 9.10. Here we shall focus on *tris*-chelate complexes, for which the *d-d* chromophore **A** has approximately D_{3d} symmetry. We therefore expect two transitions close together in energy: a *z*-polarized A_2 symmetry transition and a degenerate *x/y*-polarized E symmetry transition.

The isotropic polarizability terms of the *CD* equations for *d-d* transitions of *tris*-chelates are:[245]

$$R(E) = \frac{15\alpha_c(\varepsilon_a)}{4R_{AC}^4} \operatorname{Im}\left[m_x^a Q_{x^2}^a - m_x^a Q_{y^2}^a - 2m_y^a Q_{xy}^a\right]\left[x\left(x^2 - 3y^2\right)\right] \quad (9.11)$$

$$R(A_2) = \frac{35\alpha_c(\varepsilon_a)}{4R_{AC}^6} \operatorname{Im} m_z^a\left[H_{x^3z}^a - 3H_{xy^2z}^a\right]\left[x\left(x^2 - 3y^2\right)\left(9z^2 - 1\right)\right] \quad (9.12)$$

where $Q_{x^2}^a$ *etc.* and $H_{x^3z}^a$ *etc.* are, respectively, quadrupole and hexadecapole transition moments. The different distance dependence of these equations would lead us to expect the E band *CD* to be larger than the A_2 band. The 'dominant E band rule' was in fact first postulated[246] as an empirical rule based on the study of a significant number of systems; it is illustrated by the spectra in Figure 1.3 and Figure 9.11.

There has been some controversy about the dominant E band rule: when the *CD* spectrum of crystalline $[Co(ethylenediamine)_3]^{3+}$ was measured it was found that the crystal *CD* for the A_2 component of the T_{1g} *d-d* band had almost the same magnitude as that of the E band and both were an order of magnitude larger than those observed in the solution spectrum—so the solution spectrum appeared to result from the cancellation between two large A_2 and E bands. This raises the question of why the E band in solution spectra is always the larger if the observed solution *CD* is only a small percentage of the total crystal signal.

Such an apparent contradiction could arise because the conformation of the ethylenediamine rings in solution is different from that in the crystal. However, $[Co(propylenediamine)_3]^{3+}$, which is forced to adopt the crystal conformation (lel) in solution due to its extra methylene group (Figure 9.9), shows a qualitatively similar solution *CD* spectrum Figure 9.11 to that of $[Co(ethylenediamine)_3]^{3+}$ (Figure 1.3).

Thus we must look for an alternative explanation. As always, to resolve an

Under C_{2v} symmetry, transitions with *x*- and *y*-polarized magnetic dipole transition moments actually have electric dipole transition moments of, respectively, *y*- and *x*-polarization. However, the eda coupled-oscillator *CD* (Chapter 8) for *d-d* transitions is very small because their electric dipole transition moments are very small. This C_{2v} chromophore is a distorted octahedron and the electric dipole transition moments would be zero under the higher symmetry of the octahedral template.

apparent contradiction between a model and experiment we must examine the assumptions underlying the model. The first assumption made in deriving equations (9.11) and (9.12)[247] was that the *d-d* chromophore is achiral. As soon as one examines the crystal coordinates one realises that the crystalline *d-d* chromophore is not achiral—there is always an element of twist. This contribution will dominate the solid state spectrum where the A_2 and E components are measured separately, though they will be almost the same magnitude and so cancel in the overlapping solution phase spectrum where both A and E components are measured simultaneously. Thus in both solid and solution the *net CD*, when the A_2 and E bands are combined, is dominated by the mechanisms of equations (9.11) and (9.12).

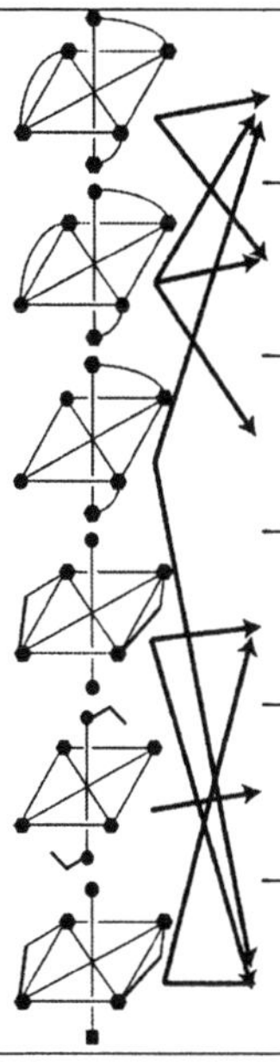

Molecule	A symmetry	R_{AC}^{-n}	Isotropic geometry factor
D_3	O_h	$n=6$	0
C_3	D_{3d}	E: $n=4$ A_2: $n=6$	x^3-3xy^2 $(x^2-y^2)(9z^2-1)$
C_2	C_{3v}	E: $n=3$ A_2: $n=5$	x^3-3xy^2 $(x^3-3xy^2)(9z^2-1)$
C_2	C_{4v}	E: $n=3$ A_2: $n=6$	xy $(x^2-y^2)xyz$
D_4	D_{4h}	E: $n=4$ A_2: $n=6$	$0, \Rightarrow n=6$ $(x^2-y^2)xyz$
C_2	C_{2v}	B: $n=3$ A_2: $n=4$	xy xyz

Figure 9.10 Distance dependence and geometry factors for the isotropic polarizability term in equations describing the *CD* induced into *d-d* chromophores of different symmetry. Circles, squares, and hexagons denote different ligating atoms.

This problem may be explained in terms of a difference between *additivity* and *coalescence* effects. Let us consider the hypothetical experiment of gradually changing the geometry of **A** so that the symmetry of the complex increases. It is intuitively clear that the magnitude of *CD* bands will continuously decrease with increasing symmetry until, when the system becomes achiral, the *CD* vanishes. Although symmetry changes are abrupt, the geometry changes involved can be thought of as continuous so there is a tendency to imagine bands belonging to transitions of different polarizations merging to form one band as transitions become degenerate. However, two distinct factors are operative, both of which may contribute to the large difference between crystal and solution *CD*.

We saw the effect of additivity when we considered the coupled-oscillator *CD* for the situation where the interaction energy, *V*, is zero (§8.3). That the octant rule does not contribute to *tris*-chelate *CD* is an example of coalescence.

- *Additivity* describes the cancellation of *CD* intensity observed when transitions occurring at different energies in low symmetry systems are degenerate in higher symmetry systems. The component band *CD* strengths, which derives from coupling with other transitions (in different chromophores), are the same for the low and high symmetry systems. In the $[Co(ethylenediamine)_3]^{3+}$ case, this is the twist contribution.
- *Coalescence* describes what occurs when increased symmetry causes *CD* mechanisms that contribute to the *CD* of lower symmetry systems to vanish, so the net *CD* decreases. This effect corresponds to the situation of an isolated chiral chromophore in which the *CD* of all transitions should sum to zero. To a reasonable approximation, a local sum rule also holds for the *CD* of, *e.g.*, the two transitions of a $\mathbf{D}_{3d}$ chromophore deriving from the T_{2g} band.

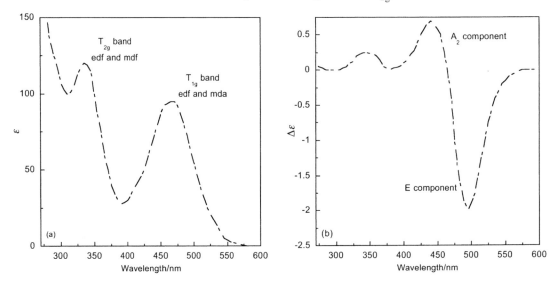

Figure 9.11 (a) Absorbance and (b) *CD* spectrum of the *d-d* transitions of Δ-[Co(R-propylenediamine)$_3$]$^{3+}$.

In deriving equations (9.11) and (9.12) for *tris*-chelate *CD* it was assumed that **A** has $\mathbf{D}_{3d}$ symmetry. This implicitly performs an additivity cancellation for the A_2 and E band *CD* intensity. In particular it cancels any *CD* due to the intrinsic chirality of the *d-d* chromophore that we have ignored. In solution this does not matter because the small energy splitting between the A_2 and E bands also performs this additivity cancellation. However, in measuring the single crystal *CD*, only the E band is measured so no automatic additivity cancellation occurs. In both instances, however, the *net CD* is dominated by the mechanisms of equations (9.11) and (9.12).

9.6   Magnetic circular dichroism

The final section of this chapter is devoted to magnetic circular dichroism, *MCD*. An *MCD* spectrum is measured by first placing the sample in a magnetic field that is aligned parallel with (by convention the direction of the magnetic field is north to south) the direction of propagation of light and then measuring the *CD* spectrum (Figure 9.12). Some permanent magnets are illustrated in Figure 9.13, though electromagnetics are more widely used.

In practice, most *MCD* signals are much smaller than intrinsic *CD* signals, so *MCD* is only ever measured for achiral molecules. *MCD* is most commonly used to see how many transitions are under one absorption envelope, since transitions of different polarizations often have differently signed *MCD* signals. *MCD* spectroscopy has never become a general chemical tool, probably because the standard approach to the theory looks very complicated.

MCD signal magnitudes are dependent on the magnitude of the magnetic field, but the large magnetic fields used for NMR and mass spectroscopy are not typically used for MCD.

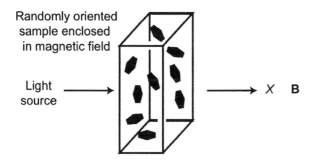

Figure 9.12 Configuration of an *MCD* experiment.

Our treatment of *MCD*,[248] which is derived in Chapter 10, is much simpler than that usually found in the literature[7, 249,250] since the latter approach traditionally explicitly includes the band shapes into derivations and calculations. The advantage of including band shapes from the beginning is that the final equations relate directly to the shapes of observed spectra, however, to a considerable extent this is at the expense of a simple physical understanding of the origin of the measured *CD*.

Our approach to deriving *MCD* equations is analogous to the coupled-oscillator analysis of Chapter 8. The main difference between the two situations is that, in the coupled-oscillator model the coupling is between an eda transition and the magnetic dipoles generated by other transitions on neighbouring chromophores, whereas with *MCD* the *CD* arises from the coupling between an eda transition and another eda transition on the same chromophore *via* the perturbation from the externally applied magnetic field.

Now consider a randomly oriented collection of molecules in a magnetic field of magnitude B that is directed along X, the direction of propagation of the incident radiation. The *MCD* of a non-degenerate transition with electric dipole transition moment $\boldsymbol{\mu}^a$ and transition energy ε_a is

$$MCD = \frac{4k\mathrm{BIm}\left(\boldsymbol{\mu}^a \times \boldsymbol{\mu}^b \cdot \boldsymbol{m}\right)}{3\left(\varepsilon_b - \varepsilon_a\right)} \qquad (9.13)$$

where k is a constant (*cf.* Chapter 10), Im denotes 'imaginary part of', $\boldsymbol{\mu}^b$ is a neighbouring transition occurring at energy ε_b, and $\boldsymbol{m}$ is the magnetic dipole transition moment between excited states from state $|a\rangle$ to state $|b\rangle$. For a non-zero *MCD* signal $\boldsymbol{\mu}^a$ and $\boldsymbol{\mu}^b$ must therefore not be parallel, as also seen from the vector product $\boldsymbol{\mu}^a \times \boldsymbol{\mu}^b$ in equation 9.13. If there is more than one state close in energy to the one for which we are measuring the *MCD*, then a sum over these states must be included in equation (9.13).

(a) (b)

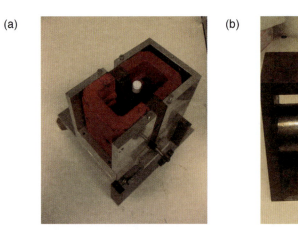

Figure 9.13 Permanent magnets with cell holders for measurement of magnetic circular dichroism (*MCD, cf.* §9.6), with (a) adjustable pole gap (B= 0.08-0.2 T) and (b) strong field (B = 1.2 T).[251]

The *MCD* for the neighbouring transition $\boldsymbol{\mu}^b$ follows from equation (9.13) upon doing three things, each of which changes the sign of the expression so the final result is an *MCD* signal of opposite sign from that of the $\boldsymbol{\mu}^a$:

- exchange $\boldsymbol{\mu}^a$ and $\boldsymbol{\mu}^b$
- exchange ε_a and ε_b
- reverse the direction of $\boldsymbol{m}$.

The last sign reversal follows from exchanging the wave functions in the expression for the magnetic transition moment (*cf.* §10.9).

To sum up, the *MCD* signal for a non-degenerate transition arises from coupling with another transition that is fairly close in energy and whose transition polarization is different. The *MCD* signal of the second transition (assuming coupling to all other transitions can be ignored) will be equal in magnitude, but opposite in sign from that of the first. Figure 9.14 shows the *MCD* spectrum for dibromothiophene, together with the corresponding resolved stretched film polarized spectra and deconvolved component spectra showing the presence of two orthogonal, polarized absorption components.

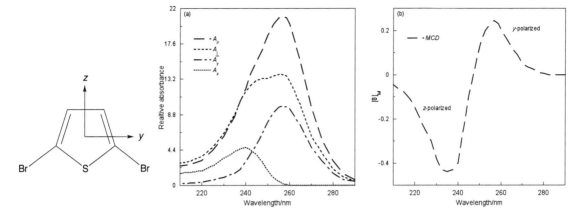

Figure 9.14 (a) Polarized absorbance spectra and (b) *MCD* spectrum of dibromothiophene showing opposite signed signals for neighbouring transitions of perpendicular polarization.[252]

If the transition of interest, with electric dipole transition moment $\boldsymbol{\mu}^a$ and energy ε, goes from a non-degenerate ground state to a degenerate excited state (of symmetry E or T), then there will be another transition, with electric dipole transition moment $\boldsymbol{\mu}^b$ occurring at the same energy. There will then be two *MCD* bands with respectively intensity

$$MCD(\pm) = \pm k\boldsymbol{\mu}^a \times \boldsymbol{\mu}^b \cdot \hat{\mathbf{B}} \qquad (9.14)$$

at energies

$$\varepsilon^{\pm} = \varepsilon \mp \mathrm{Im}(\mathbf{B}\cdot\mathbf{m}) \qquad (9.15)$$

In equation (9.14) the *MCD magnitude* does not formally depend on the magnitude of the magnetic field, but equation (9.15), which gives the energy splitting between the transitions, does. This is similar to the situation with the degenerate coupled-oscillator model, and an analogous exciton-like spectrum is therefore expected for such a situation. To a good approximation the cancellation of the overlapping oppositely signed bands will be inversely proportional to the energy splitting and therfore make the net *MCD* intensity scale linearly with the field strength. The energy splitting at normal field strengths is so small that only the *MCD* intensity not the shape of the emerging *MCD* spectrum is independent of field strength.

Although the underlying mechanisms of all *MCD* spectra are in principle the same, the equations for different cases have by convention been given specific labels: the *MCD* equation for a non-degenerate transition is known as the 'B term' while that for a transition to a degenerate excited state is the 'A term'. The situation with a degenerate ground state, where the magnetic field lifts the degeneracy and there is an asymmetric thermal population distribution, is known as the 'C term'.

Note: We defined $\hat{\mathbf{B}} = (1,0,0)_b$. The choice of sense for $\boldsymbol{\mu}^a$ and $\boldsymbol{\mu}^b$ then defines the sense of $\boldsymbol{m}$.

Extra sensitive measurement of *MCD* measurements may be made by exploiting a permanent magnet with a hatched pole-gap that produces strong but extremely heterogeneous fields. Despite this field heterogeity, the recorded *MCD* spectra are well resolved as a result of

$$MCD = \int C\ell\mathrm{B}(\tau)\Delta\varepsilon_\mathrm{B}d\tau$$
$$= \Delta\varepsilon_\mathrm{B}C\ell\int\mathrm{B}(\tau)\,d\tau$$

corrsponding to the *MCD* at the effective field <**B**>, with **B**(τ) the field in volume element τ, C the sample concentration, ℓ the path length and

$\Delta\varepsilon_\mathrm{B}/(\mathrm{mol}^{-1}\mathrm{cm}^{-1}\mathrm{dm}^3\mathrm{T}^{-1})$

the specific molar magnetic circular dichroism at unit field.[251]

10 Circular dichroism formalism

The derivation of a number of fundamental equations has been postponed until this chapter, including the CD equation of Rosenfeld, the coupled-oscillator CD and the MCD for sets of non-degenerate and non-parallel electric-dipole allowed transitions.

Linear Dichroism and Circular Dichroism: A Textbook on Polarized-Light Spectroscopy
By Bengt Nordén, Alison Rodger and Timothy Dafforn
© B. Nordén, A. Rodger and T. Dafforn, 2010
Published by the Royal Society of Chemistry, www.rsc.org

10.1   Introduction

This chapter is designed to enable readers to understand the underlying basis of the equations used in previous chapters and to enable the *CD* theory literature to be read. Much of the interpretation of the final equations has already been covered in the previous chapters. This chapter therefore has very few pictures, comparatively little discussion, but many equations.

The level of mathematical sophistication required for this chapter is comparatively low, but the necessity of labelling transition moments and often transition energies with the states they connect and the chromophore to which they belong makes many of the equations appear complicated. The mathematics required includes: a basic acceptance of quantum theory (*i.e.* molecules have wave functions that describe their electron distribution), the use of *bra-ket* notation, vector products, trigonometry, complex numbers, and simple determinants. Brief textual and marginal notes endeavour to provide sufficient information on these topics to remind readers who have previously covered them but have forgotten the details. For readers who find these notes too brief, most physical chemistry textbooks or mathematics-for-chemists books will give more details.

The chapter is structured so the next three sections cover general aspects of the interaction of polarized radiation and matter. The emphasis of the final sections of the chapter is on determining wave functions and hence *CD* expressions for a transition in chromophore **A** when perturbed by the rest of the molecule or by an external magnetic field. Each type of transition gains the helical character required for a *CD* signal by perturbation from another part of the system. In §10.6 and §10.7 the focus is on transitions that to a first approximation only have linear charge displacements (*i.e.* eda and mdf transitions). Circular charge displacements are covered in §10.8. In the final section we shall see the effect of an external static magnetic field on eda transitions of achiral molecules when we derive the *MCD* equations quoted in §9.6.

10.2   Beer-Lambert law

The Beer-Lambert law (equation (1.5)) is arguably the single most useful spectroscopy equation. It follows from considering what happens to a beam of photons propagating along the X direction as it passes through a uniform sample of absorbing molecules. If light of intensity $I(X)$ enters a slice of the sample at position X then the intensity $I(X+\delta X)$ after an infinitesimal distance δX is

$$I(X + \delta X) = I(X) - (\text{number of molecules hit by a photon}) \times \Pi \quad (10.1)$$

where Π is the probability that a photon that hits a molecule is absorbed. Now

$$\text{number of molecules hit by a photon} = I(X)\frac{N_{slice} \times \sigma_{molecule}}{\text{volume of slice}}\delta X \quad (10.2)$$

$$= I(X)C\sigma_{molar}\delta X$$

where N_{slice} is the number of molecules in the slice of the sample of width δX, $\sigma_{molecule}$ is the cross-sectional area of the molecules facing the incident light beam and σ_{molar} is the corresponding molar quantity, and C is concentration of the analyte. Thus,

$$\frac{I(X+\delta X) - I(X)}{\delta X} = -I(X)C\sigma_{molar}\Pi \quad (10.3)$$

In the limit of small δX, this becomes

$$\frac{dI(X)}{dX} = -I(X)C\sigma_{molar}\Pi = -I(X)\varepsilon C \quad (10.4)$$

where ε is the wavelength dependent extinction coefficient. From this it follows that

$$\frac{dI(X)}{I(X)} = -\varepsilon C dX \quad (10.5)$$

Upon integrating across a sample of path length ℓ this gives the Beer-Lambert law.

As discussed in Chapter 2, the Beer Lambert law is not valid if too few photons are emitted from the sample for the photomultiplier tube to accurately count them. It also breaks down when there are too many molecules in some parts of the sample and too few in other parts, or if there are concentration dependent intermolecular interactions that affect the spectroscopy.

10.3   Polarized light and spectropolarimeters

Polarized light

As discussed in §1.2, linearly polarized light beams have all photons with their electric fields oscillating in the same plane (Figure 1.2). When two linearly polarized beams of different polarization but propagating in the same direction are combined (Figure 10.1), this will only result in a new linearly polarized light beam if the phases (and wavelengths) of both beams are the same—*i.e.* if both light beams have zero amplitude at the same points in space. If they are out of phase then elliptically polarized light results. If we take X to be the direction of propagation of the light, then the unit vectors describing the polarization (by convention the direction of the electric field component of the light) of two in-phase perpendicularly polarized light beams may be written[3,6]

$$\hat{\mathbf{e}}_1 = \hat{\mathbf{e}}_Y = (0,1,0) \quad (10.6)$$

and

$$\hat{\mathbf{e}}_2 = \hat{\mathbf{e}}_Z = (0,0,1) \quad (10.7)$$

Throughout this book $\{X,Y,Z\}$ are the laboratory fixed axes systems; $\{x,y,z\}$ are the molecule fixed axis systems.

In the special case where two equal magnitude linearly polarized beams of orthogonal polarizations are combined with a phase difference of $\pi/2$ (quarter of a wavelength) we get circularly polarized light. The electric field polarization vector for left circularly polarized light may be written:[3,6]

$$\hat{\mathbf{e}}_\ell = \frac{1}{\sqrt{2}}\left\{(0,1,-i)\exp\left(\frac{2\pi i X}{\lambda} - i\omega t\right)\right\}\tag{10.8}$$

where i is the square root of -1. The oscillation of the electric field vector in time is described by the factor $\exp(-i\omega t)$ where ω is the angular frequency ($2\pi\nu$) and t is time. The magnitude of the electric field vector remains constant in time, but its direction rotates about X with frequency ω. By taking the real part of equation (10.8) we see that for an observer viewing oncoming radiation, the electric field vector of left circularly polarized light rotates about X in an anticlockwise manner. In what follows we use the polarization vectors

$$\hat{\mathbf{e}}_\ell = \frac{1}{\sqrt{2}}(0,1,-i)$$

$$\hat{\mathbf{e}}_r = \frac{1}{\sqrt{2}}(0,1,i)\tag{10.9}$$

In using this definition of left (and hence right) circularly polarized light we are following the optics convention.[3,6] For an observer facing an oncoming left circularly polarized light beam, the rotation is anti-clockwise, forming a right handed helix in space and a left handed helix in time.

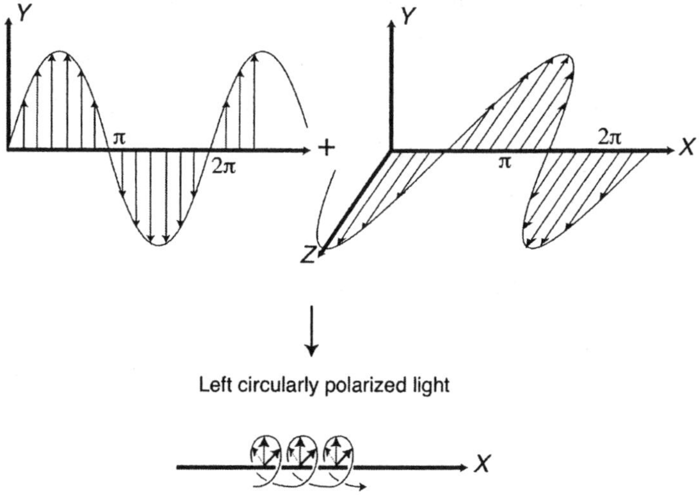

Figure 10.1 Production of circularly polarized light from two linearly polarized light beams. Arrows indicate electric field polarizations.

Circular dichroism spectropolarimeter

As *CD* is the difference in absorption of left and right circularly polarized light, the key feature of a *CD* spectropolarimeter is a means of producing both polarizations of light with exactly equal intensities. The light source in most instruments is a xenon arc lamp for UV and visible *CD* measurements.

The optics of a typical *CD* machine are illustrated in Figure 10.2. A series of mirrors and prisms and slits are used to produce (relatively) collimated monochromatic radiation (in reality, it is not monochromatic but has a well-defined wavelength range, typically of 0.5–2 nm, as discussed in §2.4). This light is then linearly polarized and subsequently circularly polarized. The conversion of linearly polarized light into circularly polarized light is achieved by the photoelastic modulator, PEM.

The PEM consists of a piece of crystalline quartz mechanically coupled (glued) to a piece of isotropic (silica) quartz, the light passing through the latter. If the incident linearly polarized light is, say, horizontally polarized, and the long axis of the PEM is oriented at 45°, then the light beam may be considered as split into equal magnitude long and short axis components (Figure 10.3). The PEM exhibits birefringence so that $(n_{\text{long}} - n_{\text{short}}) \neq 0$, *i.e.* the refractive indices n_{long} and n_{short} for orthogonal polarizations of light are different. During passage through the PEM the phase difference between the two light beam components will amount to:

$$\delta = \frac{2\pi D}{\lambda}$$
$$= \frac{2\pi d\left(n_{long} - n_{short}\right)}{\lambda}$$

(10.10)

where d is the thickness of the PEM, λ is the wavelength of the light, and D is the path difference between the light components. At a certain value of $(n_{\text{long}} - n_{\text{short}})$, δ will be equal to $\pi/2$ and D equal to $\lambda/4$, making the so-called 'quarter-wave' plate required to produce circularly polarized light.

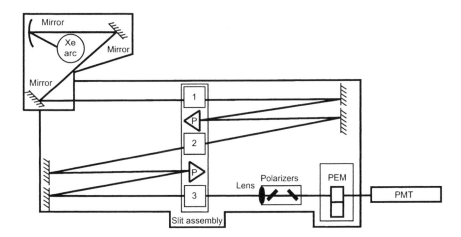

Figure 10.2 Schematic diagram of the optics of a *CD* spectropolarimeter. In this instrument the monochromator prisms (P) are also the polarizers. To perfect the polarization a series of tilted fused-silica plates are inserted in the light path preceding the PEM. Sample and photomultiplier tube are after the PEM. The xenon source means that the instrument has most sensitivity in the 300–400 nm region of the spectrum. A decrease in lamp efficiency occurs with time and can be observed as a decreased light intensity below 200 nm and above 700 nm.

An AC voltage is applied to the crystalline part of the PEM, causing the whole PEM assembly to oscillate (at 50 kHz). By adjusting the voltage amplitude so that the birefringence amplitude corresponds to the quarter-wave condition at each λ, the time dependence of ($n_{\text{long}}-n_{\text{short}}$) as it oscillates between $+|n_{\text{long}}-n_{\text{short}}|$ and $-|n_{\text{long}}-n_{\text{short}}|$ gives light that is alternately left and right circularly polarized. The polarized light thus produced then passes through the sample compartment (and the sample) and what is not absorbed is detected by the photomultiplier tube.

If the sample has no *CD*, the photomultiplier current will be constant (direct current, DC, only) with a magnitude determined by the normal absorbance of the sample. If the sample exhibits *CD* the photomultiplier current will also show an oscillating component (alternating current, AC). The *CD* is obtained as the ratio between the AC and DC components. Its sign is determined from the phase of the AC component using a lock-in amplifier that has the AC voltage of the PEM as a time reference.

A factor of 2 increase in voltage across the PEM produces alternating polarizations of linearly polarized light rather than alternating left and right circularly polarized light. So the PEM becomes a half wave plate and produces alternating pulses of orthogonal beams of linearly polarized light for *LD*. Since the linear polarization of light hitting the PEM will appear twice (when birefringence crosses the zero line) during the oscillation period, the lock-in amplifier has to be tuned to look for *LD* at twice the oscillation frequency of the PEM. An alternative that avoids the extra stress on the PEM is to insert an achromatic quarter-wave plate in the light beam (§2.2).[8,253]

> In all commercial *CD* instruments, instead of forming the <AC>/DC ratio, the DC current is kept constant by a servo adjusting the PM voltage: in this way the *CD* becomes proportional to the recorded <AC> signal (*cf.* §2.6).

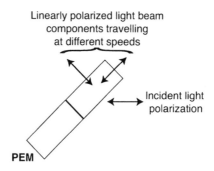

Linearly polarized light beam components travelling at different speeds

Incident light polarization

PEM

Figure 10.3 Schematic of a photoelastic modulator.[254]

10.4 Interaction of radiation with matter

Using classical mechanics, the energy of the interaction, H^{int}, between the electric, **E,** and magnetic, **B,** fields of electromagnetic radiation and the molecule is given by

$$H^{\text{int}} = -\boldsymbol{\mu} \cdot \mathbf{E} - \boldsymbol{m} \cdot \mathbf{B} + \text{higher order multipole terms} \qquad (10.11)$$

where $\boldsymbol{\mu}$ is the electric dipole moment of the molecule and $\boldsymbol{m}$ is its magnetic dipole moment. $\boldsymbol{\mu}$ and $\boldsymbol{m}$ are determined by summing over the electron distribution of the molecule:

> The higher order multipole terms include the electric quadrupole term:[6]
>
> $$-Q_{ij}\nabla_i E_j$$

$$\boldsymbol{\mu} = e\sum_j \boldsymbol{r}_j$$

$$\boldsymbol{m} = \frac{e}{2m_e}\sum_j \{\boldsymbol{r}_j \times \boldsymbol{p}_j\}$$

(10.12)

where e is the (negative) unit charge on an electron, $\boldsymbol{r}_j$ denotes the position vector of the jth electron, $\boldsymbol{p}_j$ is the momentum vector for the jth electron, and m_e is the mass of an electron.

m is a circulation of charge about an axis, hence the appearance of a cross product (*cf.* §11.1) in equation (10.12).

When we are interested in spectroscopic phenomena such as absorbance, *CD*, and *LD*, we have to move beyond classical mechanics to quantum mechanics since the transitions we study involve discrete amounts of energy—*i.e.* the energy is quantized. Thus, instead of an expression for the energy of the interaction between the radiation and the molecule, we need the interaction Hamiltonian. This follows from equation (10.11) by converting $\boldsymbol{\mu}$ and $\boldsymbol{m}$ to operators and the interaction energy to the Hamiltonian operator. Operators are denoted by using tilde 'hats' to indicate operators for the whole system:

$$\tilde{H}^{\text{int}} = -\tilde{\boldsymbol{\mu}}\cdot\mathbf{E} - \tilde{\boldsymbol{m}}\cdot\mathbf{B} + \text{higher order multipole terms} \qquad (10.13)$$

When we compare the relative strengths of different transitions experimentally we measure the intensity of the transition, which is the area under the absorption band. The related theoretical quantity is the probability per unit time, P, that a transition will occur from the initial state $|i\rangle$ (which is usually the ground state) to the final state $|f\rangle$ [6,17]

With the notation known as bra-ket notation, the 'bra' $\langle j|$ state is the complex conjugate of the 'ket' $|j\rangle$, and when they are coupled together either directly or on either side of an operator, such as $\tilde{H}^{\text{int}}$ in equation (10.14), then we imply that an integral over all space is being performed. We shall use round brackets $|j)$ for the unperturbed states of chromophores **A** and **C** (see below), and pointed ones $|j\rangle$ for the states of the whole system.

$$P(|i\rangle \rightarrow |f\rangle) = k'\left|\langle f|\tilde{H}^{\text{int}}|i\rangle\right|^2$$

$$= k'\langle f|\tilde{H}^{\text{int}}|i\rangle^*\langle f|\tilde{H}^{\text{int}}|i\rangle \qquad (10.14)$$

$$= k'\langle i|\tilde{H}^{\text{int}}|f\rangle\langle f|\tilde{H}^{\text{int}}|i\rangle$$

where the asterisk denotes the complex conjugate, the integral is over molecular position coordinates, and k' is a constant. Equation (10.14) is the Fermi Golden Rule. Upon substituting equation (10.13) into equation (10.14)

$$P(|i\rangle \rightarrow |f\rangle) = k\{\langle i|\tilde{\boldsymbol{\mu}}|f\rangle\cdot\hat{\mathbf{e}}^* + \langle i|\tilde{\boldsymbol{m}}|f\rangle\cdot\hat{\mathbf{b}}^*\}\{\langle f|\tilde{\boldsymbol{\mu}}|i\rangle\cdot\hat{\mathbf{e}} + \langle f|\tilde{\boldsymbol{m}}|i\rangle\cdot\hat{\mathbf{b}}\}$$

(10.15)

where $\hat{\mathbf{b}}$ is the unit vector along $\mathbf{B}$, $|i\rangle^* = \langle i|$ and the magnitudes of the electric and magnetic fields are absorbed into k. $\langle i|\tilde{\boldsymbol{\mu}}|f\rangle$ is the electric dipole transition moment *from* state i *to* state f, and $\langle f|\tilde{\boldsymbol{m}}|i\rangle$ is the corresponding magnetic dipole transition moment. Upon expansion and simplification, it follows that

Here we are assuming that we only need consider dipolar contributions to $\tilde{H}^{\text{int}}$. In some situations, however, for example when measuring *CD* of oriented samples, the quadrupolar terms may be as important as the electric dipole-magnetic dipole terms.

$$P(|i\rangle \rightarrow |f\rangle) = k\{\boldsymbol{\mu}^{fi}\cdot\hat{\mathbf{e}}^*\boldsymbol{\mu}^{if}\cdot\hat{\mathbf{e}} + \mathbf{m}^{fi}\cdot\hat{\mathbf{b}}^*\mathbf{m}^{if}\cdot\hat{\mathbf{b}}\}_I$$

$$+ k\{\boldsymbol{\mu}^{fi}\cdot\hat{\mathbf{e}}^*\mathbf{m}^{if}\cdot\hat{\mathbf{b}} + \boldsymbol{\mu}^{if}\cdot\hat{\mathbf{e}}\,\mathbf{m}^{fi}\cdot\hat{\mathbf{b}}^*\}_{II}$$

(10.16)

Note inversion of order of states in the notation.

where $\boldsymbol{\mu}^{if} = \boldsymbol{\mu}^{i\rightarrow f} = \langle f|\tilde{\boldsymbol{\mu}}|i\rangle$ etc.

In the next subsections we consider a number of particular cases where a molecule interacts with radiation of different polarizations.

Interaction of a molecule with linearly polarized light: A_{iso} and LD

If no external magnetic field is present (except that due to the radiation) we can choose $|i\rangle$ and $|f\rangle$ to be real wavefunctions; thus $\boldsymbol{\mu}^{if} = \boldsymbol{\mu}^{fi}$ but $\boldsymbol{m}^{if} = -\boldsymbol{m}^{fi}$ since $\hat{\boldsymbol{m}}$ is imaginary. The polarization vectors for **E**, and hence **B**, are also real for linearly polarized light. It follows that the two parts of term *II* in equation (10.16) are equal in magnitude but opposite in sign for linearly polarized light, so term *II* is zero. As magnetic terms are invariably small compared with electric ones, we can ignore the second term within term *I* and write the probability of the transition occurring from $|i\rangle$ to $|f\rangle$ with linearly polarized light to be:

See §11.5 for the relationships between these theoretical equations and the experimental quantities.

$$P\big(|i\rangle \rightarrow |f\rangle\big) = A_{\text{linearly polarised}}\big(|i\rangle \rightarrow |f\rangle\big) = k\big|\boldsymbol{\mu}^{if} \cdot \hat{\mathbf{e}}\big|^2 \qquad (10.17)$$

Isotropic absorption of linearly polarized light

For the usual situation where the system is a collection of randomly oriented randomly placed molecules, equation (10.17) becomes:

$$A_{iso} = A = \frac{k}{3}\big|\boldsymbol{\mu}^{if}\big|^2 \qquad (10.18)$$

Linear dichroism

Linear dichroism *(LD)* was defined in Chapter 1 to be the difference in absorption of two linearly polarized light beams propagating in the same direction with perpendicular polarizations. Let the parallel direction in the laboratory-fixed $\{X,Y,Z\}$ be $(0,0,1)$ and the perpendicular direction be $(0,1,0)$. Then from equation (10.17)

$$LD_{molecule} = \big(A_{//} - A_{\perp}\big)_{molecule}$$
$$= k\left\{\big|\mu_Z^{if}\big|^2 - \big|\mu_Y^{if}\big|^2\right\} \qquad (10.19)$$

where μ_Y^{if} is the Y component of the $|i\rangle$ to $|f\rangle$ transition dipole moment, *etc*. For a collection of N molecules we sum over the LD for each molecule

$$LD_{total} = \sum_{molecules}\big(A_{//} - A_{\perp}\big)_{total}$$
$$= Nk\left\{\big\langle\big|\mu_Z^{if}\big|^2\big\rangle - \big\langle\big|\mu_Y^{if}\big|^2\big\rangle\right\} \qquad (10.20)$$

where $\langle\ \rangle$ denotes average. The direction of Z within the molecule-fixed axis systems is different for different molecules since the molecules are seldom perfectly aligned. The net LD thus depends on the method and extent of sample orientation. In the absence of any molecular orientation the LD in equation (10.20) vanishes because

$$\big\langle\big|\mu_Z^{if}\big|^2\big\rangle = \big\langle\big|\mu_Y^{if}\big|^2\big\rangle = \frac{1}{3}\big|\boldsymbol{\mu}^{if}\big|^2 \qquad (10.21)$$

Interaction of a molecule with circularly polarized light: *CD*

CD is the difference in absorption of left and right circularly polarized light by a molecule. As for *LD*, with *CD* there is no external magnetic field present so we can again choose to use real wave functions so $\boldsymbol{\mu}^{if} = \boldsymbol{\mu}^{fi}$ and $\boldsymbol{m}^{if} = -\boldsymbol{m}^{fi}$. To evaluate equation (1.2) we substitute first the expression for left circularly polarized light into equation (10.16) then subtract from it the result of substituting the expression for right circularly polarized light. We also make use of the equality $\hat{\mathbf{e}}_r = \hat{\mathbf{e}}_\ell^*$ (equation (10.9)) and note that (§1.2)

$$\hat{\mathbf{b}} = \mathbf{k} \times \hat{\mathbf{e}} \tag{10.22}$$

so

$$\hat{\mathbf{b}}_\ell = \frac{1}{\sqrt{2}}(1,0,0) \times (0,1,-i)$$

$$= \frac{1}{\sqrt{2}}(0,i,1) = \hat{\mathbf{b}}_r^* \tag{10.23}$$

Thus from equation (10.16),

$$CD = A_\ell - A_r$$

$$= k \left[\begin{array}{l} \left(\boldsymbol{\mu}^{fi}.\hat{\mathbf{e}}_\ell^* \boldsymbol{\mu}^{if}.\hat{\mathbf{e}}_\ell + \boldsymbol{m}^{fi} \cdot \hat{\mathbf{b}}_\ell^* \boldsymbol{m}^{if} \cdot \hat{\mathbf{b}}_\ell \right) \\ -\left(\boldsymbol{\mu}^{fi}.\hat{\mathbf{e}}_\ell \boldsymbol{\mu}^{if}.\hat{\mathbf{e}}_\ell^* + \boldsymbol{m}^{fi} \cdot \hat{\mathbf{b}}_\ell \boldsymbol{m}^{if} \cdot \hat{\mathbf{b}}_\ell^* \right) \end{array} \right]_I \tag{10.24}$$

$$+ k \left[\begin{array}{l} \left(\boldsymbol{\mu}^{fi}.\hat{\mathbf{e}}_\ell^* \boldsymbol{m}^{if}.\hat{\mathbf{b}}_\ell + \boldsymbol{\mu}^{if} \cdot \hat{\mathbf{e}}_\ell \boldsymbol{m}^{fi} \cdot \hat{\mathbf{b}}_\ell^* \right) \\ -\left(\boldsymbol{\mu}^{fi}.\hat{\mathbf{e}}_\ell \boldsymbol{m}^{if}.\hat{\mathbf{b}}_\ell^* + \boldsymbol{\mu}^{if} \cdot \hat{\mathbf{e}}_\ell^* \boldsymbol{m}^{fi} \cdot \hat{\mathbf{b}}_\ell \right) \end{array} \right]_{II}$$

Because $\boldsymbol{\mu}^{fi}.\hat{\mathbf{e}}_\ell^* \boldsymbol{\mu}^{if}.\hat{\mathbf{e}}_\ell = \boldsymbol{\mu}^{fi}.\hat{\mathbf{e}}_\ell \boldsymbol{\mu}^{if}.\hat{\mathbf{e}}_\ell^*$ *etc.* (the wave functions are real), the first term (*I*) vanishes. Upon rearranging term (*II*) so that the electric dipole transition moments are all from $|f\rangle$ to $|i\rangle$ and the magnetic moments are the reverse, equation (10.24) then becomes

$$CD = A_\ell - A_r$$

$$= 2k \left(\boldsymbol{\mu}^{fi}.\hat{\mathbf{e}}_\ell \boldsymbol{m}^{if} \cdot \hat{\mathbf{b}}_\ell - \boldsymbol{\mu}^{fi}.\hat{\mathbf{e}}_\ell \boldsymbol{m}^{if} \cdot \hat{\mathbf{b}}_\ell^* \right) \tag{10.25}$$

Upon substituting explicit forms for the electric (equation (10.9)) and magnetic field polarizations for the circularly polarized radiation we get

$$CD = 2ik \left(\mu_Y^{fi} m_Y^{if} + \mu_Z^{fi} m_Z^{if} \right)$$

$$= 2k \, \mathrm{Im} \left(\mu_Y^{fi} m_Y^{if} + \mu_Z^{fi} m_Z^{if} \right) \tag{10.26}$$

where 'Im' denotes imaginary part of what follows.

CD of a collection of randomly oriented molecules

CD experiments are usually performed on collections of randomly oriented molecules (*i.e.* solutions). In that case the laboratory (radiation)-defined axis system does not relate in any fixed way to a molecular axis system, so what is measured is a rotational average of equation (10.26):

$$CD = \frac{4}{3} k \, \text{Im}\left[\boldsymbol{\mu}^{fi} \cdot \boldsymbol{m}^{if} \right] = \frac{4}{3} k \langle i | \tilde{\boldsymbol{\mu}} | f \rangle \cdot \langle f | \tilde{m} | i \rangle \qquad (10.27)$$

Interaction of a molecule with circularly polarized light in the presence of an external static magnetic field: magnetic circular dichroism (*MCD*)

In §9.6 we looked at the *CD* signal, the so-called magnetic *CD*, that can be measured for achiral molecules when an external static magnetic field gives eda transitions some helical character. In deriving the equations for such a situation where an external magnetic field is imposed, the wave functions can no longer be taken to be real. Thus $\boldsymbol{\mu}^{fi} . \hat{\boldsymbol{e}}_\ell^* \boldsymbol{\mu}^{if} . \hat{\boldsymbol{e}}_\ell$ may differ from $\boldsymbol{\mu}^{fi} . \hat{\boldsymbol{e}}_\ell \boldsymbol{\mu}^{if} . \hat{\boldsymbol{e}}_\ell^*$ and term (*I*) in equation (10.24) does not vanish. However, if, as in Chapter 9, we limit consideration to achiral molecules then term (*II*) vanishes. Further, we ignore the terms that involve squares of magnetic dipole transition moments as they will be much smaller than any electric dipole transition moment terms. The *MCD* may then be written

$$MCD = k\left[\left(\mu_Y^{fi} + i\mu_Z^{fi}\right)\left(\mu_Y^{if} - i\mu_Z^{if}\right) - \left(\mu_Y^{fi} - i\mu_Z^{fi}\right)\left(\mu_Y^{if} + i\mu_Z^{if}\right)\right]$$
$$= 2ik\left(\mu_Z^{fi}\mu_Y^{if} - \mu_Y^{fi}\mu_Z^{if}\right) \qquad (10.28)$$

In §10.9 equation (10.28) is used to derive the equations used in §9.6 and to see how to determine spectroscopic and geometric information about a molecule from *MCD* spectra.

10.5   *CD*, transition moment operators, and transition moments

What happens when we measure a *CD* spectrum is summarized in Figure 10.4. The Rosenfeld equation for the rotatory or *CD* strength is

$$R = CD = \text{Im}\langle 0 | \tilde{\boldsymbol{\mu}} | 1 \rangle \cdot \langle 1 | \tilde{m} | 0 \rangle = \text{Im}\left(\boldsymbol{\mu}^{10} \cdot \boldsymbol{m}^{01} \right) \qquad (10.29)$$

where we use the notation defined above but use $|0\rangle$ and $|1\rangle$ rather than $|i\rangle$ and $|f\rangle$ for the initial and final state of the transition. $\boldsymbol{\mu}^{10}$ is thus the electric dipole transition moment for the transition from the excited to the ground state and $\boldsymbol{m}^{01}$ is the magnetic dipole transition moment for the reverse transition. In this section we shall see how to express transition moments of a system in terms of the moments of its (usually achiral) sub-units.

We shall work within the chromophoric approach to *CD* theory (§8.2), and will limit our consideration to transitions in achiral chromophores. A chromophore is strictly a part of a system whose wave functions have no overlap with the rest of the system; electronic wave functions in a chromophore therefore have no electron exchange with the rest of the system. In practice, a chromophore is usually identified for a given electronic transition if that transition seems to be more or less dependent on only the identity of a subset of the molecule (such as the carbonyl functional group). We shall use perturbation theory to develop equations for the *CD* resulting from the coupling of independent chromophores.[119, 245, 255, 256]

The Rosenfeld equation for a transition within an isolated achiral chromophore **A** gives no *CD*:

$$CD(\text{achiral}) = \text{Im}\left(0|\hat{\boldsymbol{\mu}}|1\right) \cdot \left(1|\hat{\boldsymbol{m}}|0\right) = \text{Im}\left(\boldsymbol{\mu}_A^{10} \cdot \boldsymbol{m}_A^{01}\right) = 0 \qquad (10.30)$$

where we use curved bra-ket notation for the states of isolated chromophores and subscripts on the transition moments to denote that they belong entirely to the chromophore **A**. Equation (10.30) vanishes because the electron displacement of a transition within an achiral chromophore is not helical: it is either linear ($\boldsymbol{m}_A^{01}$=0), or circular ($\boldsymbol{\mu}_A^{10}$=0), or a planar spiral if the two moments are perpendicular (§4.2).

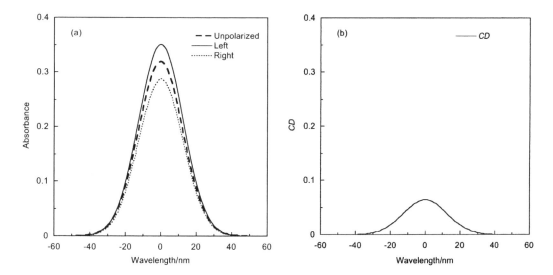

Figure 10.4 (a) Spectra from absorption of unpolarized, left circularly polarized, and right circularly polarized light by a randomly oriented sample of chiral molecules. (b) The resulting *CD* spectrum. The differences in absorption have been exaggerated for illustrative purposes.

As soon as **A** is allowed to interact with other chromophores, $\{C_j\}$, unless the whole system is still achiral, we should expect to be able to see a *CD* signal induced into the **A** transitions. Since the *CD* strength depends crucially on the geometry of the whole system, the *CD* spectrum should be able to tell us about the arrangement of the $\mathbf{C}_j$ about **A**. The way we extract that information is different depending whether the **A** transition is electric dipole allowed (eda) or electric dipole forbidden (edf) and magnetic dipole allowed (mda). For the rest of this chapter we shall generally consider two chromophores, **A** and **C**. Including more $\mathbf{C}_j$ simply requires the inclusion of a sum over *j*.

We also need to be able to write an equation for the interaction between **A** and **C**. Since there is no electron exchange between **A** and **C** (by definition), their interaction, *V*, is purely Coulombic

$$V = \sum_{a,c} \frac{q_a q_c}{\left|R_{AC} + r_c - r_a\right|} \qquad (10.31)$$

where q_a is the charge of a particle in **A** located at the end of the vector $\boldsymbol{r}_a$, which begins at the **A** origin, similarly q_c and $\boldsymbol{r}_c$. $\boldsymbol{R}_{AC}$ is the vector from the origin of chromophore **A** to the origin of chromophore **C** (Figure 10.5). $\boldsymbol{R}_{AC} = \boldsymbol{R}_c - \boldsymbol{R}_a$ where $\boldsymbol{R}_c$ is the vector from the system origin to the **C** origin *etc.*

Now, from equation (10.12), the electric dipole moment operator $\tilde{\boldsymbol{\mu}}$ (where ˜ is used to denote the operators for the whole system) is

$$\tilde{\boldsymbol{\mu}} = e\sum_j \tilde{r}_j \tag{10.32}$$

where $\tilde{r}_j$ is the position operator in the global coordinate system for a particle in **A** or **C**. We may rewrite equation (10.32) in terms of **A** and **C** operators using curved 'hats' to indicate operators of chromophores as

$$\tilde{\boldsymbol{\mu}} = e\left\{\sum_a \left(\widehat{\boldsymbol{R}}_A + \hat{r}_a\right) + \sum_c \left(\widehat{\boldsymbol{R}}_C + \hat{r}_c\right)\right\} \tag{10.33}$$

We shall usually end up defining the system origin to be the **A** chromophore origin so

$$\widehat{\boldsymbol{R}}_A = \boldsymbol{R}_A = (0,0,0)$$

and

$$\boldsymbol{R}_{AC} = \boldsymbol{R}_C - \boldsymbol{R}_A = \boldsymbol{R}_C$$

where $\widehat{\boldsymbol{R}}_A = \boldsymbol{R}_A$ is the position operator for the **A** origin in the global coordinate system, and $\hat{r}_a = r_a$ is the position operator within **A** for particle a (Figure 10.5). We consider only *transition* moments, so since $\boldsymbol{R}_A$ is constant and since all non-degenerate states are orthogonal, it follows that

$$\tilde{\boldsymbol{\mu}} = e\left\{\sum_a \hat{r}_a + \sum_c \hat{r}_c\right\} \tag{10.34}$$

$$= \hat{\boldsymbol{\mu}}_A + \hat{\boldsymbol{\mu}}_C$$

where $\hat{\boldsymbol{\mu}}_A$ is the electric dipole moment operator for **A** and operates only on wave functions of **A**.

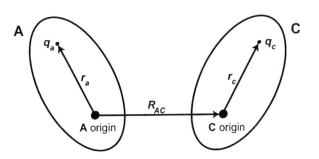

Figure 10.5 **A** and **C** chromophores.

To evaluate the Rosenfeld equation for two interacting chromophores we need to know $\boldsymbol{\mu}^{10}$ and $\boldsymbol{m}^{10}$ for the whole system, *i.e.* for the $|0\rangle \rightarrow |1\rangle$ transition, which is the $|0\rangle \rightarrow |1\rangle$ transition on **A** in the presence of any perturbation due to **C**. We can write the wavefunctions for the **A/C** system when their interaction is switched off as the product wavefunctions $|a\rangle|c\rangle \rightarrow |ac\rangle$ where $|a\rangle$ is the wavefunction of **A** and $|c\rangle$ is a wavefunction of **C**. We always write the **A** function first, even in the ket form, thus

$$\left\{\left|a\right\rangle\left|c\right\rangle\right\}^{*} = \left\{\left|ac\right\rangle\right\}^{*} = \left(a\right|\left(c\right| = \left(ac\right| \tag{10.35}$$

It follows, for $\delta_{cc'} = 0$ unless $c' = c$, that

$$\left(a'c'\left|\hat{\boldsymbol{\mu}}_{A}\right|ac\right) = \left(a'\left|\hat{\boldsymbol{\mu}}_{A}\right|a\right)\left(c'\left|c\right) \tag{10.36}$$

$$= \boldsymbol{\mu}_{A}^{aa'}\delta_{cc'}$$

Similarly it follows from equation (10.12) that the magnetic dipole moment operator for the whole system is

$$\tilde{\boldsymbol{m}} = \frac{e}{2m_{e}}\sum_{j}\left\{\tilde{\boldsymbol{r}}_{j} \times \tilde{\boldsymbol{p}}_{j}\right\} \tag{10.37}$$

Equation (10.37) may then be rewritten

$$\tilde{\boldsymbol{m}} = \tilde{\boldsymbol{m}}_{A} + \tilde{\boldsymbol{m}}_{C}$$

$$= \sum_{a}\left(\hat{\boldsymbol{R}}_{A} + \hat{\boldsymbol{r}}_{a}\right) \times \hat{\boldsymbol{p}}_{a} + \sum_{c}\left(\hat{\boldsymbol{R}}_{C} + \hat{\boldsymbol{r}}_{c}\right) \times \hat{\boldsymbol{p}}_{c}$$

$$= \frac{e}{2m_{e}}\hat{\boldsymbol{R}}_{A} \times \sum_{a}\hat{\boldsymbol{p}}_{a} + \hat{\boldsymbol{m}}_{A} + \frac{e}{2m_{e}}\hat{\boldsymbol{R}}_{C} \times \sum_{c}\hat{\boldsymbol{p}}_{c} + \hat{\boldsymbol{m}}_{C} \tag{10.38}$$

$$= \frac{e}{2m_{e}}\hat{\boldsymbol{R}}_{A} \times \hat{\boldsymbol{p}}_{A} + \hat{\boldsymbol{m}}_{A} + \frac{e}{2m_{e}}\hat{\boldsymbol{R}}_{C} \times \hat{\boldsymbol{p}}_{C} + \hat{\boldsymbol{m}}_{C}$$

where $\tilde{\boldsymbol{m}}_{A}$ is the total magnetic moment for chromophore $\mathbf{A}$, $\hat{\boldsymbol{p}}_{a}$ is the linear momentum operator within chromophore $\mathbf{A}$, $\hat{\boldsymbol{m}}_{A}$ is the intrinsic magnetic dipole moment operator within $\mathbf{A}$ *etc.*

It is convenient to re-express magnetic transition momenta in terms of electric dipole transition moments for the appropriate transition exploiting a very useful relationship that is derived in §11.4:

$$\left(k\left|\hat{\boldsymbol{p}}_{C}\right|j\right) = \frac{im_{e}}{e\hbar}\left(\varepsilon_{k} - \varepsilon_{j}\right)\boldsymbol{\mu}_{C}^{jk} \tag{10.39}$$

where $\left|k\right)$ is at energy ε_{k} and $\left|j\right)$ at energy ε_{j}, just as in the transition moment integrals. If there is no intrinsic magnetic moment for the transition within $\mathbf{C}$ (*i.e.* it is eda and mdf) it follows

$$m_{C}^{jk} = \left(k\left|\hat{\boldsymbol{m}}_{C}\right|j\right)$$

$$= \frac{e}{2m_{e}}\left(k\left|\hat{\boldsymbol{R}}_{C} \times \hat{\boldsymbol{p}}_{C}\right|j\right) \tag{10.40}$$

$$= \frac{i}{2\hbar}\boldsymbol{R}_{C} \times \left(\varepsilon_{k} - \varepsilon_{j}\right)\boldsymbol{\mu}_{C}^{jk}$$

This is the magnetic moment at the system origin created by the tangential eda transition moment $\boldsymbol{\mu}_{C}^{jk}$ located within $\mathbf{C}$.

10.6   CD from the coupling of degenerate electric dipole transition moments in identical chromophores: the degenerate coupled-oscillator model

If an eda transition, of an achiral chromophore, $\mathbf{A}$, with electric dipole transition moment $\boldsymbol{\mu}_A^{01}$ and transition energy ε, has a measurable *CD* spectrum, then it must have 'borrowed' some magnetic character, to give a net helical electron rearrangement. The degenerate coupled-oscillator model describes the situation where the magnetic character comes from the coupling of $\boldsymbol{\mu}_A^{01}$ with another electric dipole transition moment in **C**, $\boldsymbol{\mu}_C^{01}$, whose transition energy is the same but whose orientation is skewed relative to that of $\boldsymbol{\mu}_A^{01}$. $\boldsymbol{\mu}_C^{01}$ induces a magnetic effect in **A** since, although it is a linear motion of charge within **C**, when viewed from **A** it moves around **C**'s origin as illustrated in Figure 8.1. $\boldsymbol{\mu}_A^{01}$ in **A** simultaneously induces a magnetic component into the transition(s) of **C** $\boldsymbol{\mu}_C^{01}$).

To write the wave functions of the interacting **A**/**C** system we first consider an **A**/**C** system where **A** and **C** do not interact. The Hamiltonian, $\tilde{H}$, of the non-interacting system is then simply the sum

$$\tilde{H} = \hat{H}_A + \hat{H}_C \tag{10.41}$$

where $\hat{H}_A$ is the Hamiltonian of isolated **A**, and similarly $\hat{H}_C$. The total energy of a state $|ac)$ is the energy of $|a)$ in **A** plus that of $|c)$ in **C**. So, for example, $|01)$ means **A** is in its ground state and **C** is in state $|1)$. The energy of the $|01)$ state of the combined system is therefore $0 + \varepsilon = \varepsilon$ (taking the zero point of energy to be when both **A** and **C** are in their ground states).

When the electrostatic interaction, V (equation (10.31)), is 'switched on' between **A** and **C** then the Hamiltonian of the interacting system is

$$\tilde{H} = \hat{H}_A + \hat{H}_C + \tilde{V} \tag{10.42}$$

This term and further terms are derived in §10.10.

If **A** and **C** are not too close together, then $\tilde{V}$ of equation (10.31) may be expanded using a double Taylor series expansion to give terms dependent on monopoles, dipoles, quadrupoles *etc.* of **A** and **C** as outlined in §10.10. For uncharged **A** and **C**, the first term in the expansion is the dipole-dipole term:

$$\tilde{V} = \frac{\hat{\boldsymbol{\mu}}_A \cdot \hat{\boldsymbol{\mu}}_C - 3\hat{\boldsymbol{R}}_{AC} \cdot \hat{\boldsymbol{\mu}}_A \hat{\boldsymbol{\mu}}_C \cdot \hat{\boldsymbol{R}}_{AC}}{R_{AC}^3} \tag{10.43}$$

where $\boldsymbol{R}_{AC}$ is the vector from the **A** origin to the **C** origin. The **A**/**C** wave functions have the form

$$|a)|c) = |ac)$$
$$= |ac) + \text{other terms} \tag{10.44}$$

Again we use the rounded bra-ket notation for unperturbed wave functions of chromophores and the pointed bra-ket notation for the wave functions of the interacting system.

The identity of the 'other terms' depends on whether or not **A** and **C** are identical. If **A** and **C** are identical then their transitions are degenerate (have the same energy) and by symmetry (or degenerate perturbation theory) the two states that result from the mixing of $|10)$ (**A** in $|1)$ and **C** in the ground state) and $|01)$ are

$$|10_\pm\rangle = \frac{1}{\sqrt{2}}\left\{|10) \pm |01)\right\} \tag{10.45}$$

As above, we use real normalized wave functions. If we also assume that permanent moments are much smaller than transition moments, then the dominant terms in the energies of these two states are

$$\varepsilon^\pm = \langle 10_\pm | \hat{H}_A + \hat{H}_C + \hat{V} | 10_\pm \rangle$$

$$= \frac{1}{2}\left\{ \left(\langle 10| \pm \langle 01| \right) \hat{H}_A + \hat{H}_C + \hat{V} \left\{ |10\rangle \pm |01\rangle \right\} \right\}$$

$$= \frac{1}{2}\left\{ \left(\langle 1|\hat{H}_A|1\rangle \right) + \left(\langle 0|\hat{H}_C|0\rangle \right) \pm \left(\langle 10|\hat{V}|01\rangle \right) + \left(\langle 0|\hat{H}_A|0\rangle \right) + \left(\langle 1|\hat{H}_C|1\rangle \right) \pm \left(\langle 01|\hat{V}|10\rangle \right) \right\}$$

$$\varepsilon^\pm = \langle 10_\pm | \hat{H}_A + \hat{H}_C + \hat{V} | 10_\pm \rangle$$

$$= \frac{1}{2}\left\{ \left(\langle 10| \pm \langle 01| \right) \hat{H}_A + \hat{H}_C + \hat{V} \left\{ |10\rangle \pm |01\rangle \right\} \right\}$$

$$= \frac{1}{2}\left\{ \left(\langle 1|\hat{H}_A|1\rangle \right) + \left(\langle 0|\hat{H}_C|0\rangle \right) \pm \left(\langle 10|\hat{V}|01\rangle \right) + \left(\langle 0|\hat{H}_A|0\rangle \right) + \left(\langle 1|\hat{H}_C|1\rangle \right) \pm \left(\langle 01|\hat{V}|10\rangle \right) \right\} \tag{10.46}$$

So

$$\varepsilon^\pm = \varepsilon^0 + \varepsilon^1 \pm \frac{1}{2}\left(V^{01,10} + V^{10,01} \right) \tag{10.47}$$

$$= \varepsilon \pm V^{11}$$

where, for example,

$$V^{ba,dc} = \left(\langle ac|\hat{V}|bd\rangle \right)$$

$$= \frac{\boldsymbol{\mu}_A^{ba} \cdot \boldsymbol{\mu}_C^{dc} - 3\hat{\boldsymbol{R}}_{AC} \cdot \boldsymbol{\mu}_A^{ba} \, \boldsymbol{\mu}_C^{dc} \cdot \hat{\boldsymbol{R}}_{AC}}{R_{AC}^3} \tag{10.48}$$

V^{11} is much smaller than ε so the two new perturbed states are close in energy.

The electric dipole transition moments for the transitions from the two excited states of equation (10.45) to the ground state $|00\rangle$ are:

$$\langle 00|\tilde{\boldsymbol{\mu}}|10_\pm \rangle = \frac{1}{\sqrt{2}}\left\{ \left(\langle 00|\tilde{\boldsymbol{\mu}}_A + \tilde{\boldsymbol{\mu}}_C \left[|10\rangle \pm |01\rangle \right] \right) \right\}$$

$$= \frac{1}{\sqrt{2}}\left\{ \boldsymbol{\mu}_A^{10} + \boldsymbol{\mu}_C^{10} \right\} \tag{10.49}$$

where the upper sign of the $\pm$ refers to the upper signs in the above equations. The $|0\rangle \rightarrow |1\rangle$ transitions in both **A** and **C** are taken to be mdf, and so have no intrinsic magnetic moment. We therefore set $\hat{m}_A = \hat{m}_C = 0$ in equation (10.38) and write the magnetic dipole transition moments for the transition from the ground states to the excited perturbed states to be

$$\langle 10_\pm|\tilde{m}|00\rangle = \frac{i\varepsilon}{2\sqrt{2}\hbar}\left\{ \boldsymbol{R}_A \times \boldsymbol{\mu}_A^{01} \pm \boldsymbol{R}_C \times \boldsymbol{\mu}_C^{01} \right\} \tag{10.50}$$

Substituting equations (10.49) and (10.50) into the Rosenfeld equation gives the *CD* strengths of bands centred at ε^+ and ε^- (as given by equation (7.41)) arising from the coupling of degenerate $|0\rangle \rightarrow |1\rangle$ transitions on **A** and **C** to be

An eda transition of an achiral chromophore may also be mda (if the polarizations are orthogonal), but any *CD* arising *via* perturbation of the magnetic moment will usually be much smaller than the coupled-oscillator *CD*.

$$R_{\pm}^{01} = \pm \frac{\varepsilon}{4\hbar} \left\{ \boldsymbol{\mu}_C^{01} \times \boldsymbol{\mu}_A^{01} \cdot (\boldsymbol{R}_C - \boldsymbol{R}_A) \right\}$$

$$= \pm \frac{\varepsilon}{4\hbar} \left\{ \boldsymbol{\mu}_C^{01} \times \boldsymbol{\mu}_A^{01} \cdot (\boldsymbol{R}_{AC}) \right\} \quad (10.51)$$

where the ± subscripts refer to the + and − states respectively in equation (10.45).

Note the + state is not necessarily higher or lower in energy than the − state.

Equation (8.11) is a notationally simplified version of equation (10.51). The physical interpretation and applications of equation (10.51) for solving geometric and spectroscopic problems is outlined in Chapter 8.

If there are $n > 2$ identical chromophores in the system arranged around an n-fold rotational symmetry axis as for *tris*-chelate in-ligand transitions (Figure 8.6) then equation (10.51) is replaced by the two symmetry adapted projections and normalization factors. The result is always a transition polarized along the n-fold axis, z, whose *CD* is always equal in magnitude but opposite in sign from a second transition polarized in the plane perpendicular to the n-fold axis. The normalization factor for the wave functions is $1/\sqrt{n}$. Assuming that each chromophore couples only with its two nearest neighbours, the *CD* signals are then[257]

The two transitions have equal magnitude and opposite signed *CD* signals so that if the energy splitting is zero then we observe no net effect.

$$R(z, n \text{ chromophores}) = \left(\frac{\sqrt{2}}{\sqrt{n}}\right)^2 nR_+(z, 2 \text{ chromophores})$$

$$= 2R_+(z, 2 \text{ chromophores}) \quad (10.52)$$

$$= -R(x/y, n \text{ chromophores})$$

10.7 *CD* from the coupling of electric dipole transition moments in non-identical chromophores: the non-degenerate coupled-oscillator model

If **A** and **C** are not identical then the wave functions of the combined system are not determined by symmetry. This makes writing an expression for the wave functions and hence the transition moments more difficult, but the compensating factor is that, because the **A** and **C** transitions occur at very different energies, we can take the energies of the states of the interacting **A**/**C** system to be simple sums of the energies of the non-interacting **A** and **C** states.

Further, we need only determine the *CD* induced into the $|0\rangle_A \rightarrow |1\rangle_A$ transition at energy ε_a of **A**. So we require expressions for the perturbed states of the interacting **A**/**C** system where both chromophores are in the ground state, $|00\rangle$, or where **A** is in an excited state and **C** is in its ground state, $|10\rangle$. For simplicity, we shall consider only these two states and the two others $|0c\rangle$ and $|1c\rangle$ where **C** is in the excited state $|c\rangle$ on **C**. The final equations are easily generalized by introducing summations at the end. From perturbation theory we may write

A little imprecision about the ordering of states with electric transition moments does not matter as long as the wave functions are real. The same is not true for magnetic dipole transition moments.

$$|00\rangle = |1c\rangle - \frac{\left(1c|\tilde{V}|00\right)}{\varepsilon_c + \varepsilon_a}|1c\rangle \quad (10.53)$$

$$|10\rangle = |10\rangle - \frac{\left(0c|\tilde{V}|10\right)}{\varepsilon_c - \varepsilon_a}|0c\rangle \tag{10.54}$$

We may also write

$$V^{1c} = \left(1c|\tilde{V}|00\right) = \left(0c|\tilde{V}|10\right)$$
$$= \frac{\boldsymbol{\mu}_A^{01} \cdot \boldsymbol{\mu}_C^{c0} - 3\hat{\boldsymbol{R}}_{AC} \cdot \boldsymbol{\mu}_A^{01} \boldsymbol{\mu}_C^{c0} \hat{\boldsymbol{R}}_{AC}}{R_{AC}^3} \tag{10.55}$$

It then follows (again assuming terms containing permanent moments are smaller than those containing transition moments) that to first order in V the electric dipole transition moment of the perturbed $|10\rangle \rightarrow |00\rangle$ transition is

$$\boldsymbol{\mu}^{10} = \langle 00|\tilde{\boldsymbol{\mu}}|10\rangle \tag{10.56}$$
$$= \langle 00|\hat{\boldsymbol{\mu}}_A + \hat{\boldsymbol{\mu}}_C|10\rangle$$

So

$$\boldsymbol{\mu}^{10} = \left(00|\hat{\boldsymbol{\mu}}_A|01\right) - \left(00|\hat{\boldsymbol{\mu}}_C|0c\right)\frac{\left(0c|\tilde{V}|10\right)}{\varepsilon_c - \varepsilon_a} - \left(1c|\hat{\boldsymbol{\mu}}_C|10\right)\frac{\left(00|\tilde{V}|1c\right)}{\varepsilon_c + \varepsilon_a}$$
$$= \boldsymbol{\mu}_A^{10} - \frac{2\varepsilon_c V^{1c}}{\varepsilon_c^2 - \varepsilon_a^2}\boldsymbol{\mu}_C^{c0} \tag{10.57}$$

Similarly,

$$\boldsymbol{m}^{01} = \langle 10|\bar{\boldsymbol{m}}|00\rangle$$
$$= \frac{e}{2m_e}\left\{\boldsymbol{R}_A \times \boldsymbol{p}_A^{01} - \frac{2\varepsilon_a V^{1c}}{\varepsilon_c^2 - \varepsilon_a^2}\boldsymbol{R}_C \times \boldsymbol{p}_C^{0c}\right\} \tag{10.58}$$

so

$$\boldsymbol{m}^{01} = \frac{i}{2\hbar}\left\{\varepsilon_a \boldsymbol{R}_A \times \boldsymbol{\mu}_A^{01} - \frac{2\varepsilon_a \varepsilon_c V^{1c}}{\varepsilon_c^2 - \varepsilon_a^2}\boldsymbol{R}_C \times \boldsymbol{\mu}_C^{0c}\right\} \tag{10.59}$$

since the transition has no intrinsic magnetic moment within **A**. The *CD* strength is therefore (introducing a sum over **C** states)

$$R(|00\rangle \rightarrow |10\rangle) = \sum_c \frac{-\varepsilon_a \varepsilon_c V^{1c}}{\hbar(\varepsilon_c^2 - \varepsilon_a^2)}\left\{\boldsymbol{\mu}_A^{10} \cdot \boldsymbol{R}_C \times \boldsymbol{\mu}_C^{0c} + \boldsymbol{\mu}_C^{c0} \cdot \boldsymbol{R}_A \times \boldsymbol{\mu}_A^{01}\right\}$$
$$= \sum_c \frac{-\varepsilon_a \varepsilon_c V^{1c}}{\hbar(\varepsilon_c^2 - \varepsilon_a^2)}\left\{\boldsymbol{\mu}_C^{0c} \times \boldsymbol{\mu}_A^{01} \cdot \boldsymbol{R}_{AC}\right\} \tag{10.60}$$

As with the degenerate coupled-oscillator equations, the application of equation (10.60) is covered in Chapter 8.

For the particular case where one chromophore, **A**, non-degenerately couples to *n*-identical chromophores arranged around it so **A** lies on the *z*-axis of the *n*-chromophore system, then the coupled-oscillator *CD* follows

Note that for example,

$$\left(00|\hat{\boldsymbol{\mu}}_C|01\right) = \left(0|0\right)\left(0|\hat{\boldsymbol{\mu}}_C|1\right)$$
$$= \bar{\boldsymbol{\mu}}_C^{10}$$

where the subscript C denotes the integral is over the coordinates of chromophore **C**. The initial state in the bra-ket is on the right-hand side, but when the states are written as superscripts to transition moments the initial state is on the left. H_A operates only on the states of **A** etc.

simply by inserting a summation over n into equation (10.60). Depending on the symmetry of **A** and the polarization of the **A** transition the equation can be further simplified as summarized in Table 10.1[257] and as was illustrated in Chapter 8 for cyclodextrin inclusion compounds.

Table 10.1 Coupled-oscillator *CD* expressions at energy ε_a for transitions of a guest molecule **A** at the centre of a host composed of n identical **C** chromophores arranged so the host has an n-fold rotational symmetry. Guest transition polarizations are indicated in parentheses and state labels have been omitted for clarity. The notation used is that:

$$H_{uvw} = \sum_c \frac{\varepsilon_a \varepsilon_c}{\hbar R_{AC}^2 \left(\varepsilon_c^2 - \varepsilon_a^2\right)}\left(\mu_{Cu}\hat{R}_{ACv} - \mu_{Cv}\hat{R}_{ACu}\right)\mu_{Cw}$$

Host	Guest	CD expression
$n=2$	non-degenerate C_{2v}, D_{2h} symmetry	$R(x) = n\mu_{Ax}\mu_{Ax}H_{yzx}$ $R(y) = n\mu_{Ay}\mu_{Ay}H_{zxy}$ $R(z) = n\mu_{Az}\mu_{Az}H_{xyz}$
	u/v-degenerate	$R(u/v) = -\frac{n}{2}\left(\mu_{Au}\mu_{Au} + \mu_{Av}\mu_{Av}\right)H_{uvw}$ $R(w) = n\mu_{Aw}\mu_{Aw}H_{uvw}$
$n>2$	non-degenerate	$R(x) = -\frac{n}{2}\mu_{Ax}\mu_{Ax}H_{xyz}$ $R(y) = -\frac{n}{2}\mu_{Ay}\mu_{Ay}H_{xyz}$ $R(z) = n\mu_{Az}\mu_{Az}H_{xyz}$
	x/y-degenerate	$R(x/y) = -\frac{n}{2}\left(\mu_{Ax}\mu_{Ax} + \mu_{Ay}\mu_{Ay}\right)H_{xyz}$ $R(z) = n\mu_{Az}\mu_{Az}H_{xyz}$
	x/z-degenerate	$R(x/y) = -\frac{n}{4}\left(\mu_{Ax}\mu_{Ax} + \mu_{Az}\mu_{Az}\right)H_{xyz}$ $R(z) = n\mu_{Ay}\mu_{Ay}H_{xyz}$

10.8 Magnetic dipole allowed transitions: the dynamic coupling model

In this section the origin of the equations used in Chapter 9 to account for the *CD* induced into mda transitions is outlined. *Ab initio* calculations[240] have been used to show that the mechanisms arising from this perturbation theory approach, first developed by Höhn and Weigang,[241] do indeed account for the *CD* in chiral adamantanones. However, only qualitative analysis of the electronic effects in planar zig-zag molecules (§9.4) is feasible within a chromophoric approach, since it breaks down for such molecules as discussed below.

An mda transition of an achiral chromophore, **A**, must 'borrow' some electric character in order to have a *CD* signal as discussed in Chapter 9. As with coupled-oscillator *CD*, the required perturbation arises as a result of the electrostatic coupling of **A** with the rest of the molecule. The general case is more difficult to deal with for mda transitions than for eda transitions since the coupling due to the first term in the Taylor series expansion of V (equation (10.43)) often vanishes. We shall therefore explicitly evaluate V^{lc} in the equations below as late as possible in the derivations.

We also again assume that transition moments give significantly stronger effects than permanent moments. This leads to the 'dynamic coupling' model for the *CD* of mda transitions. Historically much more effort has been put into what is known as the 'static coupling' or 'one electron' mechanism (since one electron on **A** is assumed to be the only one to move) where the molecular framework provides a static perturbation *via* permanent moments. This mechanism certainly contributes to the net observed *CD*, however, its contribution is almost always smaller than that due to the dynamic coupling mechanism. More details about the static coupling model may be found in references [242, 247, 255].

The formalism for dealing with mda transitions is very similar to the one we used for the non-degenerate coupled-oscillator *CD* in the preceding section, we just need to remember that here the $|0\rangle \to |1\rangle$ transition of **A** is now mda and edf so $m_A^{01} \neq 0$ but $\mu_A^{01} = 0$. The perturbed wavefunctions $|i\rangle = |00\rangle$ and $|f\rangle = |10\rangle$ are given to first order in V by equations (10.53) and (10.54), from which it follows that[242,245]

$$\mu^{10} = \langle 00|\tilde{\mu}_A|10\rangle = \langle 00|\hat{\mu}_A + \hat{\mu}_C|10\rangle$$

$$= \left(00|\hat{\mu}_A|10\right) - \frac{\left(00|\tilde{V}|1c\right)}{\varepsilon_c + \varepsilon_a}\left(1c|\hat{\mu}_C|10\right) - \frac{\left(0c|\tilde{V}|10\right)}{\varepsilon_c - \varepsilon_a}\left(0c|\hat{\mu}_C|0c\right) \quad (10.61)$$

$$= -\frac{\left(00|\tilde{V}|1c\right)}{\varepsilon_c + \varepsilon_a}\mu_C^{0c} - \frac{\left(0c|\tilde{V}|10\right)}{\varepsilon_c - \varepsilon_a}\mu_C^{c0} = -\frac{2\varepsilon_c V^{lc}}{\varepsilon_c^2 - \varepsilon_a^2}\mu_C^{c0}$$

since the **A** and **C** wave functions are real; a sum over c is implied. Similarly,

$$m^{01} = \langle 10|\tilde{m}|00\rangle = \langle 10|\hat{m}_A + \hat{m}_C|00\rangle$$

$$= \left(10|\hat{m}_A|00\right) - \frac{\left(1c|\tilde{V}|00\right)}{\varepsilon_c + \varepsilon_a}\left(10|\hat{m}_C|1c\right) - \frac{\left(10|\tilde{V}|0c\right)}{\varepsilon_c - \varepsilon_a}\left(0c|\hat{m}_C|00\right)$$

$$= m_A^{01} - \frac{\left(1c|\tilde{V}|00\right)}{\varepsilon_c + \varepsilon_a}m_C^{c0} - \frac{\left(10|\tilde{V}|0c\right)}{\varepsilon_c - \varepsilon_a}m_C^{0c} \qquad (10.62)$$

$$= m_A^{01} - \frac{2\varepsilon_a V^{lc}}{\varepsilon_c^2 - \varepsilon_a^2}m_C^{0c}$$

Analogous dynamic coupling expressions may be derived for eda transition, however, they are expected to be much smaller than the coupled-oscillator terms as they have higher inverse R_{AC} dependence.

Substitution into the Rosenfeld equation gives the so-called dynamic coupling *CD* expressions to first order in *V*

$$R = \text{Im} \left\{ \sum_c \frac{-2\varepsilon_c V^{1c}}{\varepsilon_c^2 - \varepsilon_a^2} \boldsymbol{\mu}_C^{c0} \cdot \boldsymbol{m}_A^{01} \right\} \tag{10.63}$$

In order to apply equation (10.63) we need to have explicit forms for V^{1c} in terms of electric transition moments of **A** and **C**. It is now convenient to use a general form for the terms in the expansion of *V* using the Kronecker delta, δ_{jk}, which is zero unless $j = k$. We rewrite equation (10.48), the dipole-dipole term in the expansion of *V* as

$$V^{1c}(\mu_A - \mu_C) = \frac{1}{R_{AC}^3} \left[\left(\mu_A^{10} \right)_j \left(\mu_C^{0c} \right)_k \right] \left[\delta_{jk} - 3\left(\hat{R}_{AC} \right)_j \left(\hat{R}_{AC} \right)_k \right] \tag{10.64}$$

where a sum over all repeated indices and *c* is implied.

When substituted into equation (10.63), the dipole-dipole term of *V* leads to a *CD* expression containing a

$$\left(\mu_A^{01} \right)_j \left(m_A^{10} \right)_k \tag{10.65}$$

term. This moment product vanishes except for chromophores of $\mathbf{C}_s$ symmetry and *x*- and *y*-polarized transitions of $\mathbf{C}_{2v}$ chromophores. For these transitions, the **A** moment product does not vanish; however, an mda transition in such a molecule is also eda so the *CD* will almost certainly be dominated by the coupled-oscillator *CD* of the electric dipole transition moment which is dependent on R_{AC}^{-3} (see above). The one case where we cannot assume that a coupled-oscillator *CD* dominates was discussed in Chapter 9: mda *d-d* transitions even of low symmetry transition metal complexes have very small electric dipole transition moments due to the intrinsic symmetry of the *d*-orbitals.

Thus the dynamic coupling mechanism first (in order of increasing chromophore symmetry) becomes important for transitions with *z*-polarized magnetic dipole transition moments of $\mathbf{C}_{2v}$ chromophores such as the carbonyl which forms the subject matter of much of Chapter 9. The quadrupole-dipole term of *V* is the first non-vanishing one for $\mathbf{C}_{2v}$ chromophores:

$$\begin{aligned} V^{1c}(Q_A - \mu_C) = \frac{3}{2R_{AC}^4} &\left[\left(Q_A^{10} \right)_{jk} \left(\mu_C^{0c} \right)_\ell \right] \\ &\times \begin{bmatrix} \delta_{jk}\left(\hat{R}_{AC} \right)_\ell + \delta_{\ell j}\left(\hat{R}_{AC} \right)_k + \delta_{k\ell}\left(\hat{R}_{AC} \right)_j \\ -5\left(\hat{R}_{AC} \right)_j \left(\hat{R}_{AC} \right)_k \left(\hat{R}_{AC} \right)_\ell \end{bmatrix} \end{aligned} \tag{10.66}$$

where $\hat{Q}_A$ is the electric quadrupole moment operator of **A** and a sum over *c* is implied. $\hat{Q}_A$ is a second rank tensor whose components are

$$\left(\hat{Q}_A \right)_{jk} = e \sum_a \left(r_j r_k \right) \tag{10.67}$$

(*cf.* equation (10.12) for notation). The quadrupole-dipole *CD* strength for an mda transition of *i*-polarization is thus

$$R = -\frac{3}{R_{AC}^4}\,\mathrm{Im}\left[\left(Q_A^{10}\right)_{jk}\left(m_A^{01}\right)_i\right]\sum_c\left[\frac{\varepsilon_c\left(\mu_C^{c0}\right)_i\left(\mu_C^{c0}\right)_\ell}{\left(\varepsilon_c^2-\varepsilon_a^2\right)}\right]$$

$$\times\begin{bmatrix}\delta_{jk}\left(\hat{R}_{AC}\right)_\ell+\delta_{\ell j}\left(\hat{R}_{AC}\right)_k+\delta_{k\ell}\left(\hat{R}_{AC}\right)_j\\-5\left(\hat{R}_{AC}\right)_j\left(\hat{R}_{AC}\right)_k\left(\hat{R}_{AC}\right)_\ell\end{bmatrix}\qquad(10.68)$$

For transitions with z-polarized magnetic dipole transition moments, $i = z$ in equation (10.68). Further, under $\mathbf{C}_{2v}$ symmetry only $\{j,k\} = \{x,y\}$ or $\{y,x\}$ is non-zero. Thus for $\mathbf{C}_{2v}$ $\mathbf{A}$, the CD for a z-polarized mda transition is:

$$R(C_{2v})_z = -\frac{6}{R_{AC}^4}\,\mathrm{Im}\left[\left(Q_A^{01}\right)_{xy}\left(m_A^{10}\right)_z\right]\sum_c\left[\frac{\varepsilon_c\left(\mu_C^{0c}\right)_z}{\left(\varepsilon_c^2-\varepsilon_a^2\right)}\right]$$

$$\times\begin{bmatrix}\left(\mu_C^{c0}\right)_x\left(\hat{R}_{AC}\right)_y+\left(\mu_C^{c0}\right)_y\left(\hat{R}_{AC}\right)_x\\-5\left(\mu_C^{c0}\right)_\ell\left(\hat{R}_{AC}\right)_x\left(\hat{R}_{AC}\right)_y\left(\hat{R}_{AC}\right)_\ell\end{bmatrix}\qquad(10.69)$$

In §9.4 we assumed that the isotropic polarizability term in equation (10.69) is dominant giving the octant rule. However, there are occasions when the octant rule is not valid because other terms in equation (10.69) (or even higher order terms in the V expansion) become dominant.

Equation (10.69) also leads to the CD expressions for degenerately x/y-polarized transitions of molecules with $\mathbf{D}_{3d}$ $\mathbf{A}$, since the $\mathbf{A}$ moment product of equation (10.69) does not vanish for this case. If z is the three-fold rotation axis and x is one of the two-fold rotation axes of $\mathbf{A}$. By symmetry,

$$R(\mathbf{D}_{3d})_{x,y} = -\mathrm{Im}\sum_c\frac{3\varepsilon_c}{4R_{AC}^4\left(\varepsilon_c^2-\varepsilon_a^2\right)}$$

$$\left\{\begin{array}{l}\left[\left(Q_A^{10}\right)_{xx}\left(m_A^{01}\right)_x-\left(Q_A^{10}\right)_{yy}\left(m_A^{01}\right)_x-2\left(Q_A^{10}\right)_{xy}\left(m_A^{01}\right)_y\right]\\[4pt]\times\begin{bmatrix}2\left(\mu_C^{c0}\right)_x\left(\mu_C^{0c}\right)_x\left(\hat{R}_{AC}\right)_x-5\left(\mu_C^{c0}\right)_\ell\left(\mu_C^{0c}\right)_x\left(\hat{R}_{AC}\right)_x\left(\hat{R}_{AC}\right)_x\left(\hat{R}_{AC}\right)_\ell\\-4\left(\mu_C^{c0}\right)_x\left(\mu_C^{0c}\right)_y\left(\hat{R}_{AC}\right)_y+5\left(\mu_C^{c0}\right)_\ell\left(\mu_C^{0c}\right)_x\left(\hat{R}_{AC}\right)_y\left(\hat{R}_{AC}\right)_y\left(\hat{R}_{AC}\right)_\ell\\-2\left(\mu_C^{c0}\right)_y\left(\mu_C^{0c}\right)_y\left(\hat{R}_{AC}\right)_x+10\left(\mu_C^{c0}\right)_\ell\left(\mu_C^{0c}\right)_x\left(\hat{R}_{AC}\right)_x\left(\hat{R}_{AC}\right)_y\left(\hat{R}_{AC}\right)_\ell\end{bmatrix}\\[4pt]-4\left[\left(Q_A^{10}\right)_{yz}\left(m_A^{01}\right)_x-\left(Q_A^{10}\right)_{xz}\left(m_A^{01}\right)_y\right]\\[4pt]\times\begin{bmatrix}\left(\mu_C^{c0}\right)_y\left(\mu_C^{0c}\right)_z\left(\hat{R}_{AC}\right)_x+5\left(\mu_C^{c0}\right)_\ell\left(\mu_C^{0c}\right)_x\left(\hat{R}_{AC}\right)_y\left(\hat{R}_{AC}\right)_z\left(\hat{R}_{AC}\right)_\ell\\-\left(\mu_C^{c0}\right)_z\left(\mu_C^{0c}\right)_x\left(\hat{R}_{AC}\right)_y-5\left(\mu_C^{c0}\right)_\ell\left(\mu_C^{0c}\right)_y\left(\hat{R}_{AC}\right)_x\left(\hat{R}_{AC}\right)_z\left(\hat{R}_{AC}\right)_\ell\\-+\left(\mu_C^{c0}\right)_x\left(\mu_C^{0c}\right)_y\left(\hat{R}_{AC}\right)_z-\left(\mu_C^{c0}\right)_y\left(\mu_C^{0c}\right)_x\left(\hat{R}_{AC}\right)_z\end{bmatrix}\end{array}\right.$$

$$(10.70)$$

The isotropic polarizability parts of equation (10.70) are

$$R(\mathbf{D}_{3d})_{x,y} = \sum_c \frac{15\varepsilon_c \alpha_C(\varepsilon_a)}{4R_{AC}^4(\varepsilon_c^2 - \varepsilon_a^2)}$$

$$\times \left\{ \begin{array}{l} \mathrm{Im}\left[\left(Q_A^{10}\right)_{xx}\left(m_A^{01}\right)_x - \left(Q_A^{10}\right)_{yy}\left(m_A^{01}\right)_x - 2\left(Q_A^{10}\right)_{xy}\left(m_A^{01}\right)_y\right] \\ \times \left(\hat{R}_{AC}\right)_x\left[\left(\hat{R}_{AC}\right)_x^2 - 3\left(\hat{R}_{AC}\right)_y^2\right] \end{array} \right\}$$

$$(10.71)$$

This equation was the one used in the discussion of *d-d* transitions of *tris*-chelate complexes in Chapter 9.

When **A** has higher than two-fold rotational symmetry, then the quadrupole-dipole term in the expansion of *V* is zero for transitions polarized along the high symmetry rotation axes. Since the octupole term in the *V* expansion also vanishes under these circumstances, the first non-zero term in the *V* expansion is the hexadecapole-dipole term which has an R_{AC}^{-6} factor in the denominator. The appropriate equation may be found in reference [247].

The above equations may be used as the basis for qualitative comparisons of magnitudes expected for the *CD* induced into different transitions, and if the isotropic polarizability terms (with their comparatively simple sector rules) are dominant, then more detailed relationships between geometry and *CD* may be deduced. The degree of parameterization or calculation required to implement the full equation for, say, a $\mathbf{D}_{3d}$ molecule means that these equations are unlikely to be used.

10.9   Magnetic circular dichroism

In §10.6, §10.7 and §10.8 we derived the equations describing the *CD* induced into transitions by a chiral molecular environment. Another way of inducing helical character into a linear electron displacement is to impose a magnetic field on the system. In this section we derive the equations of §9.6 using a formalism consistent with the rest of this book.[248] They describe the effect of a permanent magnetic field on an eda transition. The presence of the external magnetic field means that we cannot work only with real wave functions in this section and it is therefore now important to be careful with the order of superscripts on the electric dipole transition moments as well as on the magnetic dipole transition moments. We proceed by taking the real wave functions of the molecule in the absence of the magnetic field and then determine the effect of the magnetic field using perturbation theory and the Zeeman perturbation

It was Michael Faraday who discovered the electromagnetic nature of light. The Faraday effect is the optical activity that is induced by a magnetic field.

$$\hat{V} = -\hat{\mathbf{k}} \cdot \hat{m}\mathrm{B}_{\mathrm{external}} \qquad (10.72)$$

where $\hat{\mathbf{k}}$ is the unit vector along the radiation propagation direction (which we take to be the *X*-axis), $\mathrm{B}_{\mathrm{external}}$ is the magnitude of the external magnetic field along the *X*-axis, and $\hat{m}$ is the molecule's magnetic dipole moment operator. Maximal *MCD* signals will be generated when the direction of the external magnetic field is also along X, so we take

$$\mathbf{B}_{external} = B(1,0,0) \tag{10.73}$$

The wavefunction for, say, the ground state of a molecule in the presence of the external magnetic field may then be expressed in terms of the unperturbed ground state wavefunction plus contributions 'mixed in' from unperturbed excited states. The coefficient describing how much any one unperturbed excited state contributes to the perturbed state is, to first order in V, purely imaginary since $\hat{m}$ is imaginary. The *MCD* then follows upon finding expressions for $|0\rangle$ and $|j\rangle$, $j = 1, 2, \ldots$, the states of the molecule in the presence of the magnetic field. The magnetic field that is imposed during an *MCD* experiment is small compared with other effects, so we can use perturbation theory to write expressions for the wavefunctions. As above, this separates into two cases: degenerate and non-degenerate.

MCD for a transition from a non-degenerate ground state to a degenerate excited state

If the transition we are probing is degenerate (as is the case for, for example, T_{1g} transitions in octahedral transition metal complexes) then degenerate perturbation theory is used to express the coupling between the degenerate excited states in the presence of the magnetic field. The effect of the magnetic field on the non-degenerate ground state is comparatively small so may be ignored. We limit consideration to two excited states, $|1\rangle$ and $|2\rangle$, that are degenerate in the absence of the magnetic field. Three or more are easily included by replacing $|1\rangle$ by $|j\rangle$ and $|2\rangle$ by $|k\rangle$ and summing over j and k in the final equations.

Degenerate perturbation theory involves solving the secular determinant for the degenerate states $|1\rangle$ and $|2\rangle$, which are by assumption real and orthonormal. The perturbation is $\hat{m}_X B$, so the secular determinant is:

> The Jahn-Teller effect precludes the existence of degenerate ground states, so we focus on degenerate excited states. If the Jahn-Teller distortion is small, then it may be valid to consider the ground state as degenerate in some examples. The analysis is similar to that presented here.

$$\begin{vmatrix} E - \varepsilon & -(1|\hat{m}_X B|2) \\ -(2|\hat{m}_X B|1) & E - \varepsilon \end{vmatrix} \tag{10.74}$$

where E is the energy of the perturbed states and $\varepsilon_1 = \varepsilon_2 = \varepsilon$ are the energies of the unperturbed states. So

> Note: many treatments of *MCD* begin by assuming we already know the functions that occur in the presence of the field.

$$(E - \varepsilon)^2 = B^2(1|\hat{m}_X|2)(2|\hat{m}_X|1) \tag{10.75}$$

from which it follows that

$$E = \varepsilon \mp iB(1|\hat{m}_X|2)(2|\hat{m}_X|1) \tag{10.76}$$

If we write the excited states in the form:

$$|f\rangle = c_1|1\rangle + c_2|2\rangle \tag{10.77}$$

the secular equation relates the coefficients c_1 and c_2 as follows

> $|1\rangle$ and $|2\rangle$ are real so
> $$(1|\hat{m}_X|2)^* = (2|\hat{m}_X|1)$$
> $$= -(1|\hat{m}_X|2)$$

$$\pm c_1 iB\left(2|\hat{m}_X|1\right) = c_2 B\left(2|\hat{m}_X|1\right)$$
$$\pm i c_1 = c_2 \tag{10.78}$$

so the two states that result from the coupling of the degenerate states $|1\rangle$ and $|2\rangle$, when the magnetic field is present are:

$$|f_+\rangle = \frac{|1\rangle + i|2\rangle}{\sqrt{2}}$$

$$E_+ = \varepsilon - iB\left(2|\hat{m}_X|1\right)$$
$$= \varepsilon + B\mathrm{Im}\left(2|\hat{m}_X|1\right) \tag{10.79}$$
$$= \varepsilon + B\mathrm{Im}\hat{m}_X^{12}$$

$$|f_-\rangle = \frac{|1\rangle - i|2\rangle}{\sqrt{2}}$$

$$E_- = \varepsilon + iB\left(2|\hat{m}_X|1\right)$$
$$= \varepsilon - B\mathrm{Im}\left(2|\hat{m}_X|1\right) \tag{10.80}$$
$$= \varepsilon - B\mathrm{Im}\hat{m}_X^{12}$$

To evaluate the *MCD* using equation (10.28) we require the electric dipole transition moments of the perturbed transitions from the ground to excited states:

$$\boldsymbol{\mu}^{0f_\pm} = \langle f_\pm|\hat{\boldsymbol{\mu}}|0\rangle = \frac{1}{\sqrt{2}}\left\{\left(\langle 1| \mp i\langle 2|\right\}\hat{\boldsymbol{\mu}}|0\rangle\right) = \frac{1}{\sqrt{2}}\left(\boldsymbol{\mu}^{01} \mp i\boldsymbol{\mu}^{02}\right) \tag{10.81}$$

The reverse transition is the complex conjugate of equation (10.81). So from equation (10.28) we write the *MCD* rotatory strengths for two coupled transitions as

$$MCD(|0\rangle \to |f_\pm\rangle) = 2ik\left\{\mu_Z^{f_\pm 0}\mu_Y^{0f_\pm} - \mu_Y^{f_\pm 0}\mu_Z^{0f_\pm}\right\}$$

$$= ik\left\{ \begin{array}{c} \left(\mu_Z^{10} \pm i\mu_Z^{20}\right)\left(\mu_Y^{01} \mp i\mu_Y^{02}\right) \\ -\left(\mu_Y^{10} \pm i\mu_Y^{20}\right)\left(\mu_Z^{01} \mp i\mu_Z^{02}\right) \end{array} \right\} \tag{10.82}$$

$$= \pm 2k\left\{\mu_Y^{20}\mu_Z^{01} - \mu_Y^{01}\mu_Z^{20}\right\}$$

The last line of equation (10.82) follows because the unperturbed states are real and all the electric dipole transition moments on its right side are in terms of unperturbed states. Equation (10.82) may alternatively be written

$$MCD(|0\rangle \to |f_\pm\rangle) = \pm 2k\left\{\mu_Y^{01}\mu_Z^{20} - \mu_Y^{02}\mu_Z^{10}\right\}$$

$$= \pm 2k\left\{\left(\boldsymbol{\mu}^{01} \times \boldsymbol{\mu}^{20}\right) \cdot \hat{\mathbf{B}}\right\} \tag{10.83}$$

$$= \pm 2k\left(\boldsymbol{\mu}^{01} \times \boldsymbol{\mu}^{20}\right)_X$$

In this case note that equations (10.79)–(10.83) are in terms of the laboratory fixed axis. Also both the energy and the final *MCD* equations are in terms of the *X*-components of axial vectors. The directions of the electric dipole transition moments may be arbitrarily chosen, but they define the direction of the magnetic dipole transition moment in the energy expression, so reversing one electric dipole transition moment inverts the signs of both the *MCD* equations and the energies so the same experimental spectrum is predicted.

Thus we expect two *MCD* bands of equal magnitude and opposite sign occurring at the energies given by equation (10.79). The order of +/− is important as it relates to the energies of equation (10.79). As with the degenerate coupled-oscillator *CD* (§8.3), the *MCD* for degenerate transitions appears to be independent of the perturbation which, in this case, is the external magnetic field. However, that is not the case since the energy separation of the two components depends on the applied magnetic field, and were we to include a complete treatment of band shapes as is usually done with *MCD* (see *e.g.* reference [7]), then the *MCD* would be obviously dependent on the applied field.

To measure an *MCD* spectrum of transitions to a degenerate excited state we therefore require two degenerate transitions with different polarizations. Further, neither of the electric dipole transition moments must be parallel to the applied magnetic field. The usual experimental situation is to measure the *MCD* of a solution of randomly oriented molecules.

If a molecule is rotated by 180° about an axis in the *Y/Z* plane, the *MCD* signal inverts; however, as the energy perturbation changes sign simultaneously, rotational averaging does not cause the *MCD* signal to vanish so we can measure *MCD* on solutions.

MCD for non-degenerate transitions

For the non-degenerate *MCD* case, again we limit our explicit consideration to three states on the unperturbed molecule (a summation can be introduced to generalize the result if needed): $|0\rangle$, the ground state, and two excited states $|1\rangle$ and $|2\rangle$, at energies ε_1 and ε_2, respectively, above the ground state. We choose the same experimental geometry as for the degenerate case and use non-degenerate perturbation theory, so

$$|0\rangle = |0\rangle + B\left\{\frac{\left(1|\hat{m}_X|0\right)}{\varepsilon_1}|1\rangle + \frac{\left(2|\hat{m}_X|0\right)}{\varepsilon_2}|2\rangle\right\}$$

$$|1\rangle = |1\rangle + B\left\{\frac{\left(0|\hat{m}_X|1\right)}{-\varepsilon_1}|0\rangle + \frac{\left(2|\hat{m}_X|1\right)}{\varepsilon_2 - \varepsilon_1}|2\rangle\right\} \qquad (10.84)$$

$$|2\rangle = |2\rangle + B\left\{\frac{\left(0|\hat{m}_X|2\right)}{-\varepsilon_2}|0\rangle + \frac{\left(1|\hat{m}_X|2\right)}{\varepsilon_1 - \varepsilon_2}|1\rangle\right\}$$

If we ignore all permanent moments, the required transition moments to evaluate the *MCD* for the perturbed system to first order in B are then:

$$\boldsymbol{\mu}(|0\rangle \rightarrow |1\rangle) = \left(1|\bar{\boldsymbol{\mu}}|0\right) + B\left\{\frac{\left(2|\bar{\boldsymbol{\mu}}|0\right)\left(1|\hat{m}_X|2\right)}{\varepsilon_2 - \varepsilon_1} + \frac{\left(1|\bar{\boldsymbol{\mu}}|2\right)\left(2|\hat{m}_X|0\right)}{\varepsilon_2}\right\}$$

$$= \boldsymbol{\mu}^{01} + B\left\{\frac{\boldsymbol{\mu}^{02}m_X^{21}}{\varepsilon_2 - \varepsilon_1} + \frac{\boldsymbol{\mu}^{21}m_X^{02}}{\varepsilon_2}\right\}$$

$$\qquad (10.85)$$

$$\boldsymbol{\mu}(|0\rangle \rightarrow |2\rangle) = \left(2|\bar{\boldsymbol{\mu}}|0\right) + B\left\{\frac{\left(1|\bar{\boldsymbol{\mu}}|0\right)\left(2|\hat{m}_X|1\right)}{\varepsilon_1 - \varepsilon_2} + \frac{\left(2|\bar{\boldsymbol{\mu}}|1\right)\left(1|\hat{m}_X|0\right)}{\varepsilon_1}\right\}$$

$$= \boldsymbol{\mu}^{02} + B\left\{\frac{\boldsymbol{\mu}^{01}m_X^{12}}{\varepsilon_1 - \varepsilon_2} + \frac{\boldsymbol{\mu}^{12}m_X^{01}}{\varepsilon_1}\right\}$$

From equations (10.28) and (10.85) it follows that the *MCD* is

$$MCD(|0\rangle \rightarrow |1\rangle) = i2kB \left\{ \begin{array}{l} \dfrac{\mu_Z^{10} \mu_Y^{02} m_X^{21}}{\varepsilon_2 - \varepsilon_1} + \dfrac{\mu_Z^{10} \mu_Y^{21} m_X^{02}}{\varepsilon_2} + \dfrac{\mu_Y^{01} \mu_Z^{20} m_X^{12}}{\varepsilon_2 - \varepsilon_1} \\[3mm] + \dfrac{\mu_Y^{01} \mu_Z^{12} m_X^{20}}{\varepsilon_2} - \dfrac{\mu_Y^{10} \mu_Z^{02} m_X^{21}}{\varepsilon_2 - \varepsilon_1} - \dfrac{\mu_Y^{10} \mu_Z^{21} m_X^{02}}{\varepsilon_2} \\[3mm] - \dfrac{\mu_Z^{01} \mu_Y^{20} m_X^{12}}{\varepsilon_2 - \varepsilon_1} - \dfrac{\mu_Z^{01} \mu_Y^{12} m_X^{20}}{\varepsilon_2} \end{array} \right\} \quad (10.86)$$

$$= i4kB \left\{ \begin{array}{l} \left(\mu_Y^{02} \mu_Z^{10} - \mu_Z^{02} \mu_Y^{10} \right) \dfrac{m_X^{21}}{\varepsilon_2 - \varepsilon_1} \\[3mm] + \left(\mu_Z^{10} \mu_Y^{21} - \mu_Y^{10} \mu_Z^{21} \right) \dfrac{m_X^{02}}{\varepsilon_2} \end{array} \right\}$$

and

$$MCD(|0\rangle \rightarrow |2\rangle) = i4kB \left\{ \begin{array}{l} \left(\mu_Y^{20} \mu_Z^{01} - \mu_Z^{20} \mu_Y^{01} \right) \dfrac{m_X^{12}}{\varepsilon_1 - \varepsilon_2} \\[3mm] + \left(\mu_Z^{20} \mu_Y^{12} - \mu_Y^{20} \mu_Z^{12} \right) \dfrac{m_X^{01}}{\varepsilon_1} \end{array} \right\} \quad (10.87)$$

The final lines in the previous two equations follow because all the electric moments are real, and all the magnetic ones are purely imaginary.

For the usual situation where the sample is randomly oriented then the *MCD* of the $|0\rangle$ to $|1\rangle$, transition is

$$MCD(|0\rangle \rightarrow |1\rangle) = \frac{-i4kB}{3} \left\{ \frac{\boldsymbol{\mu}^{01} \times \boldsymbol{\mu}^{20} \cdot \boldsymbol{m}^{21}}{\varepsilon_2 - \varepsilon_1} + \frac{\boldsymbol{\mu}^{01} \times \boldsymbol{\mu}^{12} \cdot \boldsymbol{m}^{02}}{\varepsilon_2} \right\} \quad (10.88)$$

In most cases $(\varepsilon_2 - \varepsilon_1) \ll \varepsilon$, so the first term in equation (10.88) is dominant. Thus

$$MCD(|0\rangle \rightarrow |1\rangle) = \frac{-i4kB}{3} \left\{ \frac{\boldsymbol{\mu}^{01} \times \boldsymbol{\mu}^{20} \cdot \boldsymbol{m}^{21}}{\varepsilon_2 - \varepsilon_1} \right\}$$

$$= \mathrm{Im} \frac{4kB}{3} \left\{ \frac{\boldsymbol{\mu}^{01} \times \boldsymbol{\mu}^{20} \cdot \boldsymbol{m}^{21}}{\varepsilon_2 - \varepsilon_1} \right\} \quad (10.89)$$

The key feature of this equation is that we can only expect to observe *MCD* if a transition can couple with one that is not parallel to it. By permuting the labels '1' and '2' in the above equation and remembering $\boldsymbol{m}^{12} = -\boldsymbol{m}^{21}$ we find the *MCD* induced into the transition at energy ε_2. If the simple three-level system we have assumed is actually appropriate, then the *MCD* signals in the two transitions will be equal in magnitude and opposite in sign. Thus, we can use *MCD* to probe the relative polarizations of transitions, and also to see how many transitions there are in a given region of the spectrum.

The equations we have derived here correspond to what are commonly referred to as the 'B' terms.[7,249] They do not look exactly the same as those commonly quoted, since, as noted above, we have made no attempt to consider the band shape.

10.10 Multipole expansion of the interaction operator

Throughout this book we have focused on dipolar interaction energies as summarized in equation (10.11). However, in some cases higher order terms must be considered. For example, in oriented samples electric dipole-electric quadrupole terms can be as large as electric dipole-magnetic dipole terms. Consider two chromophores **A** and **C** with sets of charges q_a and q_c respectively as illustrated in Figure 10.5. The electrostatic interaction operator between the two collections of charges is then

$$\tilde{V} = \sum_{a,c} \frac{q_a q_c}{|R_{AC} + r_c - r_a|} \tag{10.90}$$

Using a Taylor series expansion equation (10.90) can be re-expressed

$$\tilde{V} = \sum_{a,c,n} \frac{q_a q_c}{n!} (r_c - r_a)^n (\cdot)^n \hat{\nabla}^n \left(|R_{AC} + r_c - r_a|^{-1} \right)_{r_c - r_a = 0}$$

$$= \sum_{a,c,n} \frac{q_a q_c}{n!} P(n,k) \left[r_c^{n-k} r_a^k \right]^n (\cdot)^n \hat{\nabla}^n \left(|R_{AC} + r_c - r_a|^{-1} \right)_{r_c - r_a = 0} \tag{10.91}$$

where $(\cdot)^n$ indicates an n-fold tensor dot product, and $P(n,k) \left[r_c^{n-k} r_a^k \right]^n$ denotes the sum of all permutations of $(n-k)$ r_c's and k r_a's. We wish to work in terms of multipoles on **A** and **C**, which requires all r_a's to be collected together and similarly all r_c's. We therefore put the permutation into $\hat{\nabla}^n \left(|R_{AC} + r_c - r_a|^{-1} \right)$ and write

$$\tilde{V} = \sum_{n,k} \frac{(-1)^k}{n! k! (n-k)!} \hat{A}^k \hat{C}^{n-k} (\cdot)^n \tilde{V}_n \tag{10.92}$$

where

$$\hat{A}^k = \sum_a q_a r_a^k \tag{10.93}$$

is the kth order multipole on **A**. So $\hat{A}^1 = \mu_A$ the electric dipole operator, $\hat{A}^2 = Q_A$ the electric quadrupole operator, *etc.* The components of $\tilde{V}_n$ are defined by

$$V_{\alpha_1 .. \alpha_n} = \sum_{P(\beta_1 ... \beta_n)} \left[\hat{\nabla}^n \left(|R_{AC} + r_c - r_a|^{-1} \right)_{r_c - r_a = 0} \right]_{\beta_1 ... \beta_n} \tag{10.94}$$

where the summation is over all components whose indices are some permutation of $\beta_1 ... \beta_n$. As a result $\tilde{V}_n$ is symmetric to any permutation of its indices. The first four $\tilde{V}_n$ are

$$\tilde{V}_o = (R_{AC})^{-1} \tag{10.95}$$

$$\tilde{V}_\alpha = (R_{AC})^{-2} \left[-\hat{R}_{AC} \right]_\alpha \tag{10.96}$$

$$\check{V}_{\alpha\beta} = 2(R_{AC})^{-3}\left[-\delta_{\alpha\beta} + 3\left[-\hat{R}_{AC}\right]_{\alpha}\left[-\hat{R}_{AC}\right]_{\beta}\right] \tag{10.97}$$

$$\check{V}_{\alpha\beta\gamma} = 18(R_{AC})^{-4}\left[\begin{array}{c}\delta_{\alpha\beta}\left[-\hat{R}_{AC}\right]_{\gamma} + \delta_{\alpha\gamma}\left[-\hat{R}_{AC}\right]_{\beta} + \delta_{\beta\gamma}\left[-\hat{R}_{AC}\right]_{\alpha} \\ -5\left[-\hat{R}_{AC}\right]_{a}\left[-\hat{R}_{AC}\right]_{\beta}\left[-\hat{R}_{AC}\right]_{\gamma}\end{array}\right] \tag{10.98}$$

where $\delta_{\alpha\beta}=0$ unless $\alpha=\beta$. Equations (10.92) and (10.97) lead to equation (10.43) for the dipole-dipole interaction term. Whereas equations (10.92) and (10.98) give the quadrupole-dipole interaction term

$$\frac{3\left[I_{\hat{Q}_A}\hat{R}_{AC}\cdot\hat{\mu}_C + 2\hat{R}_{AC}\cdot\hat{Q}_A\cdot\hat{\mu}_C - 5\hat{R}_{AC}\cdot\hat{Q}_A\cdot\hat{R}_{AC}\hat{R}_{AC}\cdot\hat{\mu}_C\right]}{2R_{AC}^4} \tag{10.99}$$

and similarly for higher order terms where $I_{\hat{Q}_A} = \left(\hat{Q}_{Axx} + \hat{Q}_{Ayy} + \hat{Q}_{Azz}\right)$ is the trace of the quadrupole tensor. To determine the quadrupole-dipole contributions to any *CD* expression insert equation (10.99) in place of $\check{V}$ in the appropriate equation.

11 Further derivations and definitions

This chapter contains some useful mathematical and physical definitions related to properties and interactions of molecules and the derivation of equations useful for analysing binding data.

Linear Dichroism and Circular Dichroism: A Textbook on Polarized-Light Spectroscopy
By Bengt Nordén, Alison Rodger and Timothy Dafforn
© B. Nordén, A. Rodger and T. Dafforn, 2010
Published by the Royal Society of Chemistry, www.rsc.org

11.1   Vectors

Vectors are used to describe any property that has both magnitude and direction. Pictorially they are represented by a straight arrow, whose length is proportional to the magnitude of the property, which points along the direction of the property. For example a car moving west with velocity 100 km/hour could be represented by an arrow 1 cm in length pointing to the left on a map oriented with north at the top. In this case the scale is 1 cm corresponding to 100 km/hour. A car moving at 200 km/hour in the same direction would be represented by a 2 cm arrow.

It is usually convenient to express vectors in terms of a right handed $\{x,y,z\}$ cartesian axis system. In this case the vector, v, is written $v = (X,Y,Z)$ where the point (X,Y,Z) is the cartesian coordinate of the tip of the arrow head when the tail is sitting at the origin (Figure 11.1). The length of the vector is then

$$|v| = v = \sqrt{X^2 + Y^2 + Z^2} \qquad (11.1)$$

We define a unit vector, $\hat{v}$, to be the vector of unit length along the direction of v

$$\hat{v} = \frac{v}{v} \qquad (11.2)$$

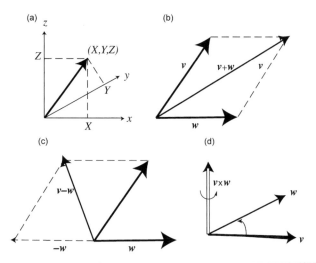

Figure 11.1 (a) Cartesian representation of a vector. (b) Sum of two vectors. (c) Difference of two vectors. (d) The cross product of two vectors.

Two vectors may be added together. Pictorially this performed by determining the diagonal of the parallelogram as illustrated in Figure 11.1b. Subtraction involves reversing the direction of the second vector (Figure 11.1c). Mathematically for two vectors $\mathbf{v} = (v_x, v_y, v_z)$ and $\mathbf{w} = (w_x, w_y, w_z)$ this may be written

$$\mathbf{v} + \mathbf{w} = \left(v_x, v_y, v_z \right) + \left(w_x, w_y, w_z \right)$$
$$= \left(v_x + w_x, v_y + w_y, v_z + w_z \right) \tag{11.3}$$

and

$$\mathbf{v} - \mathbf{w} = \left(v_x, v_y, v_z \right) - \left(w_x, w_y, w_z \right)$$
$$= \left(v_x - w_x, v_y - w_y, v_z - w_z \right) \tag{11.4}$$

Three vector products are used in the text: the scalar product $\mathbf{v} \cdot \mathbf{w}$ the cross product $\mathbf{v} \times \mathbf{w}$, and the scalar triple product which is a combination of the other two: $\mathbf{u} \cdot \mathbf{v} \times \mathbf{w}$. They are evaluated as follows:

$$\mathbf{v} \cdot \mathbf{w} = \mathbf{w} \cdot \mathbf{v} = \left(v_x, v_y, v_z \right) \cdot \left(w_x, w_y, w_z \right)$$
$$= v_x w_x + v_y w_y + v_z w_z \tag{11.5}$$

and

$$\mathbf{v} \times \mathbf{w} = -\mathbf{w} \times \mathbf{v} = \left(v_x, v_y, v_z \right) \times \left(w_x, w_y, w_z \right)$$
$$= \begin{vmatrix} \mathbf{i} & \mathbf{j} & \mathbf{k} \\ v_x & v_y & v_z \\ w_x & w_y & w_z \end{vmatrix} \tag{11.6}$$
$$= \left(v_y w_z - v_z w_y, v_z w_x - v_x w_z, v_x w_y - v_y w_x \right)$$

where $\boldsymbol{i}, \boldsymbol{j}$, and $\boldsymbol{k}$ are, respectively, the unit vectors along the x-, y-, and z-axes. Thus,

$$\mathbf{u} \cdot \mathbf{v} \times \mathbf{w} = \mathbf{v} \times \mathbf{w} \cdot \mathbf{u} = \mathbf{u} \times \mathbf{v}.\mathbf{w}$$
$$= -\mathbf{u} \cdot \mathbf{w} \times \mathbf{v} = -\mathbf{u} \times \mathbf{w}.\mathbf{v}$$
$$= \left(u_x, u_y, u_z \right) \cdot \left(v_x, v_y, v_z \right) \times \left(w_x, w_y, w_z \right) \tag{11.7}$$
$$= \begin{vmatrix} u_x & u_y & u_z \\ v_x & v_y & v_z \\ w_x & w_y & w_z \end{vmatrix}$$
$$= u_x \left(v_y w_z - v_z w_y \right) - u_y \left(v_x w_z - v_z w_x \right) + u_z \left(v_x w_y - v_y w_x \right)$$

The scalar or dot product of two vectors is a measure of the projection of one vector onto the other. If they are parallel it is simply a product of their magnitudes, if they are antiparallel, then it has the same magnitdue but is negative in sign, and if the vectors are perpendicular then the dot product is zero. The angle between the two vectors, τ is given by

$$\cos\tau = \frac{v \cdot w}{vw} \tag{11.8}$$

The cross or vector product of two vectors (Figure 11.1d) is a vector that is perpendicular to both of the original vectors and whose direction is such that the first vector, the second vector, and the new vector make a right-handed axis system (Figure 1.2). The magnitude of the new vector is the area of the parallelogram that the original two vectors define, so it is zero if those vectors are parallel.

The scalar triple product, as its name implies, is a number that measures the volume of the parallelepiped defined by the three vectors. So if any of the vectors are coplanar, the scalar triple product vanishes.

In the context of *CD* theory we often refer to axial vectors, which are as described above, and polar vectors. A polar vector is a vector that corresponds to a physical property, *e.g.* angular momentum or magnetic dipole, that is defined as the cross product of two axial vectors. Because it is helpful to think of these vectors in terms of the original two, we think of it as a circulation about the vector perpendicular to the plane defined by the two original vectors with the direction defined to make a right-handed system as described above.

11.2 Relationship between isotropic and unpolarized absorbance

The reduced linear dichroism is LD/A_{iso}. In many instances one can measure LD and A_{iso} from which LD^r follows. However, in some cases our sample (such as a film) cannot be readily randomized to measure A_{iso}. In such a situation it is convenient to have a way to determine A_{iso} from the absorbance, A, of the sample measured with unpolarized light at normal incidence in a normal absorption spectrophotometer.

A relation between A_{iso} and A may be derived for a uniaxial sample, with radiation propagating along the X and Z being the unique axis of the sample. For a uniaxial sample $A_x = A_y$. So

$$A_{iso} = \frac{\left(A_X + A_Y + A_Z\right)}{3} = \frac{\left(2A_Y + A_Z\right)}{3}$$

$$= \frac{1}{3}\left(2\log\left(\frac{I_{oY}}{I_Y}\right) + \log\left(\frac{I_{oZ}}{I_Z}\right)\right) \qquad (11.9)$$

$$= \left(\log\left(\frac{I_{oZ}}{I_Y^{2/3}I_Z^{1/3}}\right)\right)$$

Note that, except at $LD = 0$, A(unpolarized) will generally differ from the average $(A_Z + A_Y)/2$ which is often incorrectly used as measure of the absorption of the sample.

and

$$A = \log\left(\frac{I_{oY} + I_{oZ}}{I_Y + I_Z}\right) = \log\left(\frac{2I_{oZ}}{I_Y + I_Z}\right) \qquad (11.10)$$

where I_{oZ} and I_Z denote, respectively, Z polarized light before and after passage of thesample *etc*. Similarly, the linear dichroism of the sample is:

$$LD = \log\left(\frac{I_{oZ}}{I_Z}\right) - \log\left(\frac{I_{oY}}{I_Y}\right) = \log\left(\frac{I_Y}{I_Z}\right) \qquad (11.11)$$

From equations (11.9), (11.10), and (11.11) it follows that

$$A_{iso} = \log\left(\frac{I_{oZ}}{I_Y^{2/3}I_Z^{1/3}}\right) = A + \log\left(\frac{I_{oZ}\left(I_Y + I_Z\right)}{2I_Y^{2/3}I_Z^{1/3}I_{oZ}}\right) \qquad (11.12)$$

From which it follows that

$$A_{iso} = A + \log\left(\frac{I_Y}{I_Z}\right)^{1/3} + \log\left(\frac{(I_Y + I_Z)}{2I_Y}\right)$$

$$= A + \frac{LD}{3} + \log\left(\frac{1}{2}\left(1 + \frac{I_Z}{I_Y}\right)\right)$$

$$= A + \frac{LD}{3} + \log\left(\frac{1}{2}\left(1 + 10^{-LD}\right)\right) \qquad (11.13)$$

$$= A + \frac{LD}{3} + \log\left(1 + 10^{-LD}\right) - \log 2$$

which provides a correction function that is only a function of A and the LD of the sample.

11.3   Determination of equilibrium binding constants

There are many methods of determining binding constants. The ones outlined below are among the simplest and indicate some of the possibilities.

The simplest measure of the binding strength between two molecules (such as a DNA-binding ligand and DNA) is the equilibrium binding constant,

$$K = \frac{L_b}{L_b S_f} \qquad (11.14)$$

for the equilibrium

free ligand + empty binding site $\rightleftharpoons$ bound ligand

where L_b is the concentration of bound ligand, L_f is the concentration of free ligands, and S_f is the free site concentration. The total site concentration

$$S_{tot} = \frac{C_M}{n} \qquad (11.15)$$

where C_M is the macromolecule concentration. For DNA we usually use the concentration of bases, in which case n is the number of bases in a binding site. For a protein, C_M is usually taken to be the concentration of protein molecules in which case

$$n' = \frac{1}{n} \qquad (11.16)$$

is the number of ligand binding sites on each protein molecule. Typically $n' = 1$ for proteins

Sometimes for macromolecules one is reduced to evaluating the apparent binding constant at a given DNA and ligand concentration, since the binding strength and/or geometry is concentration dependent. However, concentration regimes either of a single binding mode or of constant proportions of a number of modes will often occur. K is indeed a constant in such uniform binding regimes, and the shapes of spectra such as that of the ligand induced CD or LD will also be constant. The four methods outlined below are particularly appropriate for use with LD and CD data.

The advantages of using spectroscopic data for determining K include the short timescale of the experiment (meaning that the system does not have to

If binding reduces or increases S, the LD orientation parameter, in a non-linear fashion then that needs to be taken into account.

be stable for any great length of time) and our ability to probe a signal due only to the bound ligand. The starting point for equilibrium constant determination from spectroscopic data is usually the following equation:

$$L_b = \alpha \rho \qquad (11.17)$$

where ρ is the *LD* or *CD* signal at a chosen wavelength and α (which is a function of wavelength) is a constant over the range of binding ratios being considered. Wavelengths of maximum magnitude *LD* or *CD* signal will provide the most accurate data.

The simplest means of determining α is usually from the low binding ratio limit where all the ligand may be assumed to be bound so L_f is assumed to be zero. If this is indeed the case then

A plot of *ρ versus L* should be a straight line if indeed all the ligand is bound.

$$\alpha = \frac{L_{tot}}{\rho} \qquad (11.18)$$

where L_{tot} is the total ligand concentration. Alternatively, as indicated below in method II, the maximum *LD* or *CD* signal may be used to determine α if the binding site size, n, is known.

Equation (11.18) restricts us not only to regimes of uniform binding, but also to situations where ligand-ligand interactions do not affect the signal. This is unlikely to be a problem for *LD*, but if, for example, ligands stack externally on DNA their *CD* signal arises at least in part from ligand-ligand exciton coupling whose magnitude will depend non-linearly on the number of molecules in the stack.

I. Scatchard plot

The method most widely used in one form or another for determining K is the one developed by Scatchard.[258] Equation (11.14) is rearranged as follows:

$$\frac{r}{L_f} = \frac{KS_f}{C_M} \qquad (11.19)$$

$$= \frac{K}{n} - rK$$

where

$$r = \frac{L_b}{C_M} \qquad (11.20)$$

since $S_f = S_{tot} - L_b$. So, a plot of r/L_f *versus* r has slope $-K$ and y-intercept K/n. The x-intercept occurs where $r = 1/n$. L_b, and hence L_f, may be determined directly from the *LD* or *CD* data if α of equation (11.18) has been determined as discussed above. A Scatchard plot is illustrated below in Figure 11.3c, where results from the intrinsic method (see below) are used to calculate the required r and L_f values.

II. Ligand number

This method was used by Nordén and Tjerneld[259] for *LD*; it has since been independently used for normal absorption studies.[260] The ligand number, a concept that was originally introduced in the context of step-wise formation of metal-ligand complexes,[261] may be defined as:

$$v = \frac{L_b}{S_{tot}} = \frac{\rho}{\rho_{max}} \tag{11.21}$$

where ρ_{max} is the high ligand concentration limiting value for the *LD* or *CD* corresponding to the situation where every binding site is occupied so the maximum possible signal is being measured. In this case equation (11.14) is rewritten

In practice ρ_{max} is often difficult to measure as it requires high total ligand concentrations which may mean that the total ligand absorbance (due to free and bound) is too large for the spectropolarimeter to measure the *LD* or *CD*.

$$K = \frac{L_b S_{tot}}{L_f S_{tot}(S_{tot} - L_b)} \tag{11.22}$$

which upon using the definition of v becomes

$$K = \frac{L_b}{L_f S_{tot}(1 - v)} \tag{11.23}$$

so

$$\frac{1}{(1-v)} = \frac{(L_{tot} - L_b)S_{tot}K}{L_b}$$

$$= \frac{L_{tot}}{v}K - \frac{C_M}{n}K \tag{11.24}$$

Thus $\dfrac{1}{(1-v)}$ *versus* $\dfrac{L_{tot}}{v}$ has slope K, and y-intercept $-\dfrac{C_M}{n}K$.

III. Intrinsic method

It is not always possible to get data for either high or low binding ratio limits. In such cases, equation (11.14) may be written

$$K = \frac{\alpha\rho}{(S_{tot} - \alpha\rho)(L_{tot} - \alpha\rho)} \tag{11.25}$$

Rearranging equation (11.25) gives

$$L_{tot} = \frac{L_{tot}S_{tot}}{\alpha\rho} - S_{tot} + \alpha\rho - \frac{1}{K} \tag{11.26}$$

so, for two different total ligand concentrations, L_{tot}^j and L_{tot}^k, but the same macromolecule concentration, *i.e.* $S_{tot}^j = S_{tot}^k$

$$\frac{L_{tot}^k - L_{tot}^j}{\rho^k - \rho^j} = \frac{S_{tot}}{\alpha}\left(\frac{\dfrac{L_{tot}^k}{\rho^k} - \dfrac{L_{tot}^j}{\rho^j}}{\rho^k - \rho^j}\right) + \alpha \tag{11.27}$$

Thus a plot of

$$y = \frac{L_{tot}^k - L_{tot}^j}{\rho^k - \rho^j} \tag{11.28}$$

versus

$$x = \frac{\dfrac{L_{tot}^k}{\rho^k} - \dfrac{L_{tot}^j}{\rho^j}}{\rho^k - \rho^j}. \qquad (11.29)$$

(where any pair of data points considered have the same C_M) should be a straight line with slope $C_M(n\alpha)^{-1}$ and intercept α. The concentration of bound molecules in any sample may then be determined from equation (11.18), as may the effective binding site size. The equilibrium binding constant, K, follows directly from equation (11.25). Alternatively, using these accurate values of n and α, a Scatchard plot (Method I) may be used to determine the best value of K using all the data points. The intrinsic method[11] is illustrated for some *CD* data in Figure 11.2 and for some *LD* data from which a Scatchard plot is then used to determine K in Figure 11.3.

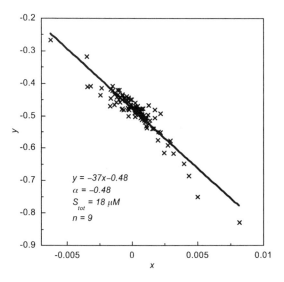

Figure 11.2 Graph illustrating application of the intrinsic method for determining α and n using the induced *CD* data given in Figure 4.8. Poly[d(G-C)$_2$] (160 µM), path length = 1 cm; anthracene-9-carbonyl-N^1-spermine:DNA ratios are: 0.008:1; 0.015:1; 0.022:1; 0.029:1; 0.037:1; 0.045:1; 0.056:1; 0.067:1; 0.083:1; 0.091:1; 0.11:1; 0.12:1; 0.14:1; 0.15:1; 0.16:1; 0.17:1. x and y are as defined in equations (11.28) and (11.29).

It is sometimes convenient to perform experiments with constant ligand and varying macromolecule concentration. In this case C_M, and hence S_{tot}, are the variables and L_{tot} is fixed. Rather than equation (11.27) we then use:

$$\frac{S_{tot}^k - S_{tot}^j}{\rho^k - \rho^j} = \frac{L_{tot}}{\alpha}\left(\frac{\dfrac{S_{tot}^k}{\rho^k} - \dfrac{S_{tot}^j}{\rho^j}}{\rho^k - \rho^j}\right) + \alpha \qquad (11.30)$$

or equivalently

$$\frac{C_M^k - C_M^j}{\rho^k - \rho^j} = \frac{L_{tot}}{\alpha} \left(\frac{\dfrac{C_M^k}{\rho^k} - \dfrac{C_M^j}{\rho^j}}{\rho^k - \rho^j} \right) + n\alpha \qquad (11.31)$$

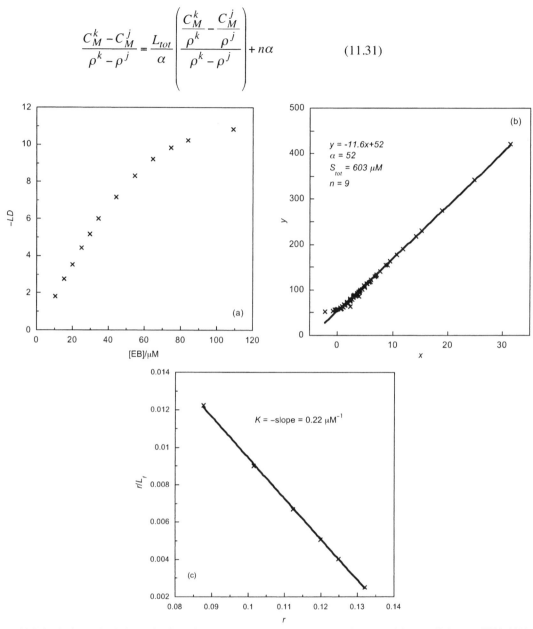

Figure 11.3 Intrinsic method determination of α and n for the binding of ethidium bromide to calf thymus DNA (426 μM in base, in 0.2 M NaCl) and subsequent Scatchard plot. *LD* data from reference[259]. (a) Experimental data, (b) intrinsic method plot, x and y as defined in equations (11.28) and (11.29), and (c) Scatchard plot using data from final six points in (a) where the relationship between *LD* and concentration is not a straight line through the origin. K agrees with the best value from reference 2. Note all concentrations are plotted in units of μM, all other units are consistent between figures, and in (b) we use only the magnitude of the *LD*, so α is positive rather than negative.

So the plot of

$$y = \frac{C_M^k - C_M^j}{\rho^k - \rho^j} \qquad (11.32)$$

versus

$$x = \frac{\dfrac{C_M^k}{\rho^k} - \dfrac{C_M^j}{\rho^j}}{\rho^k - \rho^j} \tag{11.33}$$

has slope L_{tot}/α and y-intercept $n\alpha$.

IV. McGhee and von Hippel Method

McGhee and von Hippel[262, 263] analyzed the binding of a ligand to a macromolecule, such as DNA, that could be represented as a lattice. Their concern was firstly to account for the fact that a binding site might span more than one lattice point, in which case defining $S = C_M/n$ is too simplistic except at low binding ratios (where this approach gives the same answer as the above methods). The revised form of equation (11.19) is then

$$\frac{r}{L_f} = K\left(1 - \frac{r}{n}\right)\left[\frac{n-r}{n-(1-n)r}\right]^{\frac{1}{n}-1} \tag{11.34}$$

so r/L_f versus r gives K as the intercept on the r/L_f axis and n on the r axis.

Further, they considered the effect of the ligand binding being cooperative, by which they meant a second ligand is more likely to bind next to the first than not. When cooperativity is taken into account the relationship between r/L_f and r is more complicated than indicated by equation (11.34) as outlined in reference [262, 263], however, the intercepts are still the same. As noted by McGhee and von Hippel, even moderately large degrees of cooperativity lead to extremely steep extrapolations and hence unreliable values for K and n.

V. Binding constant determined assuming one site per DNA base pair

A widely used method, especially in the bioinorganic literature, for determining binding constants is derived from the work of Schmechel and Crothers,[264] though the usual reference given is reference [265]. It is based on the original work of Benesi and Hildebrand[266] and it can be adopted for *LD* and *CD* data. The equation is:

$$\frac{[DNA]}{\varepsilon_{ap} - \varepsilon_f} = \frac{[DNA]}{(\varepsilon_b - \varepsilon_f)} + \frac{(\varepsilon_b - \varepsilon_f)}{K} \tag{11.35}$$

where ε_{ap} is the average extinction coefficient of the ligand in the sample (*i.e.* absorbance divided by total ligand concentration and path length), ε_b is the extinction coefficient of the bound ligand, ε_f is the extinction coefficient of the free ligand, [DNA] is the concentration of the DNA *in basepairs*, and K is the binding constant at a given mixing ratio. Thus, equation (11.35) is only valid when there is one potential binding site per base pair, and the experiment is performed at low binding ratio r. Many applications in the literature satisfy neither of these requirements, so the resulting values of K must be treated with suspicion.

11.4   Momentum-dipole equivalence

One of the key equations (equation (10.39)) used in deriving the coupled-oscillator *CD* is the equivalence between the expectation value between two states, $\left|j\right\rangle$ and $\left|k\right\rangle$ of the momentum operator for a chromophore, say **C**, and the electric dipole transition moment from $\left|k\right\rangle$ to $\left|j\right\rangle$:

$$\left(k\middle|\widehat{\boldsymbol{p}}_C\middle|j\right) = \frac{im_e}{e\hbar}\left(\varepsilon_k - \varepsilon_j\right)\left(k\middle|\widehat{\boldsymbol{\mu}}_C\middle|j\right)$$

$$= \frac{im_e}{e\hbar}\left(\varepsilon_k - \varepsilon_j\right)\boldsymbol{\mu}_C^{jk} \tag{11.36}$$

Note: the order of *j* and *k* in equation (11.36) is very important.

Equation (11.36) follows from first noting that the momentum operator

$$\widehat{\boldsymbol{p}} = -i\hbar\left(\frac{\partial}{\partial x}, \frac{\partial}{\partial y}, \frac{\partial}{\partial z}\right) = -i\hbar\nabla \tag{11.37}$$

Using the definition of $\boldsymbol{\mu}$ (equation (10.12)), this means what we are required to show is that

$$\left(k\middle|\frac{\partial}{\partial x}\middle|j\right) = -\frac{m_e}{\hbar^2}\left(\varepsilon_k - \varepsilon_j\right)\left(k\middle|x\middle|j\right) \tag{11.38}$$

and similarly for y and z where the coordinates belong to chromophore **C**. The derivation given here follows that of reference [267].

$\left|j\right\rangle$ and $\left|k\right\rangle$ satisfy the time independent Schrödinger equation, so in the x-direction:

$$\frac{\partial^2}{\partial x^2}\middle|k\right\rangle^* + \frac{2m_e}{\hbar^2}\left[\varepsilon_k - V(x)\right]\middle|k\right\rangle^* = 0 \tag{11.39}$$

$$\frac{\partial^2}{\partial x^2}\middle|j\right\rangle + \frac{2m_e}{\hbar^2}\left[\varepsilon_j - V(x)\right]\middle|j\right\rangle = 0 \tag{11.40}$$

Note: the notation $\left(k\right|$ denotes the complex conjugate of $\left|k\right\rangle$, and when a bra and a ket are multiplied together integration is implied.

If we multiply equation (11.39) by $\left(j\right|^*x$ and equation (11.40) by $\left(k\right|x$ and then subtract, we obtain

$$\left\{\left(\left(j\middle|x\frac{\partial^2}{\partial x^2}\middle|k\right)\right)^* - \left(k\middle|x\frac{\partial^2}{\partial x^2}\middle|j\right)\right\} = +\frac{2m_e}{\hbar^2}\left(\varepsilon_j - \varepsilon_k\right)\left\{\left(k\middle|x\middle|j\right)\right\} \tag{11.41}$$

The left two terms may now be integrated by parts; since the wave functions vanish at infinity, we have

$$\left\{\left(\frac{\partial}{\partial x}\left[\left(j\middle|x\right]\frac{\partial}{\partial x}\middle|k\right)\right)^* - \left(\frac{\partial}{\partial x}\left[\left(k\middle|x\right]\frac{\partial}{\partial x}\middle|j\right)\right)\right\} = +\frac{2m_e}{\hbar^2}\left(\varepsilon_k - \varepsilon_j\right)\left\{\left(k\middle|x\middle|j\right)\right\} \tag{11.42}$$

from which it follows that

$$\left\{\left(j\middle|\frac{\partial}{\partial x}\middle|k\right)^* - \left(k\middle|\frac{\partial}{\partial x}\middle|j\right)\right\} = +\frac{2m_e}{\hbar^2}\left(\varepsilon_k - \varepsilon_j\right)\left(k\middle|x\middle|j\right) \tag{11.43}$$

Integration by parts can be used to show that the two terms on the left-hand side of equation (11.43) are equivalent. So

$$\left(k\middle|\frac{\partial}{\partial x}\middle|j\right) = -\frac{m_e}{\hbar^2}\left(\varepsilon_k - \varepsilon_j\right)\left\{\left(k\middle|x\middle|j\right)\right\} \tag{11.44}$$

which is equation (11.36).

11.5   Definitions and units

Linearly polarized light

We have defined the direction of propagation of light to be X. Light linearly polarized along the Y or Z axes has its electric field parallel respectively to the unit vectors

$$\hat{\mathbf{e}}_Y = (0,1,0) \tag{11.45}$$

or

$$\hat{\mathbf{e}}_Z = (0,0,1) \tag{11.46}$$

Circularly polarized light

Right circularly polarized light describes a left-handed helix in space and rotates clockwise in time when viewed at a given point along X *towards* the source (§1.2.) Its electric field is parallel to

$$\hat{\mathbf{e}}_r = \frac{1}{\sqrt{2}}\left\{(0,1,i)\exp\left(\frac{2\pi iX}{\lambda} - i\omega t\right)\right\} \tag{11.47}$$

using the notation of equation (10.8). Similarly, left circularly polarized light describes a right-handed helix in space and rotates anticlockwise in time when viewed towards the light source

$$\hat{\mathbf{e}}_\ell = \frac{1}{\sqrt{2}}\left\{(0,1,-i)\exp\left(\frac{2\pi iX}{\lambda} - i\omega t\right)\right\} \tag{11.48}$$

Absorbance

The absorbance of a sample is defined to be the logarithm (usually to base 10) of the ratio of the incident and transmitted radiation

$$A = A_{10} = \log_{10}\left(\frac{I_o}{I}\right) \tag{11.49}$$

Circular dichroism

CD in absorbance units is

$$CD = A_\ell - A_r \tag{11.50}$$

Alternatively in molar units it is

$$CD = \varepsilon_\ell - \varepsilon_r \tag{11.51}$$

where ε_r is the extinction coefficient for the absorption of right circularly polarized light. The units used for extinction coefficients are almost always: $mol^{-1}\,dm^3\,cm^{-1}$ which gives values that are ten times greater than those in SI units. We retain this set of units for extinction coefficients. The Beer-Lambert law relates absorbance units to the extinction coefficient:

$$(A_\ell - A_r) = (\varepsilon_\ell - \varepsilon_r)C\ell \tag{11.52}$$

where C is in units of $mol\,dm^{-3}$ and ℓ is the path length in cm.

Ellipticity (old measure of CD)

The ellipticity, θ, is obtained from the ratio of the minor and major axes of the ellipse traced out by the electric field vector of the elliptically polarized light when it emerges from the chiral sample onto which linearly polarized light was incident

$$\tan\theta = \tanh\left(\frac{\pi\ell}{\lambda}\left(n'_\ell - n'_r\right)\right) \tag{11.53}$$

where n'_ℓ is the absorption index for left circularly polarized light. The wavelength, λ, must be in the same units as ℓ. For small ellipticities

$$\theta/\text{radians} \approx \frac{\pi\ell}{\lambda}\left(n'_\ell - n'_r\right)$$

$$= \frac{2.303\ell C}{4}\left(\varepsilon_\ell - \varepsilon_r\right) \tag{11.54}$$

so

$$\theta/\text{degrees} = 32.98C\ell\left(\varepsilon_\ell - \varepsilon_r\right)$$

$$= 32.98\left(A_\ell - A_r\right) \tag{11.55}$$

or equivalently

$$\theta/\text{millidegrees} = 32{,}980C\ell\left(\varepsilon_\ell - \varepsilon_r\right)$$

$$= 32{,}980\left(A_\ell - A_r\right) \tag{11.56}$$

which is useful since the output from most *CD* machines is in millidegrees.

Specific ellipticity

Another old measure of *CD* is the specific ellipticity

$$[\theta] = \frac{\theta}{C_g d} \tag{11.57}$$

where C_g is sample concentration in $g\,cm^{-3}$ and d is path length in dm.

Molar ellipticity

The related molar ellipticity is

$$M_\theta = \frac{[\theta]M}{100} = \frac{100\theta}{C\ell} \tag{11.58}$$

where M is molar mass in g, θ is in degrees, C is in moles dm^{-3}, and ℓ is in cm. Thus

$$M_\theta = 3298(\varepsilon_\ell - \varepsilon_r) \tag{11.59}$$

$$= 3298\Delta\varepsilon$$

Optical activity (optical rotation, *OR*)

Optical rotation, *OR*, is the difference in refractive indices of left and right circularly polarized light upon passing through the medium:

$$OR = (n_\ell - n_r)\frac{\pi d}{\lambda} \tag{11.60}$$

where $(n_\ell - n_r)$ is called the *circular birefringence* and λ and d have the same units (usually dm are used). *OR* is measured in radians.

   OR is usually measured by determining the rotation of linearly polarized light upon passing through the solution. If the linearly polarized light is rotated clockwise when viewed into the light source, the *OR* is called a positive or right (dextro) *OR*.

Optical rotation as a function of λ is called *optical rotatory dispersion* (*ORD*). *ORD* makes an S-shaped curve centred at the *CD* maximum (so-called anomalous optical rotatory dispersion). The *ORD* is also non-zero away from an absorption band, hence α_D values (the *ORD* at the sodium D line) may be used to characterize the enantiomeric excess of a solution. For a positive *CD* band the long-wavelength side of the *ORD* curve shows a larger (positive) *ORD* contribution. For a negative *CD* band a smaller (negative) *ORD* contribution.

11.6   Dipole moments

Electric dipole moment

The electric dipole moment as defined in equation (10.12) is

$$\boldsymbol{\mu} = \sum_i \left(q_i \boldsymbol{r}_i \right) \tag{11.61}$$

where q_i is the charge of the *i*th particle located at the end of vector $\boldsymbol{r}_i$. The charge on an electron is $e = -1.602 \times 10^{-19}$ C in S.I. units. The units for the electric dipole moment are: C m or Debye. These are related by:

$$1 \text{ Debye (D)} = 3.336 \times 10^{-30} \text{ C m} \tag{11.62}$$

Dipole moments of small molecules are typically about 1 D.

Magnetic dipole moment

In the absence of a magnetic field, the magnetic dipole moment of a collection of charges is:

$$\boldsymbol{m} = \sum_i \left\{ \frac{q_i}{2m_i} \boldsymbol{r}_i \times \boldsymbol{p}_i \right\} \tag{11.63}$$

where m_i is the mass of the particle. If the particles are electrons, the factor $e/(2m_i)$ may be replaced by $-\mu_B/\hbar$ where

$$\mu_B = \frac{|e|\hbar}{2m_e} \tag{11.64}$$

$$= 9.273 \times 10^{-24} \text{ m}^2 \text{ s}^{-1} \text{ C}$$

is the Bohr magneton. The Bohr magneton is often used as the unit for magnetic moments.

Normal absorption, dipole strength, and oscillator strength

The integrated absorption coefficient over an entire absorption band is

$$\varepsilon_I = \int \varepsilon(v) \mathrm{d}v \tag{11.65}$$

It may be related to the dipole strength, D^{01}, and the dimensionless oscillator strength, f^{01}, for a transition from state $|0\rangle$ to state $|1\rangle$ of a solution of randomly oriented molecules as follows.

$$D^{01} = \left| \boldsymbol{\mu}^{01} \right|^2 \tag{11.66}$$

Since experimental reality is that no transition occurs at a precisely defined frequency, but over a range, we write the dipole strength per molecule as

$$D^{01} = \frac{3 \ln 10 \, c \varepsilon_0 \hbar}{\pi N_A} \int \frac{\varepsilon(v)}{10v} \, dv$$

$$= 1.022 \times 10^{-61} \int \frac{\varepsilon(v)}{v} \, dv \tag{11.67}$$

which has units $C^2 \, m^2$ if $\varepsilon_0 = 8.854 \times 10^{-12} \, kg^{-1} \, m^{-3} \, s^2 \, C^2$ is the permittivity of a vacuum, N_A is Avagadro's number, and ε is left in its usual non-SI units (hence the factor of 10 in the denominator). The dimensionless oscillator strength is then

$$f^{01} = \frac{4 \pi m_e}{3 \hbar e^2} v \left| \boldsymbol{\mu}^{01} \right|^2$$

$$= \frac{4 m_e c \varepsilon_0 \ln 10}{N_A e^2} \int \frac{\varepsilon(v)}{10} \, dv \tag{11.68}$$

Rotatory (rotational or *CD*) strength

The Rosenfeld equation (equation (8.1)) for the *CD* strength of the transition from state $|0\rangle$ to state $|1\rangle$ of a solution of randomly oriented molecules is

$$R^{01} = \text{Im}\langle 0 | \tilde{\boldsymbol{\mu}} | 1 \rangle \cdot \langle 1 | \tilde{\boldsymbol{m}} | 0 \rangle \tag{11.69}$$

Analogously to the dipole strength, the *CD* strength is

$$R^{01} = \frac{3 \ln 10 \, c^2 \varepsilon_0 \hbar}{4 \pi N_A} \int \frac{\varepsilon_\ell(v) - \varepsilon_r(v)}{10v} \, dv$$

$$= 7.659 \times 10^{-54} \int \frac{\varepsilon_\ell(v) - \varepsilon_r(v)}{v} \, dv \tag{11.70}$$

which has units of $C^2 \, m^3 \, s^{-1}$ if ε is used in its normal unit system.

The relative intensities of the *CD* and dipole strengths gives a convenient estimate of how easy it will be to measure a *CD* spectrum. The ratio is known as the dissymmetry factor. When SI units are used it is

$$g^{01} = \frac{4 R^{01}}{CD^{01}} \tag{11.71}$$

Interaction energies

Coulomb's Law for the interaction energy between two charged particles separated by a distance R_{12} in SI units (J) is

$$V = \frac{q_1 q_2 \mu_0 c^2}{4 R_{12}}$$

$$= \frac{q_1 q_2}{4 \pi \varepsilon_0 R_{12}} \tag{11.72}$$

where $\mu_0 = 4 \times 10^{-7} \, kg \, m \, C^{-2}$ is the permeability of free space, R_{12} is in m, the charges in C, and $c = 2.9979 \times 10^8 \, m \, s^{-1}$. To express the dipole-dipole interaction energy in SI units we write (*cf.* equation (10.43))

$$V = \frac{\boldsymbol{\mu}_1 \cdot \boldsymbol{\mu}_2 - 3\hat{\mathbf{R}}_{12} \cdot \boldsymbol{\mu}_1 \boldsymbol{\mu}_2 \cdot \hat{\mathbf{R}}_{12}}{4\pi\varepsilon_{\rm o} R_{12}^3} \tag{11.73}$$

where R_{12} is in m, $\boldsymbol{\mu}_1$ and $\boldsymbol{\mu}_2$ in C m, so V is in units of J/molecule.

For example, two parallel dipoles (*i.e.* perpendicular to the line connecting them) of strength 1 D that are separated by a distance of 1 nm have a dipole-dipole interaction energy of

$$V = \frac{(3.34 \times 10^{-30})^2}{4\pi \times 8.85 \times 10^{-12} \times 10^{-27}}$$

$$= 1.00 \times 10^{-22} \text{ J/molecule} \tag{11.74}$$

12 References

1.   O'Donoghue, M., *Gems* 6th ed.; Butterworth-Heinemann: Amsterdam, 2006.
2.   Soanes, C.; Stevenson, A., *Oxford Dictionary of English, Revised Edition*. Oxford University Press: Oxford, 2005.
3.   Barron, L. D., *Molecular light scattering*. 2nd ed.; Cambridge University Press: Cambridge, 2004.
4.   Berova, N.; Nakanishi, K.; Woody, R. W., *Circular dichroism principles and applications*. 2nd ed.; Wiley-VCH: New York, 2000.
5.   Cantor, C. R.; Schimmel, P. R., *Biophysical chemistry. Part II. Techniques for the study of biological structure and function*. Freeman and Co.: San Fransisco, 1980.
6.   Craig, D. P.; Thirunamachanrdan, T., *Molecular quantum electrodynamics: An introduction to radiation-molecule interactions*. Academic Press: London, 1984.
7.   Michl, J.; Thulstrup, E. W., *Spectroscopy with polarized light*. VCH: New York, 1986.
8.   Nordén, B., Applications of linear dichroism spectroscopy. *Appl. Spectroscopy Rev.* 1978, 14, 157–248.
9.   Nordén, B.; Elvingson, C.; Jonsson, M.; Åkerman, B., Microscopic behaviour of DNA during electrophoresis: electrophoretic orientation. *Quart. Rev. Biophys.* 1991, 24, 103–164.
10.   Nordén, B.; Kubista, M.; Kurucsev, T., Linear dichroism spectroscopy of nucleic acids. *Quart. Rev. Biophysics* 1992, 25, 51–170.
11.   Rodger, A., Linear dichroism. In *Methods in Enzymology*, Riordan, J. F.; Vallee, B. L., Eds. Academic Press: San Diego, 1993; Vol. 226, pp 232–258.
12.   Rodger, A.; Nordén, B., *Circular dichroism and linear dichroism*. Oxford University Press: Oxford, 1997.
13.   Samori, B.; Thulstrup, E. W., *Polarized spectroscopy of ordered systems*. Kluwer Academic Publishers: The Netherlands, 1988.
14.   Thulstrup, E. W.; Michl, J.; Eggers, J. H. J., Polarization spectra in stretched polymer sheets. II. Separation of pi-pi* absorption of symmetrical molecules into components. *Phys. Chem.* 1970, 74, 3878–3884.
15.   Wallace, B. A.; Janes, R., *Modern Techniques for Circular Dichroism Spectroscopy*. IOS Press: Amsterdam, 2009.
16.   Thulstrup, E. W.; Michl, J., *Elementary polarization spectroscopy*. Wiley-VCH: New York, 1989.
17.   Atkins, P. W., *Molecular quantum mechanics*. Oxford University Press: Oxford, 1983.
18.   Atkins, P. W., *Physical chemistry*. 4th ed.; Oxford University Press: Oxford, 1991.
19.   Hollas, J. M., *Modern Spectroscopy*. 2nd ed.; John Wiley and Sons: Chichester, 1992.
20.   Chou, P. J.; W. C. Johnson, J., Base inclinations in natural and synthetic DNAs *J. Am. Chem. Soc.* 1993, 115, 1205–1214.

Linear Dichroism and Circular Dichroism: A Textbook on Polarized-Light Spectroscopy
By Bengt Nordén, Alison Rodger and Timothy Dafforn
© B. Nordén, A. Rodger and T. Dafforn, 2010
Published by the Royal Society of Chemistry, www.rsc.org

21. Holmén, A.; Broo, A.; Albinsson, B.; Nordén, B., Electronic transition moments of 2-aminopurine. *J. Am. Chem. Soc.* 1997, 119, 12240–12250.

22. Clark, L. B., Electronic spectrum of the adenine chromophore. *J. Phys. Chem.* 1990, 94, 2973–2879.

23. Clark, L. B., Electronic spectra of crystalline 9-ethylguanine and guanine hydrochloride. *J. Am. Chem. Soc.* 1977, 99, 3934–3938.

24. Zaloudek, F.; Novros, J. S.; Clark, L. B., The electronic spectrum of cytosine. *J. Am. Chem. Soc.* 1985, 107, 7344–7351.

25. A. L. Williams, J.; Cheong, C.; I. Tinoco, J.; Clark, L. B., *Nucleic Acids Res.* 1986, 14, 6649–6659.

26. Matsuoka, Y.; Nordén, B., Linear dichroism studies of nucleic acid bases in stretched poly(vinyl alcohol) film. Molecular orientation and electronic transition moment directions. *J. Phys. Chem.* 1982, 86, 1378–1386.

27. Kerckhoffs, J. M. C. A.; Peberdy, J. C.; Meistermann, I.; Childs, L. J.; Isaac, C. J.; Pearmund, C. R.; Reudegger, V.; Alcock, N. W.; Hannon, M. J.; Rodger, A., Enantiomeric resolution of supramolecular helicates with different surface topographies. *Dalton T* 2007, 734–742.

28. Nordén, B., Linear dichroism technique on small molecules dissolved and oriented in a polymer matrix I. Polarisations for electronic transitions in SO_2, CS_2 and NO_2. *Chemica Scripta* 1975, 7, 167–172.

29. Davidsson, A.; Nordén, B., Conversion of Legrand-Grosjean circular dichroism spectrometers to linear dichroism detection. *Chemica Scripta* 1976, 9, 49–53.

30. Nordén, B.; Seth, S., Critical aspects on measurement of circular and linear dichroism. A device for absolute calibration. *Appl. Spectr.* 1985, 39, 647–455.

31. Kahr, B.; Claborn, K., The lives of Malus and his bicentennial law. *ChemPhysChem* 2008, 9, 43–58.

32. Nordén, B.; Seth, S., Critical aspects on measurement of circular and linear dichroism. A device for absolute calibration. *Applied Spectroscopy* 1985, 39, 647–655.

33. Jablonski, A., Polarized photoluminescence of adsorbed molecules of dyes. *Nature* 1934, 133, 140–140.

34. Nordén, B.; Seth, S., The structure of strand-separated DNA in different environments studied by linear dichroism. *Biopolymers* 1979, 18, 2323–2339.

35. Matsuoka, Y.; Nordén, B., Linear dichroism studies of nucleic acids. III. Reduced dichroism curves of DNA in ethanol-water and in poly(vinyl alcohol) films. *Biopolymers* 1982, 22, 1731–1746.

36. Rizzo, V.; Schellman, J., Flow dichroism of T7 DNA as a function of salt concentration. *Biopolymers* 1981, 20, 2143–2163.

37. Wada, A., Chain regularity and flow dichroism of deoxyribonucleic acids in solution. *Biopolymers* 1964, 2, 361–380.

38. Wada, A., Dichroic spectra of biopolymers oriented by flow. *App. Spectros. Rev.* 1972, 6, 1–30.

39. Marrington, R.; Dafforn, T. R.; Halsall, D. J.; Hicks, M.; Rodger, A., Validation of new microvolume Couette flow linear dichroism cells. *Analyst* 2005, 130, 1608–1616.

40. Marrington, R.; Dafforn, T. R.; Halsall, D. J.; Rodger, A., Micro volume Couette flow sample orientation for absorbance and fluorescence linear dichroism. *Biophys. J.* 2004, 87, 2002–2012.

41.   Ardhammar, M.; Mikati, N.; Nordén, B., Chromophore orientation in liposome membranes probed with flow linear dichroism. *J. Am. Chem. Soc.* 1998, 120, 9957–9958.

42.   Rodger, A.; Rajendra, J.; Marrington, R.; Ardhammar, M.; Nordén, B.; Hirst, J. D.; Gilbert, A. T. B.; Dafforn, T. R.; Halsall, D. J.; Woolhead, C. A.; Robinson, C.; Pinheiro, T. J.; Kazlauskaite, J.; Seymour, M.; Perez, N.; Hannon, M. J., Flow oriented linear dichroism to probe protein orientation in membrane environments. *Phys. Chem. Chem. Phys.* 2002, 4, 4051–4057.

43.   Rittman, M.; Gilroy, E.; Koohy, H.; Rodger, A.; Richards, A., Is DNA a worm-like chain in Couette flow? In search of persistence length, a critical review. *Science Progress* 2009, 92, 163–204.

44.   Dafforn, T. R.; Halsall, D. J.; Rodger, A., The detection of single base pair mutations using a novel spectroscopic technique. *Chem. Commun.* 2001, 2410–2411.

45.   Hiort, C.; Nordén, B.; Rodger, A., Enantioselective DNA binding of [Ru(1,10-phenanthroline)$_3$]$^{2+}$ studied with linear dichroism. *J. Am. Chem. Soc.* 1990, 112, 1971–1982.

46.   Gilroy, E., Second year report. *University of Warwick* 2009.

47.   Tumpane, J.; Karousis, N.; Tagmatarchis, N.; Nordén, B., Alignment of carbon nanotubes in weak magnetic fields. *Angew. Chem. Int. Ed.* 2008, 47, 5148–5152.

48.   van Amerongen, H.; van Grondelle, R., Orientation of the bases of the single-stranded DNA and polynucleotides in complexes formed with the gene 32 protein of bacteriophage T4. A linear dichroism study. *J. Mol. Biol.* 1989, 433–445.

49.   Abdourakhmanov, I. A.; Ganago, A. O.; Erokhin, Y. E.; Solov'ev, A. A.; Chugunov, V. A., Orientation and linear dichroism of the reaction centers from *Rhodopseudomonas sphaeroides* R-26. *Biochim. Biophys. Acta* 1979, 546, 183–186.

50.   Nordén, B.; Eriksson, T.; Lundahl, J., *Patent Pending* 2010.

51.   Jonsson, M.; Jacobsson, U.; Takahashi, M.; Nordén, B., Orientation of large DNA during free solution electrophoresis studied by linear dichroism. *J. Chem. Soc. Far. Trans.* 1993, 89, 2791–2798.

52.   Åkerman, B.; Johnsson, M.; Nordén, B., Electrophoretic orientation of DNA detected by linear dichroism spectroscopy. *Chem. Commun.* 1985, 422–423.

53.   Hofrichter, J.; Eaton, W. A., Linear dichroism of biological chromophores. *Ann. Rev. Biophys. Bioeng.* 1976, 5, 511–560.

54.   Nordén, B.; Lindblom, G.; Jonás, I., Linear dichroism spectroscopy as a tool for studying molecular orientation in model membrane systems. *J. Phys. Chem.* 1977, 81, 2086–2093.

55.   Ekwall, P.; Mandell, L., Effect of solubilized decanol on some properties of aqueous sodium octanoate solutions. *J. Colloid Interface Sci.* 1975, 69, 384–397.

56.   Carlsson, N.; Åkerman, B.; Nordén, B., Personal communication. 2010.

57.   Marsh, D.; Muller, M.; Schmitt, F. J., Orientation of the infrared transition moments for an alpha-helix. *Biophys. J.* 2000, 78, 2499–2510.

58.   O'Brien, F. E. M., The control of humidity by saturated salt solutions. *J. Sci. Instrum.* 1948, 25, 73–76.

59. Marshall, K. E.; Hicks, M. R.; Williams, T. L.; Hoffmann, S. V.; Rodger, A.; Dafforn, T. R.; Serpell, L. C., Characterising the assembly of the Sup35 yeast prion fragment, GNNQQNY: structural changes accompany a fibre to crystal switch. *Biophys. J.* 2010, 98, 330–338.

60. Kelly, S. M.; Jess, T. J.; Price, N. C., How to study proteins by circular dichroism. *Biochim. Biophys. Acta* 2005, 1751, 119–139.

61. Billardon, M.; Badoz, J., Modulateur de birefringence. *Comptes Rendus Hebdomadaires des Seances de l'Academie des Sciences Serie B* 1966, 263, 1672–1700.

62. Damianoglou, A.; Crust, E. J.; Hicks, M. J.; Howson, S. E.; Knight, A. E.; Ravi, J.; Scott, P.; Rodger, A., A new reference material for UV-visible circular dichroism spectroscopy. *Chirality* 2008, 20, 1029–1038.

63. Miles, A. J.; Wien, F.; Lees, J. G.; Rodger, A.; Janes, R. W.; Wallace, B. A., Calibration and standardisation of synchrotron radiation circular dichroism and conventional circular dichroism spectrophotometers. *Spectroscopy* 2003, 17, 653–661.

64. Rodger, A.; Marrington, R.; Geeves, M. A.; Hicks, M.; de Alwis, L.; Halsall, D. J.; Dafforn, T. R., Looking at long molecules in solution: what happens when they are subjected to Couette flow? *Phys. Chem. Chem. Phys.* 2006, 8, 3131–3171.

65. Nordh, J.; Deinum, J.; Nordén, B., Flow orientation of brain microtubules studied by linear dichroism,. *Eur. Biophys. J* 1986, 14, 113–122.

66. Rodger, A.; Marrington, R.; Geeves, M. A.; Hicks, M.; de Alwis, L.; Halsall, D. J.; Dafforn, T. R., Looking at long molecules in solution: what happens when they are subjected to Couette flow? *Physical Chemistry Chemical Physics* 2006, 8, 3161–3171.

67. Marrington, R.; Seymour, M.; Rodger, A., A new method for fibrous protein analysis illustrated by application to tubulin microtubule polymerisation and depolymerisation. *Chirality* 2006, 18, 680–690.

68. Gordon, D. J.; Holzwarth, G., Artifacts in the measured optical activity of membrane suspensions. *Arch. Biochem. Bophys.* 1971, 142, 481–488.

69. Nordén, B., Absorption statistics in linear dichroism. *Spectroscopy Lett.* 1977, 10, 483–488.

70. Rittman, M.; Vrønning Hoffmann, S.; Gilroy, E.; Jones, N.; Hicks, M.; Rodger, A., Probing the structure of long DNA molecules in solution using synchrotron radiation linear dichroism. 2010.

71. Duysens, L. M. N., Reversible changes in bacteriochlorophyll in purple bacteria upon illumination. *Biochim. Biophys. Acta* 1956, 19, 1–12.

72. Castiglioni, E., Artifacts from absorption flattening. *Jasco Europe, Technical Reports* 2003, 86, 1.

73. Waldron, D. E.; Marrington, R.; Grant, M. C.; Hicks, M. R.; Rodger, A., Capillary circular dichroism. *Chirality* 2010.

74. Mao, D.; Wallace, B. A., Differential light scattering and absorption flattening optical effects are minimal in the circular dichroism spectra of small unilamellar vesicles. *Biochemistry* 1984, 23, 2667–2673.

75. Davidsson, Å.; Nordén, B., On the problem of obtaining accurate circular dichroism. Calibration of circular dichroism spectrometers. *Spectrochimica Acta* 1976, 32A, 717–722.

76. Davidsson, Å.; Nordén, B.; Seth, S., Measurement of oriented circular dichroism. *Chem. Phys. Lett.* 1980, 70, 313–316.

77. Nordén, B., Calibration of circular dichroism spectrometers. *Acta Chem. Scand.* 1973, 27, 4021–4024.

78. Chiesa, M.; Domini, I.; Samori, B.; Eriksson, S.; Kubista, M.; Nordén, B., Multisite distribution of solute molecules in micelles: benzene in a nematic phase of potassium laurate/decanol/water. *Gazz. Ital.* 1990, 120, 667–670.

79. Snowdon, R., *Personal communication* 2009.

80. Saenger, W., *Principles of nucleic acid structure.* Springer Verlag: New York, 1984.

81. Wang, A. H. J.; Quigley, G. J.; Kolpak, F. J.; Crawford, J. L.; van Boom, J. H.; van der Marel, G.; Rich, A., Molecular structure of a left-handed double helical DNA fragment at atomic resolution. *Nature* 1979, 282, 680–686.

82. Matsuoka, Y.; Nordén, B., Linear dichroism studies of nucleic acids. II. Calculation of reduced dichroism curves of A and B form DNA. *Biopolymers* 1982, 21, 2433–2452.

83. Patel, K. K.; Plummer, E. A.; Darwish, M.; Rodger, A.; Hannon, M. J., Aryl substituted ruthenium bisterpyridine complexes: intercalation and groove binding with DNA. *J. Inorg. Biochem.* 2002, 91, 220–229.

84. Nordén, B.; Tjerneld, F., Structure of methylene blue DNA complexes studied by linear and circular dichroism spectroscopy. *Biopolymers* 1982, 21, 1713–1734.

85. Rodger, A.; Parkinson, A.; Best, S., Molecular features of Co(III) tetra and pentammines affect their influence on DNA structure. *Eur. J. Inorg. Chem.* 2001, 9, 2311–2316.

86. Rodger, A.; Sanders, K. J.; Hannon, M. J.; Meistermann, I.; Parkinson, A.; Vidler, D. S.; Haworth, I. S., DNA structure control by polycationic species: polyamines, cobalt ammines, and di-metallo transition metal chelates. *Chirality* 2000, 12, 221–236.

87. Hannon, M. J.; Moreno, V.; Prieto, M. J.; Molderheim, E.; Sletten, E.; Meistermann, I.; Isaac, C. J.; Sanders, K. J.; Rodger, A., Intramolecular DNA coiling mediated by a metallo supramolecular cylinder. *Angew. Chem.* 2001, 40, 879–884.

88. Meistermann, I.; Moreno, V.; Prieto, M. J.; Molderheim, E.; Sletten, E.; Khalid, S.; Rodger, P. M.; Peberdy, J.; Isaac, C. J.; Rodger, A.; Hannon, M. J., Intramolecular DNA coiling mediated by metallo-supramolecular cylinders: differential binding of P and M helical enantiomers. *Proc. Nat. Acad. Sci.* 2002, 99, 5069–5074.

89. Nordén, B.; Matsuoka, Y.; Kurucsev, T., Nucleic acid-metal interactions. 5. The effect of silver(I) ions on the structures of A- and B-DNA forms. *Biopolymers* 1986, 25, 1531–1545.

90. Malina, J.; Hofr, C.; Maresca, L.; Natile, G.; Brabec, V., DNA interactions of antitumor cisplatin analogs containing enantiomeric amine ligands. *Biophys. J.* 2000, 78, 2008–2021.

91. Hannon, M. J.; Moreno, V.; Prieto, M. J.; Molderheim, E.; Sletten, E.; Meistermann, I.; Isaac, C. J.; Sanders, K. J.; Rodger, A., Intramolecular DNA coiling mediated by a metallo supramolecular cylinder. *Angewandte Chemie,* 2001. 40. 879–884.

92. Tsai, C. C.; Jain, S. C.; Sobell, H. M., X-ray crystallographic visualization of drug-nucleic acid intercalative binding: structure of an ethidium-dinucleoside monophosphate crystalline complex, Ethidium: 5-iodouridylyl (3'-5') adenosine *Proc. Nat. Acad. Sci.* 1975, 72, 628–632.

93. Tuite, E.; Nordén, B., Intercalative interactions of ethidium dyes with triplex structures. *Biorg. Med. Chem.* 1995, 3, 701–711.

94. Nordén, B.; Tjerneld, F.; Palm, E., Linear dichroism studies of binding site structures in solution. Complexes between DNA and basic arylmethane dyes. *Biophys. Chem.* 1978, 8, 1–15.

95. Bulheller, B. PhD thesis. University of Nottingham, 2009.

96. Albinsson, B.; Kubista, M.; Thulstrup, E.; Nordén, B., Near-ultraviolet electronic transitions of the tryptophan chromophore: linear dichroism, fluorescence anisotropy, and magnetic circular dichroism spectra of some indole derivatives. *J. Phys. Chem.* 1989, 93, 6646–6654.

97. Woody, R. W., Aromatic side-chain contributions to protein circular dichroism. In *Methods in protein structure and stability analysis*, Uversky, V.; Permyakov, E., Eds. Nova Science Publishers, Inc.: New York, 2007; pp 291–344.

98. Griebenow, K.; Holzwarth, A. R.; van Mourik, F.; van Grondelle, R., Pigment organization and energy transfer in green bacteria. 2 Circular and linear dichroism of protein-containing and protein-free chlorosomes isolated from *Chloroflexus aurantiacus* strain OK-70-fl*. *Biochim. Biophys. Acta* 1991, 1058, 194–202.

99. Wan, C.; Qian, J.; Johnson, C. K., Conformational motion in bacteriorhodopsin: the K to L transition. *Biochemistry* 1991, 30, 394–400.

100. Dennison, S.R.; Hicks, M.R.; Probert, F., *Personal Communication* 2009.

101. Turner, D., *Personal Communication* 2009.

102. Manolios, N.; Collier, S.; Taylor, J.; Pollard, J.; Harrison, L.; Bender, V., T-cell antigen receptor transmembrane peptides modulate T-cell function and T cell-mediated disease *Nat. Med.* 1997, 3, 84–88.

103. Marrington, R.; Small, E.; Rodger, A.; Dafforn, T. R.; Addinall, S., FtsZ fibre bundling is triggered by a calcium-induced conformational change in bound GTP. *J. of Biol. Chem.* 2004, 47, 48821–48829.

104. Li, H.; DeRosier, D. J.; Nicholson, W. V.; Nogales, E.; Downing, K. H., Microtubule structure at 8 Å resolution. *Structure* 2002, 10, 1317–1328.

105. Benevides, J. M.; Wang, A. H.-J.; van der Marel, G. A.; van Boom, J. H.; Thomas, G. J., Crystal and solution structures of the B-DNA dodecamer d(CGCAAATTTGCG) probed by Raman spectroscopy: heterogeneity in the crystal structure does not persist in the solution structure. *Biochemistry* 1988, 27, 931–938.

106. Lipanov, A.; Kopka, M. L.; Kaczor-Grzeskowiak, M.; Quintana, J.; Dickerson, R. E., Structure of the B-DNA decamer C-C-A-A-C-I-T-T-G-G in two different space groups: conformational flexibility of B-DNA. *Biochemistry* 1993, 32, 1373–1389.

107. Rich, A., The era of RNA awakening: structural biology of RNA in the early years. *Quart. Rev. Biophys.* 2009, 42, 117–137.

108. Williams, A. L. J.; Cheong, C.; Tinoco, I. J.; Clark, L. B., Vacuum ultraviolet circular dichroism as an indicator of helical handedness in nucleic acids. *Nucl. Acids Res* 1986, 6649–6659.

109. Arscott, P. G.; Ma, C.; Wenner, J. R.; Bloomfield, V. A., DNA condensation by cobalt hexaammine(III) in alcohol-water mixtures: dielectric constant and other solvent effects. *Biopolymers* 1995, 36, 345–365.

110. Fant, K.; Esbjörner, E. K.; Lincoln, P.; Nordén, B., DNA condensation by PAMAM dendrimers: self-assembly characteristics and effect on transcription. *Biochemistry* 2008, 47, 1732–1740.

111. Rizzo, V.; Schellman, J. A., Matrix-method calculation of linear and circular dichroism spectra of nucleic acids and polynucleotides. *Biopolymers* 1984, 23, 435–470.

112. Bulheller, B.; Rodger, A.; Hirst, J. D., Circular and linear dichroism of proteins. *Phys. Chem. Chem. Phys.* 2007, 9, 2020–2035.

113. Bulheller, B. M.; Rodger, A.; Hicks, M. R.; Dafforn, T. R.; Serpell, L. C.; Marshall, K.; Bromley, E. H. C.; King, P. J. S.; Channon, K. J.; Woolfson, D. N.; Hirst, J. D., Linear dichroism of some prototypical proteins. *J. Am. Chem. Soc.* 2009, 131, 13305–13314.

114. Garbett, N. C.; Ragazzon, P. A.; Chaires, J. B., Circular dichroism to determine binding mode and affinity of ligand-DNA interactions. *Nat. Protoc.* 2007, 2, 3166–3172.

115. Rodger, A.; Blagbrough, I. S.; Adlam, G.; Carpenter, M. L., DNA binding of a spermine derivative: spectroscopic study of anthracene-9-carbonyl-N^1-spermine with poly(dG-dC)$_2$ and poly(dA-dT)$_2$. *Biopolymers* 1994, 34, 1583–1593.

116. Rodger, A.; Taylor, S.; Adlam, G.; Blagbrough, I. S.; Haworth., I. S., Multiple DNA binding modes of anthracene-9-carbonyl-N^1-spermine. *Bioorg. Med. Chem.* 1995, 3, 861–872.

117. McDonnell, U.; Hicks, M. R.; Hannon, M. J.; Rodger, A., DNA binding and bending by dinuclear complexes comprising ruthenium polypyridyl centres linked by a bis(pyridylimine) ligand. *J. Bioinorg. Chem.* 2008, 102, 2052–2059.

118. Schipper, P. E.; Nordén, B.; Tjerneld, F., Determination of binding geometry of DNA-adduct systems through induced circular dichroism. *Chem. Phys. Lett.* 1980, 70, 17–21.

119. Schipper, P. E.; Rodger, A., Symmetry rules for the determination of the intercalation geometry of host / guest systems using circular dichroism: A symmetry adapted coupled-oscillator model. *J. Am. Chem. Soc.* 1983, 105, 4541–4550.

120. Palumbo, M.; Capasso, L.; Palù, G.; Marciani Magno, S., DNA-binding of water-soluble furocoumarins: a thermodynamic and conformational approach to understanding different biological effects. *Nuc. Acids Res.* 1984, 12, 8567–8578.

121. Nordén, B.; Wirth, M.; Ygge, B.; Buchardt, O.; Nielsen, P., Interactions between DNA and psoraleneamines studied with dichroism techniques. *Photochem. Photobiol.* 1986, 44, 587–594.

122. Ismail, M. A.; Sanders, K. J.; Fennel, G. C.; Latham, H. C.; Wormell, P.; Rodger, A., Spectroscopic studies of 9-hydroxyellipticine binding to DNA. *Biopolymers* 1998, 46, 127–143.

123. Kubista, M.; Åkerman, B.; Nordén, B., Induced circular dichroism in non-intercaltive DNA-drug complexes. Sector rules for structural applications. *J. Phys. Chem.* 1988, 92, 2352–2356.

124. Lyng, R.; Rodger, A.; Nordén, B., The circular dichroism of drug-DNA systems.

125. Lyng, R.; Rodger, A.; Nordén, B., The circular dichroism of drug-DNA systems. 2. Poly(dA-dT)$_2$ B-DNA. *Biopolymers* 1992, 32, 1201–1214.

126. Brahms, S.; Brahms, J., Determination of protein secondary structure in solution by vacuum ultraviolet circular dichroism. *J. Mol. Biol.* 1980, 138, 149–78.

127. Woody, R. W., Circular Dichroism: Principles and Applications. In Nakanishi, K.; Berova, N.; Woody, R. W., Eds. VCH: New York, 1994.

128. Johnson, W. C., Analyzing protein circular dichroism spectra for accurate secondary structures. *Proteins Struct. Funct. Genet.* 1999, 35, 307–312.

129. Anderluh, G.; Go, I.; Lakey, J. H., A natively unfolded toxin domain uses its receptor as a folding template. *J. Biol. Chem* 2004, 279, 22002–22009.

130. Cooper, T. M.; Woody, R. W., The effect of conformation on the *CD* of interacting helices: a theoretical study of tropomyosin. *Biopolymers* 1990, 30, 657–676.

131. Chen, Y. H.; Yang, J. T.; Chau, K. H., Determination of the helix and β form of proteins in aqueous solution by circular dichroism. *Biochemistry* 1974, 13, 3350–3359.

132. Zhong, L.; Johnson, W. C., Environment affects amino acid preference for secondary structure. *Proc. Natl Acad. Sci. USA* 1992, 89, 253–261.

133. Woody, R. W., Circular dichroism of peptides and proteins. In *Circular dichroism principles and applications*, Nakanishi, K.; Berova, N.; Woody, R. W., Eds. VCH: New York, 1994.

134. Dyson, H. J.; Wright, P. E., Coupling of folding and binding for unstructured proteins *Curr. Opin. Struct. Biol.* 2002, 12, 54–60.

135. Kapitan, J.; Baumruk, V.; Bour, P., Demonstration of the ring conformation in polyproline by the raman optical activity. *J. Am. Chem. Soc.* 2006, 128, 2438–2443.

136. Woody, R. W., Circular dichroism spectrum of peptides in the poly(Pro)II conformation. *J. Am. Chem. Soc.* 2009, 131, 8234–8245.

137. Gokce, I.; Woody, R. W.; Anderluh, G.; Lakey, J. H., Single peptide bonds exhibit poly(pro)II ("random coil") circular dichroism spectra. *J. Am. Chem. Soc.* 2005, 127, 9700–9701.

138. Sreerama, N.; Woody, R. W., Structural composition of I- and II-proteins. *Protein Science* 2003, 12, 384–388.

139. Whitmore, L.; Wallace, B. A., DICHROWEB: an online server for protein secondary structure analyses from circular dichroism spectroscopic data. *Nuc. Acids Res.* 2004, 32, W668–673.

140. Miguel, M. S.; Marrington, R.; Rodger, P. M.; Rodger, A.; Robinson, C., An Escherichia coli twin-arginine signal peptide switches between helical and unstructured conformations depending on hydrophobicity of the environment. *Eur. J. Biochem.* 2003, 270, 3345–3352.

141. Damianoglou, A.; Rodger, A., Unpublished work. 2010.

142. Juul, H.; Vrønning Hoffmann, S., *Personal Communication* 2009.

143. Damianoglou, A.; Rodger, A.; Pridmore, C.; Dafforn, T. R.; Mosely, J. A.; Sanderson, J. M.; Hicks, M. R., The synergistic action of melittin and phospholipase A2 with lipid membranes: development of linear dichroism for membrane-insertion kinetics. *Protein and peptide letters* 2010.

144. Ardhammar, M.; Lincoln, P.; Rodger, A.; Nordén, B., Absolute configuration and electronic state properties of light-switch complex [Ru(phen)$_2$dppz]$^{2+}$ deduced

from oriented circular dichroism in a lamellar liquid crystal host. *Chem. Phys. Lett.* 2002, 354, 44–50.

145.   Johansson, L. B. Å.; Lindblom, G.; Nordén, B., Micelle studies by high-sensitivity linear dichroism. Benzene solubilisation in rod-shaped micelles of cetyltrimethylammonium bromide in water. *Chem. Phys. Lett.* 1976, 39, 128–133.

146.   Rajendra, J.; Damianoglou, A.; Hicks, M.; Booth, P.; Rodger, P. M.; Rodger, A., Quantitation of protein orientation in flow-oriented unilamellar liposomes by linear dichroism. *Chem. Phys.* 2006, 326, 210–220.

147.   Rajendra, J.; Damianoglou, A.; Hicks, M.; Booth, P.; Rodger, P. M.; Rodger, A., Quantitation of protein orientation in flow-oriented unilamellar liposomes by linear dichroism. *Chemical Physics* 2006, 326, 210–220.

148.   Svensson, F. R.; Lincoln, P.; Nordén, B.; Esbjörner, E. K., Retinoid chromophres as probes of membrane lipid order. *J. Phys. Chem.* 2007, 111, 10839–10848.

149.   Esbjörner, E. K.; Oglecka, K.; Lincoln, P.; Gräslund, A.; Nordén, B., Membrane binding of pH-sensitive influenza fusion peptides. Positioning, configuration, and induced leakage in a lipid vesicle model. *Biochemistry* 2007, 46, 13490–13504.

150.   Oesterhelt, D.; Stoeckenius, W., Rhodopsin-like protein from the *purple* membrane of Halobacterium halobium. *Nature New Biol.* 1971, 233, 149–152.

151.   Henderson, R.; Baldwin, J. M.; Ceska, T. A.; Zemlin, F.; Beckmann, E.; Downing, K. H., Model for the structure of bacteriorhodopsin based on high-resolution electron cryo-microscopy. *J. Mol. Biol.* 1990, 213, 899–929.

152.   Hicks, M. R.; Dafforn, T. R.; Damianoglou, A.; Wormell, P.; Rodger, A.; Hoffmann, S. V., Synchrotron radiation linear dichroism spectroscopy of the antibiotic peptide gramicidin in lipid membranes. *The Analyst* 2009, 134, 1623–1628.

153.   Hicks, M. R.; Damianoglou, A.; Rodger, A.; Dafforn, T. R., Folding and membrane insertion of the pore-forming peptide gramicidin occur as a concerted process. *J. Mol. Biol.* 2008, 383, 358–366.

154.   Dicko, C.; Hicks, M. R.; Dafforn, T. R.; Vollrath, F.; Rodger, A.; Hoffmann, S. V., Breaking the 200 nm limit for routine flow linear dichroism measurements using UV synchrotron radiation. *Biophys. J.* 2008, 95, 5974–5977.

155.   Mabrey, S.; Sturtevant, J. M., Investigation of phase transitions of lipids and lipid mixtures by sensitivity differential scanning calorimetry. *Proc. Natl. Acad. Sci. USA* 1976, 73, 3862–3866.

156.   Urry, D. W.; Trapane, T. L.; Prasad, K. U., Is the gramicidin A transmembrane channel single-stranded or double-stranded helix? A simple unequivocal determination. *Science* 1983, 221, 1064–1067.

157.   Esbjorner, E. K.; Caesar, C. E. B.; Albinsson, B.; Lincoln, P.; Nordén, B., Tryptophan orientation in model lipid membranes. *Biochem. Biophys. Res. Comm.* 2007, 361, 645–650.

158.   Svensson, F. R.; Lincoln, P.; Nordén, B.; Esbjörner, E. K., Unpublished work.

159.   Gårding, L.; Nordén, B., Simple formulas for rotation averages of spectroscopic Intensities. *Chem. Phys.* 1979, 41, 431–437.

160.   Hicks, M. R.; Rodger, A.; Thomas, C. M.; Batt, S. M.; Dafforn, T. R., Restriction enzyme kinetics monitored by UV linear dichroism. *Biochemistry* 2006, 45, 8912–7.

161. Pingoud, A.; Fuxreiter, M.; Pingoud, V.; Wende, W., Type II restriction endonucleases: structure and mechanism. *Cell Mol. Life Sci.* 2005, 62, 685–707.

162. Kettling, U.; Koltermann, A.; Schwille, P.; Eigen, M., Real-time enzyme kinetics monitored by dual-color fluorescence cross-correlation spectroscopy. *Proc. Natl. Acad. Sci. USA* 1998, 95, 1416–1420.

163. Waters, T. R.; Connolly, B. A., Continuous spectrophotometric assay for restriction endonucleases using synthetic oligodeoxynucleotides and based on the hyperchromic effect. *Anal. Biochem.* 1992, 204, 204–209.

164. Schalch, T.; Duda, S.; Sargent, D. F.; Richmond, T. J., X-ray structure of a tetranucleosome and its implications for the chromatin fibre. *Nature* 2005, 436, 138–140.

165. Tjerneld, F.; Nordén, B.; Wallin, H., Chromatin structure studied by linear dichroism at different salt concentrations. *Biopolymers* 1982, 21, 343–358.

166. Kubista, M.; Hagmar, P.; Nielsen, P. E.; Nordén, B., Reinterpretation of linear dichroism of chromatin supports a perpendicular linker orientation in the folded state. *J. Biomol. Struct. Dynam.* 1990, 8, 37–54.

167. Hagmar, P.; Nordén, B.; Baty, D.; Chartier, M.; Takahashi, M., Structure of DNA-RecA complexes studied by residue differential linear dichroism and fluorescence spectroscopy for a genetically engineered RecA protein. *J. Mol. Biol.* 1992, 226, 1193–1205.

168. Takahashi, M.; Nordén, B., Structure of RecA-DNA complexes and mechanism of DNA strand exchange reaction in homologous recombination. *Adv. Biophys.* 1994, 30, 1–35.

169. Hagmar, P.; Nordén, B.; Baty, D.; Chartier, M.; Takahashi, M., Structure of DNA-RecA complexes studied by residue differential linear dichroism and fluorescence spectroscopy for a genetically engineered RecA protein. *J. Mol. Biol.* 1992, 226, 1193–1205.

170. Morimatsu, K.; Takahashi, M.; Nordén, B., Arrangement of RecA protein in its active filament determined by polarized-light spectroscopy. *Proc. Natl. Acad. Sci. USA* 2002, 99, 11688–11693.

171. Chen, Z.; Yang, H.; Pavletich, N. P., Mechanism of homologous recombination from the RecA-ssDNA/dsDNA structures. *Nature* 2008, 453, 489–494.

172. Story, R. M.; Weber, I. T.; Steitz, T. A., The structure of the E. coli recA protein monomer and polymer. *Nature* 1992, 355, 318–325.

173. Nordén, B.; Elvingson, C.; Kubista, M.; Sjöberg, B.; Ryberg, H.; Ryberg, M.; Mortensen, K.; Takahashi, M., Structure of RecA-DNA complexes studied by combination of linear dichroism and small angle neutron scattering measurements on flow-oriented samples. *J. Mol. Biol.* 1992, 226, 1175–1192.

174. Reymer, A.; Frykholm, K.; Morimatsu, K.; Takahashi, M.; Nordén, B., Structure of human Rad51 protein filament from molecular modeling and site-specific linear dichroism spectroscopy. *Proc. Natl. Acad. Sci. USA* 2009, 106, 13248–13253.

175. Nordén, B.; Tjerneld, F., Binding of Inert metal complexes to DNA detected by linear dichroism. *FEBS Letters* 1976, 67, 368–370.

176. Lincoln, P.; Broo, A.; Nordén, B., Diastereomeric DNA-Binding Geometries of Intercalated Ruthenium(II) Trischelates Probed by Linear Dichroism: [Ru(phen)₂DPPZ]²⁺ and [Ru(phen)₂BDPPZ]²⁺. *J. Am. Chem. Soc.* 1996, 118.

177. Lincoln, P.; Nordén, B., DNA binding geometries of ruthenium(II) complexes with 1,10-phenanthroline and 2,2'-bipyridyl ligands studied with linear dichroism spectra. Borderline cases of intercalation. *J. Phys. Chem.* 1998, 102, 9583–9594.

178. Khalil, A. S.; Ferrer, J. M.; Brau, R. R.; Kottmann, S. T.; Noren, C. J.; Lang, M. J.; Belcher, A. M., Single M13 bacteriophage tethering and stretching. *Proc. Nat. Acad. Sci. USA* 2007, 104, 4892–4897.

179. Clack, B. A.; Gray, D. M., Flow Linear Dichroism Spectra of Four Filamentous Bacteriophages: DNA and Coat Protein Contributions. *Biopolymers* 1992, 32, 795–810.

180. Wang, Y. A.; Yu, X.; Overman, S.; Tsuboi, M.; Thomas, G. J.; Egelman, E. H., The structure of a filamentous bacteriophage. *J. Mol. Biol.* 2006, 361, 209–215.

181. Pacheco-Gomez, R.; Dafforn, T. R., *Personal Communication* 2010.

182. Miles, A. J.; Wallace, B. A., Synchrotron radiation circular dichroism spectroscopy of proteins and applications in structural and functional genomics. *Chem. Soc. Rev.* 2006, 35, 39–51.

183. Miles, A. J.; Hoffmann, S. V.; Tao, Y.; Janes, R. W.; Wallace, B. A., Synchrotron radiation circular dichroism (SRCD) spectroscopy: new beamlines and new applications in biology. *Spectroscopy* 2007, 21, 245–255.

184. Åkerman, B.; Tuite, E., Single- and double-strand photocleavage of DNA by YO, YOYO and TOTO. *Nuc. Acids Res.* 1996, 24, 1080–1090.

185. Carlsson, C.; Larsson, A.; Jonsson, M.; Albinsson, B.; Nordén, B., Optical and photophysical properties of the oxazole yellow DNA probes. *J. Phys. Chem.* 1994, 98, 10313–10321.

186. Kowalski, J., LD of single-walled carbon nantobubes. 2010.

187. Monthioux, M.; Kuznetsov, V. L., Who should be given the credit for the discovery of carbon nanotubes? *Carbon* 2006, 44, 1621–1623.

188. Rajendra, J.; Baxendale, M.; Dit Rap, L. G.; Rodger, A., Flow linear dichroism to probe binding of aromatic molecules and DNA to single walled carbon nanotubes. *J. Am. Chem. Soc.* 2004, 126, 11182–11188.

189. Rajendra, J.; Rodger, A., The binding of single stranded DNA and PNA to single walled carbon nanotubes probed by flow linear dichroism. *Chem. - A Eur. J.* 2005, 11, 4841–48.

190. Cathcart, H.; Nicolosi, V.; Hughes, J. M.; W. J. Blau; Kelly, J. M.; Quinn, S. J.; Coleman, J. N., Ordered DNA wrapping switches on luminescence in single-walled nanotube dispersions. *J. Am. Chem. Soc.* 2008, 130, 12734–12744.

191. Vollmer, W.; Höltje, J. V., The architecture of the murein (peptidoglycan) in Gram-negative bacteria: vertical scaffold or horizontal layer(s)? *J. Bacteriol.* 2009, 186, 5978-5987.

192. Cordes, D. B.; Miller, A.; Gamsey, S.; Sharrett, Z.; Thoniyot, P.; Wessling, R.; Singaram, B., Optical glucose detection across the visible spectrum using anionic fluorescent dyes and a viologen quencher in a two-component saccharide sensing system. *Org. and Biomol. Chem. - A Eur. J.* 2005, 3, 1708–1713.

193. Chou, P. J.; W. C. Johnson, J., Base inclinations in natural and synthetic DNAs *J. Amer. Chem. Soc.* 1993, 115, 1205–1214.

194. Arnott, S.; Hukins, D. W., Optimized parameters for A-DNA and B-DNA. *Biophys. Res. Commun.* 1972, 47, 1504–1510.

195. Edmondson, S.; Johnson, W. C. J., Base tilt of DNA in various conformations from flow linear dichroism. *Biochemistry* 1985, 24, 4802–4806.

196. Wilson, R. W.; Schellman, J. A., The flow linear dichroism of DNA: comparison with the bead-string theory. Biopolymers 1978, 17, 1235–1248.

197. Nordén, B., Detection of n-> π* transitions in pyridine and pyrazine in polyethylene matrix by linear dichroism. *Chem. Phys. Lett.* 1973, 23, 200–202.

198. Novakova, O.; Chen, H.; Vrana, O.; Rodger, A.; Sadler, P. J.; Brabec, V., *Biochemistry* 2003, 42, 11544–11554.

199. Matsuoka, Y.; Nordén, B., Linear dichroism studies of flavins in stretched poly(vinyl alcohol) films. Molecular orientation and electronic transition moment directions. *J. Phys. Chem.* 1983, 87, 220–225.

200. Davidsson, Å.; Nordén, B., Vibronic polarized spectrum of naphthalene by means of linear dichroism studies in oriented polymer matrices. *Chem. Phys. Lett.* 1974, 28, 221–224.

201. Nordén, B., General aspects on linear dichroism spectroscopy and its application. *Spectroscopy Lett.* 1977, 10, 381–400.

202. Besley, N. A.; Hirst, J. D., Theoretical studies toward quantitative protein circular dichroism calculations *J. Am. Chem. Soc.* 1999, 121, 9636–9644.

203. Rodger, A.; Patel, K. K.; Sanders, K. J.; Datt, M.; Sacht, C.; Hannon, M. J., Anti-tumour platinum acylthiourea complexes and their interactions with DNA. *Dalton* 2002, 3656–3663.

204. Johansson, L. B. Å.; Davidsson, Å.; Lindblom, G.; Nordén, B., Linear dichroism as a tool for studying molecular orientation in membrane systems. II. Order parameters of guest Molecules from LD and NMR. *J. Phys. Chem.* 1978, 82, 260–269.

205. Thulstrup, E. W.; Michl, J.; Eggers, J. H., Polarised light spectra of aromatic molecules oriented in stretched polyethylene sheets. *J. Phys. Chem.* 1970, 74, 3868–3877.

206. Albinsson, B.; Nordén, B., Excited-state properties of the indole chromophore: electronic transition moment directions from linear dichroism measurements: effect of methyl and methoxy substituents. *J. Phys. Chem.* 1992, 96, 6204–6212.

207. Albinsson, B.; Kubista, M.; Sandros, K.; Nordén, B., Electronic linear dichroism spectrum and transition moment directions of the hypermodified nucleic acid base Wye. *J. Phys. Chem.* 1990, 94, 4006–4011.

208. Lakowicz, J. R., *Principles of Fluorescence Spectroscopy.* 3rd ed.; Berlin/Heidelberg, 2006.

209. Undeman, O.; Lycksell, P. O.; Gräslund, A.; Astlind, T.; Ehrenberg, A.; Jernström, B.; Tjerneld, F.; Nordén, B., Covalent Complexes of DNA and Two Diastereomers of Benzo(a)Pyrene-7,8,Dihydrodiol-9,10-Epoxide Studied by Fluorescence and Linear Dichroism. *Cancer Res.* 1982, 43, 1851–1860.

210. Eriksson, M.; Jernstrom, B.; Graslund, A.; Nordén, B., Binding geometries of benzo(a)pyrene diol epoxide isomers covalently bound to DNA. Orientational distribution. *Biochemistry* 1988, 27, 1213–1221.

211. Kerr, J., Personal communication. 2009.

212. Miakti, N.; Nordh, J.; Nordén, B., Scattering anisotropy of partially oriented samples: turbidity flow dichroism ('conservative dichroism') of reshaped macromolecules. *J. Phys. Chem.* 1987, 91, 6048–6055.

213. Nordh, J.; Deinum, J.; Nordén, B., Flow orientation of brain microtubules studied by linear dichroism. *Eur. Biophys. J.* 1986, 14, 113–122.

214. Nordén, B., Simple formulas for dichroism analysis. *J. Chem. Phys.* 1980, 72, 5032–5038.

215. Tanizaki, Y., Dichroism of dyes in the stretched PVA Sheet. II. The relation between the optical density ratio and the stretch ratio, and an attempt to analyse relative directions of absorption bands. *Bull. Chem Soc. Jpn.* 1959, 32, 75–80.

216. Andersson, L.; Nordén, B., On the use of moments for describing the molecular orientation distribution. *Chem. Phys. Lett.* 1980, 70, 398–402.

217. Lightner, D. A.; Bouman, T. D.; Wijekoon, W. M. D.; Hansen, A., The octant rule. 18. Mechanism of ketone n to pi* optical activity. Experimental and computed chiroptical properties of 4-axial and 4-equatorial alkyladamantan-2-ones. *J. Am. Chem. Soc.* 1986, 108, 4484–4497.

218. Rubio, M.; Mechan, M.; Orti, E.; Roos, B. O., A theoretical study of the electronic spectrum of biphenyl *Chem. Phys. Lett.* 1995, 234, 373–381.

219. Weigang, O. E. J., An amplified sector rule for electric dipole-allowed transitions. *J. Am. Chem. Soc.* 1979, 101, 1965–1975.

220. Harada, N.; arada, N.; Takuma, Y.; Uda, H., The absolute stereochemistries of 6,15-dihydro-6,15-ethanonaphtho[2.3-c]pentaphene and related homologs as determined by both exciton chirality and x-ray Bijvoet methods. *J. Am. Chem. Soc.* 1976, 98, 5408–5409.

221. Bosnich, B., Application of exciton theory to the determination of the absolute configurations of inorganic complexes. *Acc. Chem. Res.* 1969, 2, 266–273.

222. Coggan, D. Z.; Haworth, I. S.; Bates, P. J.; Robinson, A.; Rodger, A., DNA binding of ruthenium tris-(1,10-phenanthroline): evidence for the dependence of binding mode on metal complex concentration. *Inorg. Chem.* 1999, 38, 4486–4497.

223. Moffitt, W.; Yang, J. T., The optical rotatory dispersion of simple polypeptides I. *Proc. Natl. Acad. Sci. U S A* 1956, 42, 596–603.

224. W. C. Johnson, J.; I. Tinoco, J., *J. Am. Chem. Soc.* 1972, 94, 4389–4390.

225. Rizzo, V.; Schellman, J., Matrix-method calculation of linear and circular dichroism spectra of nucleic acids and polynucleotides. *Biopolymers* 1984, 23, 435–470.

226. Hidaka, K.; Douglas, B. E., Circular dichroism of coordination compounds. II. Some metal complexes of 2,2'-dipyridyl and 1,10-phenanthroline. *Inorg. Chem.* 1964, 3, 1180–1184.

227. Khalid, S.; Rodger, P. M.; Rodger, A., Theoretical aspects of the enantiomeric resolution of dimetallo helicates with different surface topologies on cellulose columns. *J. Liq. Chromat.* 2005, 28, 2995–3003.

228. Rodger, A.; Sanders, K. J.; Hannon, M. J.; Meistermann, I.; Parkinson, A.; Vidler, D. S.; Haworth, I. S., DNA structure control by polycationic species: polyamines, cobalt ammines, and di-metallo transition metal chelates. *Chirality* 2000, 12, 221–236.

229. Hannon, M. J.; Meistermann, I.; Isaac, C. J.; Blomme, C.; Aldrich-Wright, J. R.; Rodger, A., Paper: a cheap yet effective chiral stationary phase for chromatographic resolution of metallo-supramolecular helicates. *Chem. Commun.* 2001, 1078–1079.

230. Moffit, W.; Woodward, R. B.; Moscowitz, A.; Klyne, W.; Djerassi, C. J., Structure and the Optical Rotatory Dispersion of Saturated Ketones. *Amer. Chem. Soc.* 1961, 83, 4013–4018.

231. Coulombeau, C.; Rassat, A., Experimental studies on the octant rule. *Bull. Soc. Chim. de France* 1971, 516–526.

232. Lightner, D. A.; Chang, T. C., Octant rule. III. Experimental proof for front octants. *J. Am. Chem. Soc.* 1974, 96, 3015–3016.

233. Snatzke, G.; Eckhardt, G., Circulardichroismus—XXVII : circulardichroismus und UV-spektren β-substituierter adamantanone. *Tetrahedron* 1968, 24, 4543–4558.

234. Hudec, J., Organic spectroscopy and its relationship to ground-state chemistry. *J. Chem. Soc. Chem. Commun.* 1970, 829–831.

235. Lightner, D. A.; Gawronski, J. K.; Bouman, T. D., The octant rule. 7. Deuterium as an octant perturber. *J. Am. Chem. Soc.* 1980, 102, 1983–1990.

236. Fidler, J.; Rodger, P. M.; Rodger, A., Circular dichroism as a probe of chiral solvent structure around chiral molecules. *Journal of the Chemical Society Perkin II* 1993, 116, 235–241.

237. Fidler, J.; Rodger, P. M.; Rodger, A., Chiral solvent structure around chiral molecules: experimental and theoretical study. *J. Am. Chem. Soc.* 1994, 116, 7266–7273.

238. Rodger, A.; Rodger, P. M.; The circular dichroism of the carbonyl n-π^* transition: An independent systems/perturbation approach. *J. Am. Chem. Soc.* 1988, 110, 2361–2368.

239. Coulombeau, C.; Rassat, A., Experimental studies on the octant rule. *Bull. Soc. Chim. de France* 1971, 516–526.

240. Lightner, D. A.; Bouman, T. D.; Wijekoon, W. M. D.; Hansen, A., The octant rule. 18. Mechanism of ketone n to π^* optical activity. Experimental and computed chiroptical properties of 4-axial and 4-equatorial alkyladamantan-2-ones. *J. Amer. Chem. Soc.* 1986, 108, 4484–4497.

241. Höhn, E. G.; Weigang, O. E. J., Electron correlation models for optical activity. *J. Chem. Phys.* 1968, 48, 1127–1136.

242. Rodger, A.; Rodger, P. M., The circular dichroism of the carbonyl n-π^* transition: An independent systems / perturbation approach. *J. Amer. Chem. Soc.* 1988, 110, 2361–2368.

243. Rodger, A.; Moloney, M. J., n-π^* Circular dichroism of planar zig-zag carbonyl compounds. *Chem. Soc. Perkin II* 1991, 919–925.

244. Kirk, D. N.; Klyne, W., Optical rotatory dispersion and circular dichroism. Part LXXXII. An empirical analysis of the circular dichroism of decalones and their analogues. *J. Chem. Soc. Perkin I* 1974, 1076–1103.

245. Schipper, P. E.; Rodger, A., Generalized selection rules for circular dichroism: a symmetry adapted perturbation model for magnetic dipole allowed transitions. *Chem. Phys.* 1986, 109, 173–193.

246. Rodger, A.; Ismail, M. A., Introduction to circular dichroism. In *Spectrometry and spectrofluorimetry: a practical approach*, Gore, M., Ed. 2000; pp 99–139.

247. Schipper, P. E.; Rodger, A., Generalized selection rules for circular dichroism: Aasymmetry adapted perturbation model for magnetic dipole allowed transitions. *Chem. Phys.* 1986, 109, 173–193.

248. Rodger, A.; Nordén, B., A coupled-oscillator type approach to magnetic circular dichroism. *Enantiomer* 1998, 3, 409–421.

249. Stephens, P. J., Magnetic circular dichroism. *Adv. Chem. Phys.* 1976, 35, 197–264.

250. Solomon, E. I.; Brunold, T. C.; Davis, M. I.; Kemsley, J. N.; Lee, S. K.; Lehnert, N.; Neese, F.; Skulan, A. J.; Yang, Y. S.; Zhou, J., Geometric and electronic structure/function correlations in non-heme iron enzymes. *Chem. Rev.* 2000, 100, 235–350.

251. Nordén, B.; Håkansson, R.; Danielsson, S., Permanent magnet interface for magnetic circular dichroism spectroscopy. *Chemical Scripta* 1977, 11, 52–56.

252. Nordén, B.; Håkansson, R.; Pedersen, P. B.; Thulstrup, E. W., The magnetic circular dichroism of five-membered ring heterocycles. *Chem. Phys.* 1978, 33, 355–366.

253. Nordén, B.; Kubista, M.; Kuruscev, T., Linear Dichroism Spectroscopy of Nucleic-Acids. *Q. Rev. Biophysics* 1992, 25, 51–170.

254. Oakberg, T. C.; Trunk, J.; Sutherland, J. C., Calibration of photoelastic modulators in the vacuum UV. *Proc. Soc. for Photo-Optical Instrumentation Engineers* 2000, 4133, 101–111.

255. Richardson, F. S., Theory of optical activity in the ligand-field transitions of chiral transition metal complexes. *Chem. Rev.* 1979, 79, 17–36.

256. Schellman, J. A., Symmetry Rules for Optical Rotation. 1966, 44, 55–63.

257. Schipper, P. E.; Rodger, A., Symmetry rules for the determination of the intercalation geometry of host / guest systems using circular dichroism: A symmetry adapted coupled-oscillator model. *J. Amer. Chem. Soc.* 1983, 105, 4541–4550.

258. Scatchard, G., The attractions of proteins for small molecules and ions. *Ann. N.Y. Acad. Sci.* 1949, 51, 660–672.

259. Nordén, B.; Tjerneld, F., Stability constant of DNA-ethidiumbromide complex. *Biophysical Chemistry* 1976, 4, 191–198.

260. Diebler, H.; Secco, F.; Venturini, M., The binding of Mg(II) and Ni(II) to synthetic polynucleotides. *Biophys. Chem.* 1987, 26, 193–205.

261. Fronaeus, S., A new principle for the investigation of complex equilibria and the determination of complexity constants. *Acta Chem. Scand.* 1950, 4, 72–87.

262. McGhee, J. D.; von Hippel, P. H., Theoretical aspects of DNA-protein interactions: cooperative and non-cooperative binding of large ligands to a one-dimensional homogeneous lattice. *J. Mol. Biol.* 1974, 86, 469–489.

263. McGhee, J. D.; von Hippel, P. H., Theoretical aspects of DNA-protein interactions: cooperative and non-cooperative binding of large ligands to a one-dimensional homogeneous lattice: correction. *J. Mol. Biol.* 1976, 103, 679.

264. Schmechel, D. E. V.; Crothers, D. M. B., Kinetic and hydrodynamic studies of the complex of proflavine with poly A·poly U. *Biolopymers* 1971, 10, 465–480.

Subject Index

Note: page numbers in *italic* refer to figures and side-notes.